Chemistry of Synthetic Drugs

Chemistry of Synthetic Drugs

Dr. Vibha Tandon
Research Scientist
Dr. B.R. Ambedkar Centre for Biomedical Research
University of Delhi
Delhi-110007

Dr. Akhilesh K. Verma
Senior Lecturer
Dr. B.R. Ambedkar Center for Biomedical Research
University of Delhi
Delhi-110007

SHREE PUBLISHERS & DISTRIBUTORS
NEW DELHI-110 002

Edition : 2009

Published by :
SHREE PUBLISHERS & DISTRIBUTORS
4735/22, Prakash Deep Building
Ansari Road, Darya Ganj,
New Delhi–110 002

ISBN : 978-81–8329–298–6

Printed by :
Tarun Offset
Delhi–110 053

Preface

In recent years, much knowledge has been gained regarding the biologically useful derivatives of the naturally occurring drugs in nature. This knowledge has allowed researchers to reduce many of the structural relationships associated with specific biological responses, and the some of this knowledge allows one to produce pharmaceutically useful products, which produce the right pharmacological response. These include hallucinogens, stimulants, analgesics, sedatives of pharmaceutical drugs.

This book in the series on pharmacology has been designed as an introductory manual which looks into the chemistry of artificial or lab-designed drugs. Right from a general introduction the book proceeds onto the techniques, principles and practices which enable one composition, bioactivity pharmacodynamics and pharmacokinetics. In addition, it also takes a look at the current discoveries and advances in the field, enabling a relevant understanding of the subject. Further, the illustrations and diagrams in the book convey visual information, which moves beyond the descriptive capacity of words. Also, the legal and ethical implications of synthetic drug use have been included, fostering an in-depth knowledge of the subject.

It is hoped that the book's erudition and insight serve well for the readers.

Dr. Vibha Tandon

Dr. Akhilesh K. Verma

ACKNOWLEDGEMENT

Book writing-whether original or compilation is a complicated job requiring academic labour.

The present work is purely a research compilation of literature from authoritative publications/readings/ excerpts/notes/reviews/researches/literature, authored by eminent scholars/distinguished writers.

During research and compilation, the author has also taken assistance/references/literature from various Websites through Internet.

The author, therefore humbly, acknowledges the contributions of all those eminent writers/scholars alongwith their respective publishers and Websites from whose learned writings/displays references/ literature have been taken while preparing this book.

Contents

Chapter 1

Synthetic Drugs and Human Health

Human health is impacted by a wide variety of chemical substances, including those essential to human life, such as vitamins and nutrients, medicines, and toxic materials. Understandably, there is a vital interest in this subject on the part of the general public. A popular view holds that natural substances are innately superior to man-made or synthetic substances with respect to their effects, good or bad, on human health. This can extend to so-called nature-identical materials that are natural substances produced synthetically in an identical molecular form.

Drug Products

Natural products continue to play an important role in the discovery and development of new pharmaceuticals, as clinically useful drugs, as starting materials to produce synthetic drugs, or as lead compounds from which a totally synthetic drug is designed. At the same time, synthetic compounds, unrelated to natural products, have played an increasingly progressive role in new drug discovery. Continuous improvements in synthetic methodology have provided practical access to a vast array of synthetic substances, most recently in the form of combinatorial synthesis.

Natural compounds and synthetic compounds are, in many respects, complementary as avenues for new drug substances since natural products often possess complex structural features not easily accessible by total synthesis. In this brief overview, some selected examples will be provided from the categories of natural, semisynthetic. Semisynthetic drugs produced by modifying natural products are man-made substances, although they may be regarded as intermediate in character between natural and synthetic substances.

Natural Products

Anti-infectives

Antibiotics are, by definition, natural products or derivatives of natural products. Since the discovery of penicillin, a large number of antibiotics have been isolated from scores of microorganisms, and several new antibiotics make it to the clinic each year. The antibiotics are important, life-saving drugs. Major classes include ß-lactams, tetracyclines, macrolides, and aminoglycosides.

Additionally, explorations into the mechanisms by which both the clinically used antibiotics and those that are not used clinically exert their action have led to an understanding of the biology of the target pathogens that would not likely have been possible without these important biochemical probes. Further advances will surely be forthcoming with the application of more advanced molecular biology techniques.

Among the most important antibiotics, both from a historical and a clinical utility perspective, are the β-lactams, which are predominated by the penicillins and cephalosporins. The naturally occurring penicillin first isolated from cultures of Penicillium in the middle of the last century were extraordinary discoveries, yet suffered certain shortcomings as therapeutic agents.

While the β-lactam nucleus is required for the desired antimicrobial action, the early natural penicillins had a very narrow spectrum of activity, poor pharmacokinetics and chemical stability, and were easily destroyed by the action of penicillinase, which conferred resistance to bacterial strains capable of producing this enzyme. These shortcomings were overcome when it was discovered that 6-aminopenicillanic acid, the biosynthetic precursor to the natural penicillins, could be synthetically modified to introduce varying side chains (R) that determined the properties of the modified product. Thus began an era of developing semisynthetic and synthetic analogs of β-lactams with much improved properties, including better stability and pharmacokinetics, extended antimicrobial spectrum, and resistance to the action of penicillinase. Presently, there are more than a dozen penicillin antibiotics in clinical use, but only two are produced without synthetic modification: penicillin G and penicillin V.

Penicillins

Cephalosporins

Similarly, the first cephalosporin (cephalosporin C) lacked potency, spectrum, and other desirable pharmaceutical properties. With the discovery of synthetic methodology to replace the side-chain of the naturally occurring cephalosporins with side-chains designed to confer more desirable properties, the door was opened to improve semisynthetic analogs. Currently, there are more than 25 cephalosporins in clinical use today, all of which are produced through semisynthesis or total synthesis.

The discovery of new important anti-infectives is not limited to searching the microbial kingdom, but includes plant and animal sources as well. Artemisinin, a sesquiterpene with an unusual endoperoxide moiety, was isolated from the Chinese medicinal plant commonly known as Qinghaosu (Artemisia annua), an herbal remedy that had been used in China for centuries for the treatment of malaria. Isolated in 1972 as the active constituent, artemisinin was discovered to be particularly effective for the deadly cerebral malaria.

Subsequent efforts to develop artemisinin derivatives with more desirable pharmaceutical properties included many synthetic and semisynthetic studies, microbial transformations, biological evaluations, mechanism of action studies, and pharmacological studies of artemisinin and a number of related analogs. From these efforts, the semisynthetic derivative artemether was developed and is now approved for the treatment of malaria in much of the world. Thus, a natural substance was transformed into a related semisynthetic substance with superior drug properties.

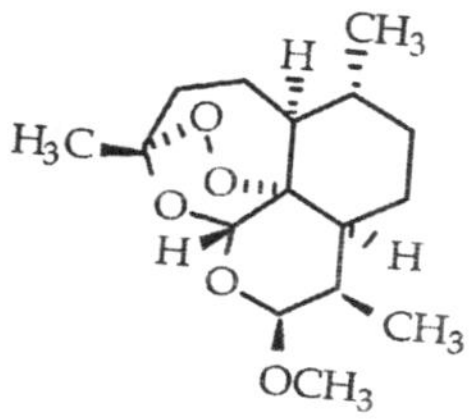

Artemether

Anticancer drugs

Natural products have provided the most important successes in the chemotherapy of cancer. Most of the major anticancer drugs are unmodified natural products obtained from plants or microorganisms, and include such important anticancer drugs as bleomycin, doxorubicin, daunorubicin, vincristine, vinblastine, mitomycin, streptozocin, and paclitaxel (Taxol™).

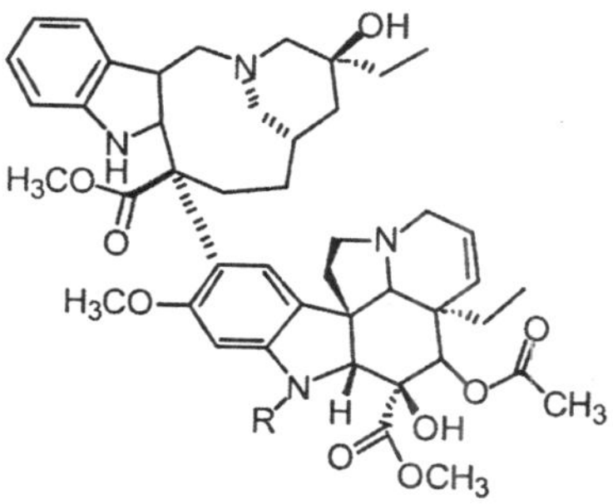

Vincristine, R = CHO
Vinblastine, R = CH_3

Ironotecan (a camptothecin derivative), and etoposide and tenoposide (podophyllotoxin derivatives) are examples of semisynthetic derivatives of natural products that are important anticancer drugs. Vincristine and vinblastine are complex, dimeric indole-indolines obtained from the rosey periwinkle *(Catharansus rosea),* and are among the most important therapies for the treatment of childhood leukemia, Hodgkin's disease, and metastatic testicular tumors. These unmodified natural products continue to be produced today by mass cultivation and processing of the plant material. Vinorelbine is a semisynthetic analog and has been reported to have decreased hematological toxicity.

As part of a programme undertaken over 40 years ago by the National Cancer Institute, which was aimed at examining higher plants as a source of anticancer agents, one of the most important discoveries of this century occurred. Wall and Wani reported that a complex diterpene, which they isolated from the bark of the Pacific yew tree and named taxol, was reported to possess significant cytotoxicity for cancer cells. Although it took some 15 years for the potential of paclitaxel to be realised, it is now recognised as a breakthrough in the treatment of ovarian breast cancers and one of the most exciting new drugs in recent history. Currently, both a semisynthetic derivative with improved water solubility, docetaxel (Taxotere®), and paclitaxel (Taxol®) are approved and used clinically.

Paclitaxel also serves as a vivid example of the utility of bioactive natural products as biochemical probes. Studies on the paclitaxel's mechanism of anticancer action revealed a previously unknown mechanism (stabilisation of microtubules). Using bioassays to detect this type of activity, two new structurally unrelated classes of natural products have now been discovered and shown to act by the same mechanism, and intense efforts are underway to identify drug candidates from these classes. Since paclitaxel was the only known compound to exhibit this activity, without its discovery and development this new target for anticancer drug discovery and development would likely not have been identified.

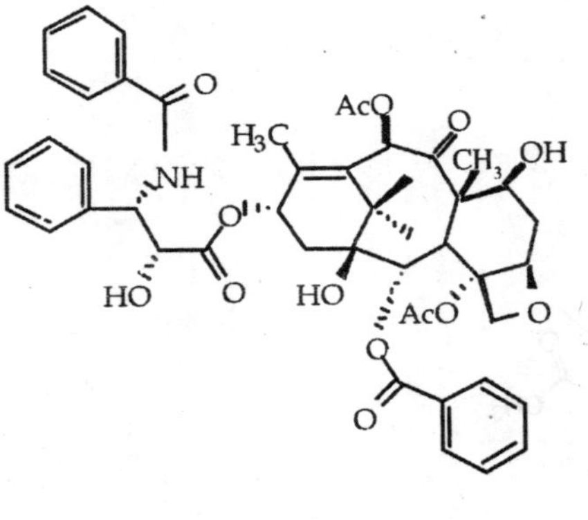

Paclitaxel

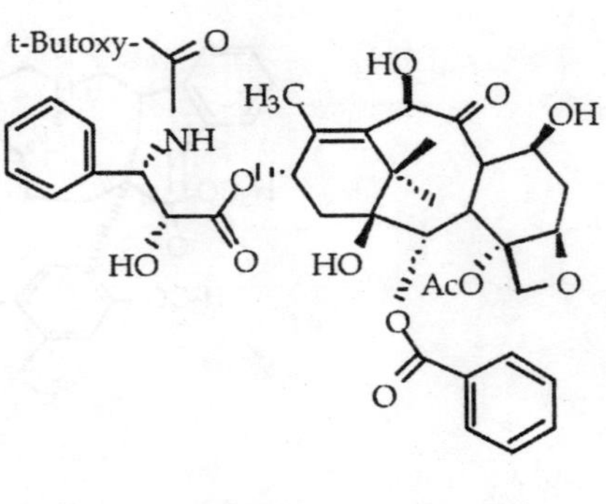

Docetaxel

Camptothecin, an alkaloid from the Chinese tree, *Camptotheca acuminata* Descne, was also discovered by Wall and Wani. Although showing promising antitumor activity, clinical studies were discontinued due to unpredictable side effects. Subsequent semisynthetic modifications led to the development of Ironotecan (Topotecin™, Campto™), a derivative of camptothecin that is now clinically available. The camptothecins act by inhibition of the topoisomerase I.

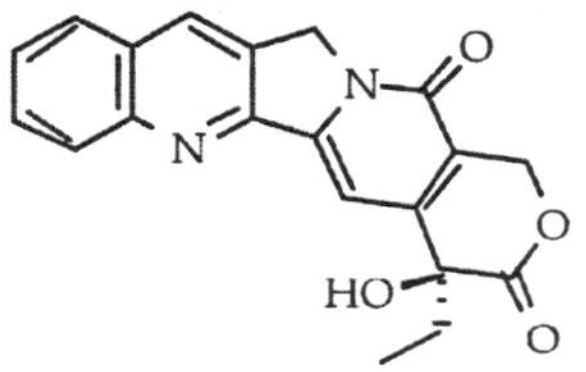

Camptothecin

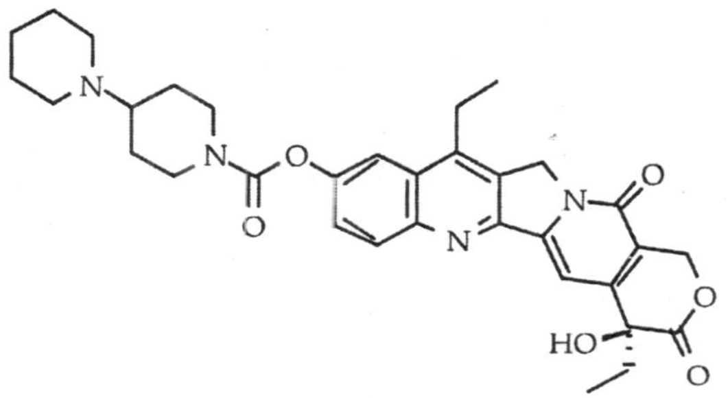

Ironotecan

Podophyllin is a crude resin derived from Podophyllum peltatum and is used topically to treat condylamata acuminata (warts). Podophyllin contains, among other things, the lignan, podophyllotoxin, which is the chemical precursor to two important semisynthetic anticancer drugs, etoposide and tenoposide. These drugs inhibit topoisomerase II, a mechanism quite different from that of podophyllotoxin (spindle poison), thus illustrating that structural similarity alone is not always a reliable predictor of similar biological effect.

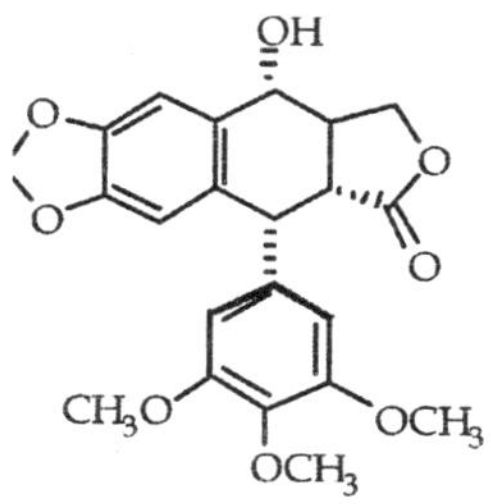

Podophyllotoxin

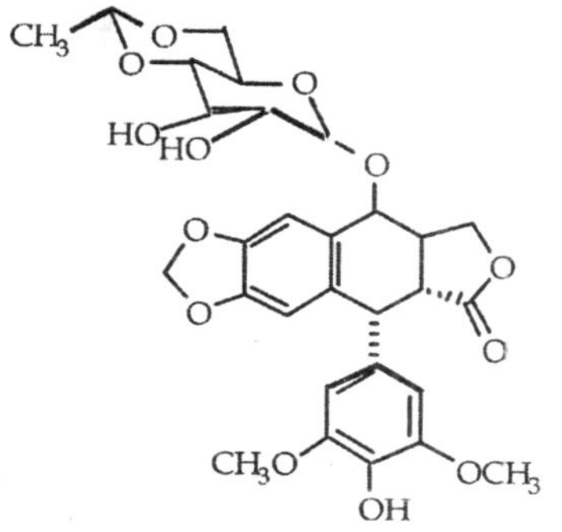

Etoposide

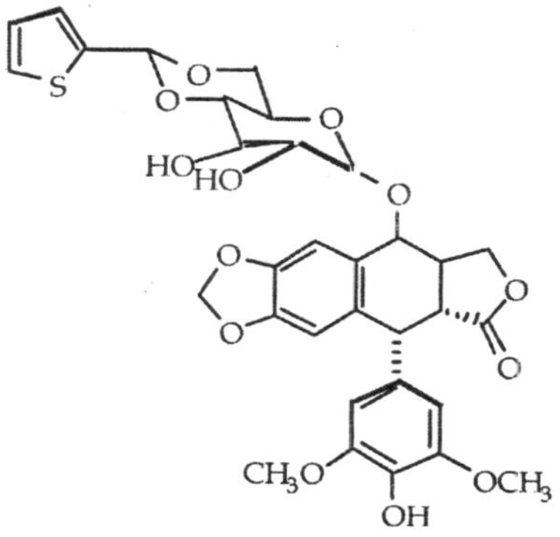

Teniposide

Cardiovascular drugs

It was known as early as 1776 that extracts of Digitalis were effective in controlling heart disease. However, the active constituents, complex glycosides such as digoxin, digitoxin, and the lanatosides, were not isolated and structurally characterised until almost a century later. These compounds exert a powerful and selective positive inotropic action on the cardiac muscle, and are important drugs in the treatment of congestive heart failure. Perhaps due to the complexity of their structures, which include numerous chiral centers, the unmodified natural products continue to be used clinically, and also continue to be produced by mass cultivation and extraction of foxglove.

In addition to the cardiac glycosides, a number of naturally occurring alkaloids are important drugs in the control of various cardiovascular conditions. For example, the alkaloid quinidine, from the bark of the Cinchona tree, is an important anti-arrhythmic drug. Other important natural products active as cardiovascular drugs include papaverine, a non-narcotic peripheral vasodilator, theophylline, an important bronchodilator used to control asthma in children, and the vasoconstrictive alkaloid ergotamine, which is obtained from a fungus that infects rye grass and is used clinically to treat migraine headaches. Cabergoline, a semisynthetic derivative of the ergot alkaloids, is a dopamine D_2 receptor agonist that is used as an antiprolactin.

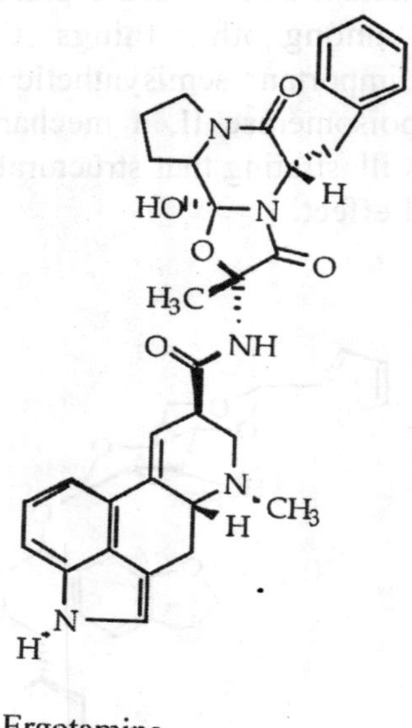

Ergotamine

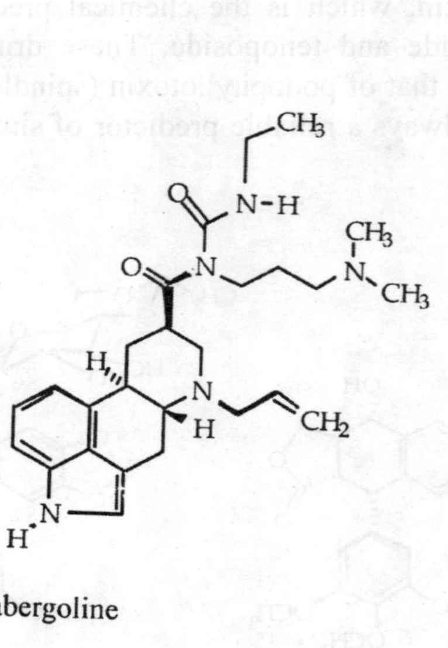

Cabergoline

CNS drugs

One of the oldest central nervous system (CNS) drugs in use is d-tubocurarine, a neuromuscular blocker derived from a plant (curare) used as an arrow poison by South

American Indians. Also, the opium alkaloids codeine and morphine are important analgesic drugs, and continue to be manufactured by processing opium exudate and extract. Delta-9-tetrahydrocannibinol (THC), the component of Cannibas sativa responsible for its CNS effects, is an important drug (Marinol®) used to reduce nausea associated with cancer chemotherapy. Although prepared commercially using synthetic methodology, Marinol® is identical to natural THC.

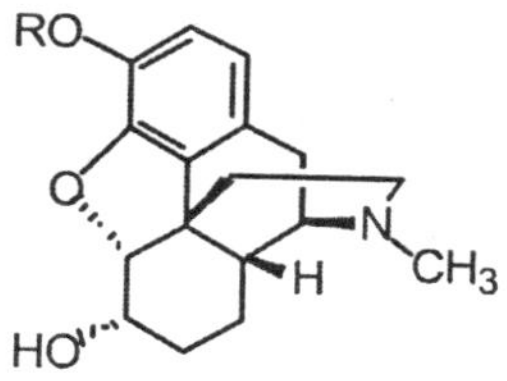

Morphine, R = H
Codeine, R = CH_3

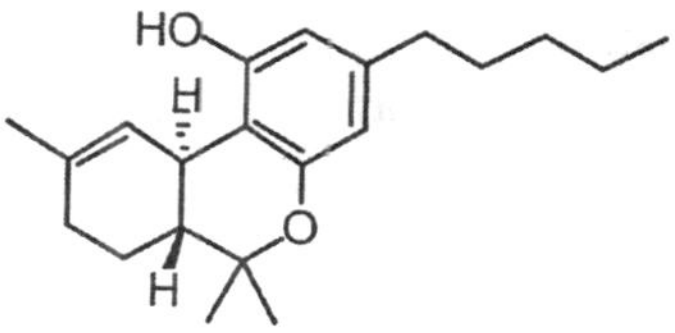

Δ^9-THC

Physostigmine, a naturally occurring alkaloid, and its carbamate ester, neostigmine, are acetylcholinesterase inhibitors used for the treatment of myasthenia gravis and as antagonists to neuromuscular blockade by nondepolarising blocking agents. Galanthamine, an alkaloid that occurs in the bulbs of daffodils, is also an acetylcholinesterase inhibitor and is currently in clinical trials as a possible therapy for cognitive impairment in Alzheimer's disease.

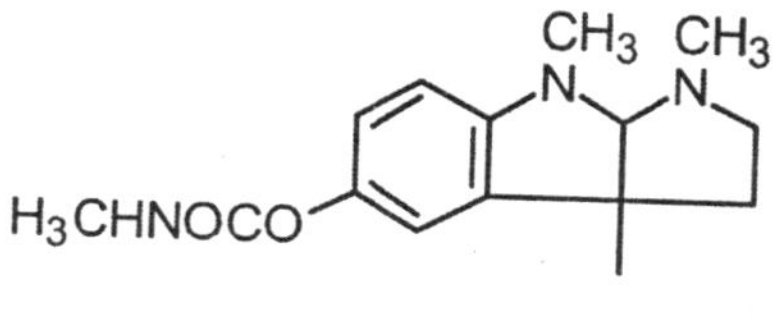

Physostigmine

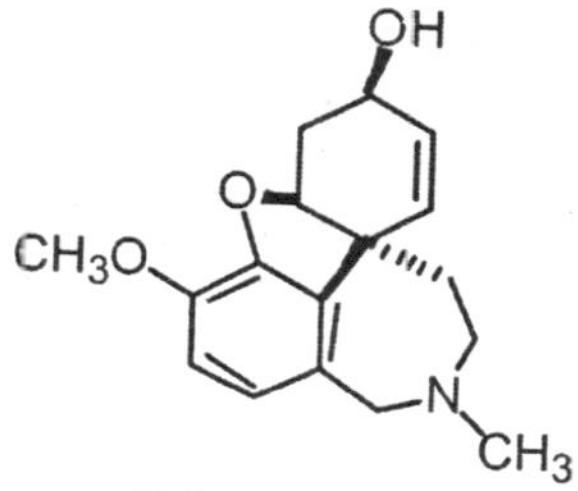

Galanthamine

Cholesterol-lowering agents

The clinically useful cholesterol-lowering agents known as the "statins" were derived from natural products isolated from a fungus. These drugs inhibit 3-hydroxy-3-methylglutaryl coenzyme A reductase (HMG-CoA reductase), an enzyme critical in the biosynthesis of cholesterol. The first such agent, compactin, was initially reported as an antifungal agent; however, once its mechanism of action was determined, a search for other naturally occurring

HMG-CoA reductase inhibitors led to the discovery of lovastatin, a metabolite of the fungus Aspergillus terreus that was first introduced to the market in 1989 followed by pravastatin in 1991. Many analogs, both semisynthetic and totally synthetic, were later prepared, and from these several have become important drugs, including simvastatin, pravastatin, and atorvastatin. The latter, a chiral totally synthetic compound, has become the drug of choice in this therapeutic category based on its superior ability to reduce cholesterol at low doses.

The development of the statins is a very good example of a natural productbased discovery of an important new drug class followed by optimisation of properties yielding improved drugs. It illustrates the interconnection between a natural product and its semisynthetic and totally synthetic analogs and the determining role of molecular structure, whether constructed by nature or humans, with respect to drug properties.

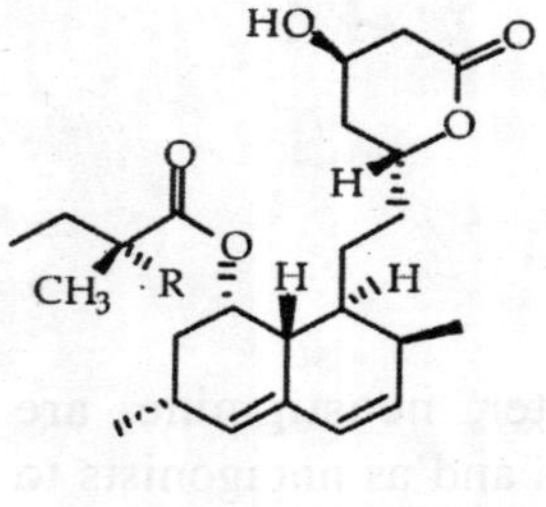

Lovastatin, R = H
Simvastatin, R = CH_3

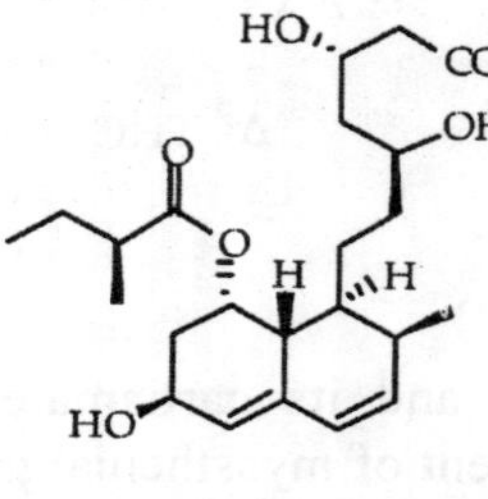

Pravastatin

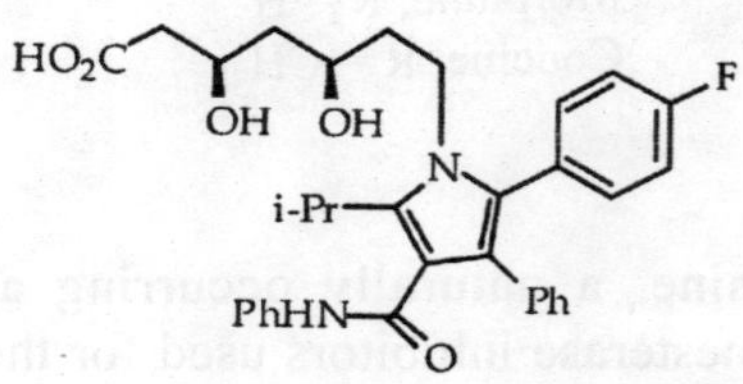

Atorvastatin

Immunomodulators

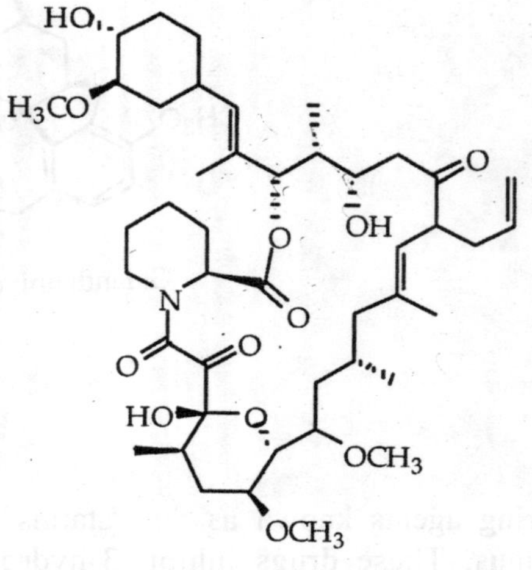

Tacrolimus

The immunomodulators cyclosporin, originally isolated from the soil fungus Trichoderma polysporum and tacrolimus (FK-506), a secondary metabolite of *Streptomyces tsukabaensis,* are used to suppress immunological rejection of the transplanted organs. These unmodified natural products represent major breakthroughs for organ transplantation.

Synthetic Origin

Anti-infectives

The important antimalarial drugs primaquine, chloroquine, and mefloquine were all patterned after the alkaloid quinine, the active constituent of the "fever tree" Cinchona succiruba, known for centuries by South American Indians to be effective in controlling malaria.

When the natural source of quinine was threatened during World Wars I and II, massive programmes to synthesize multitudes of quinoline derivatives based on the quinine prototype ensued, and the aminoquinolines cited above were among some of the successful drugs to emerge from this extensive effort. The effort to design better antimalarial agents has also led to the discovery of other important anti-infectives, including a class of synthetic antibacterials that is among the most prescribed in clinical use today-the fluoroquinolones.

The first of the quinolones to be used clinically, nalidixic acid, was synthesized as part of a large programme by the Sterling Drug Company to synthesize new antimalarials based on the quinine nucleus. The 1,8-naphthyridines were observed to be antibacterial, and in 1964, nalidixic acid became available for use in the United Kingdom for urinary tract infections. However, its limited spectrum of activity and poor pharmaceutical properties led to programmes to prepare thousands of other quinolone derivatives with improved features. More than 15 years later, in 1980, Koga and coworkers reported that structural analogs modified at C6 and C7 demonstrated certain improved properties.

In particular, it was noted that analogs carrying a fluorine at position 6 showed broader and more potent antimicrobial activity, and the class of fluoroquinolones subsequently became one of the most studied structural classes of antimicrobials. Ciprofloxacin and norfloxacin are examples of currently available fluoroquinolone antibacterials that are used extensively in both community and hospital settings.

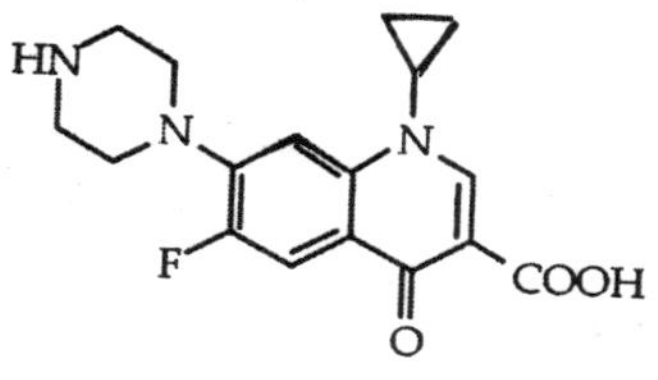

HN N F N O COOH

Ciprofloxacin

HN N F N CH_3 O COOH

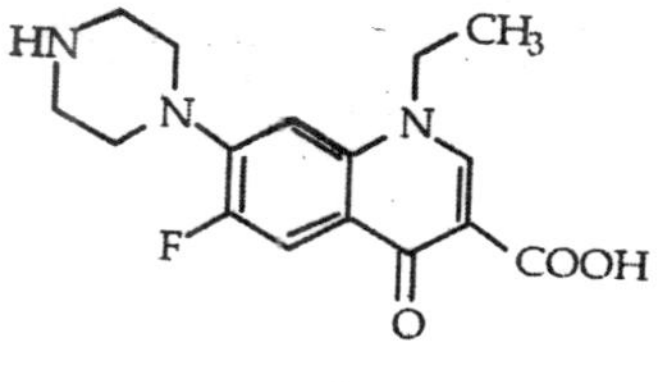

Norfloxacin

The HIV-1 protease inhibitors are another example of synthetic anti-infectives, but these agents have resulted from the rational design of inhibitors based on an understanding of the role of the aspartic protease enzyme in the life cycle of the virus. Results of early studies to design inhibitors of human renin (also an aspartic protease) for potential use in antihypertensive therapy were applied to the design of inhibitors of HIV protease. Saquinavir was the first HIV protease inhibitor to be introduced for clinical use.

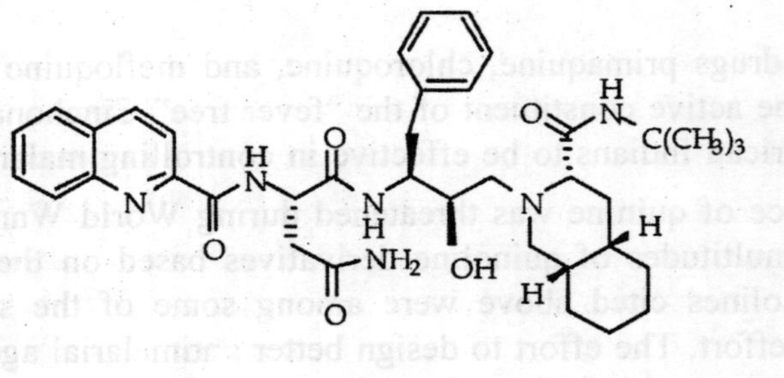

Saquinavir

Anticancer drugs

Included among the most important anticancer drugs in use today are tamoxifen and methotrexate, which are both synthetic drugs. Tamoxifen was derived from the diethylstilbesterol nucleus, which itself was patterned after estradiol. First reported for its contraceptive activity in rats, tamoxifen was later found to bind to human estrogen receptors, thus paving the way for its use in the treatment of breast cancer. Tamoxifen is now one of the most widely used and successful drugs in the treatment of breast cancer. Methotrexate is an antifolate, patterned after physiological folate, is one of the most widely used antineoplastic drugs available, and shows efficacy in the treatment of a variety of neoplasms.

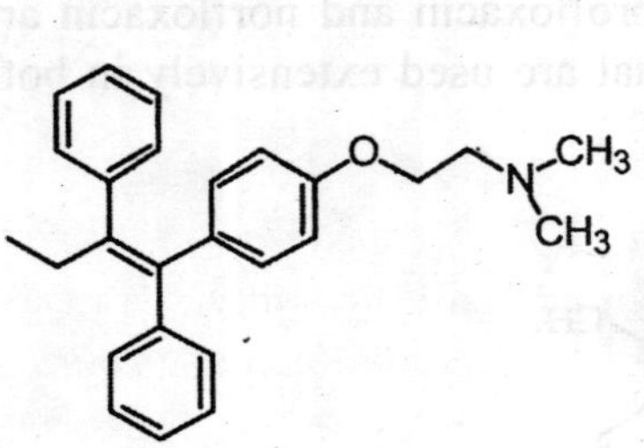

Tamoxifen

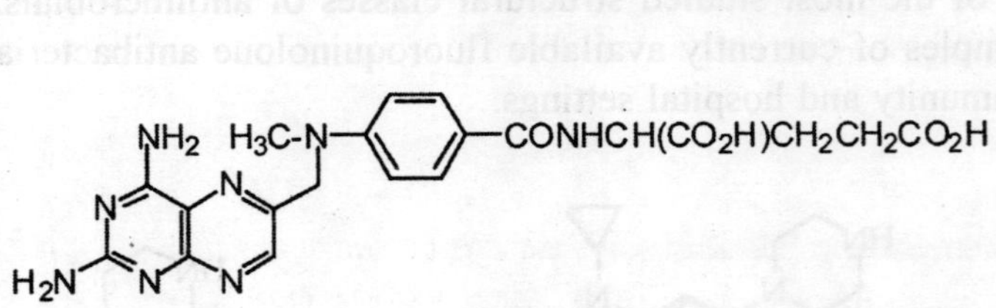

Methotrexate

Cardiovascular drugs

An examination of the pharmacological activity of tetropide, a component of the venom of the pit viper *(Bothrops jararaca)*, led to the discovery of the role of angiotensin-converting enzyme (ACE) in hypertension. Using a model system for the interaction of small peptides with the enzyme, captopril was designed as a specific, orally effective ACE inhibitor. Following the success of captopril for the control of hypertension, many additional ACE inhibitors, such as enalapril, were designed and synthesized. The ACE inhibitors now constitute one of the most important classes of cardiovascular drugs.

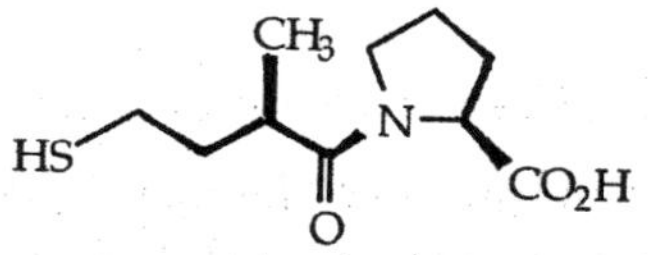

Captopril

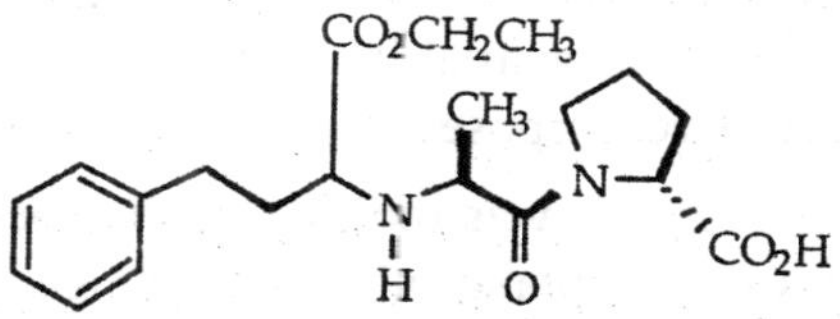

Enalapril

The development of specific agents that interfere with the renin-angiotensin system have defined the contribution of this system to blood pressure regulation and to the pathogenesis of hypertension, congestive heart failure, and chronic renal failure. This was first realised in the 1970s with the discovery of saralasin, a peptide antagonist of angiotensin II receptors. However, saralasin lacked oral activity, and at higher doses, exhibited partial angiotensin II agonist profiles.

The development of losartan, the first orally active angiotensin II receptor antagonist, derived from 1-benzylimidazole-5-acetic acid pharmacophores, represented a breakthrough in angiotensin II blockade. To date, six orally angiotensin II receptor antagonists are marketed in the United States, including, valsartan, irbesartan, candesartan, telmisartan, and eprosartan. All of these agents have identical mechanisms of action but differ in their pharmacokinetic profiles, which account for their differences in efficacy.

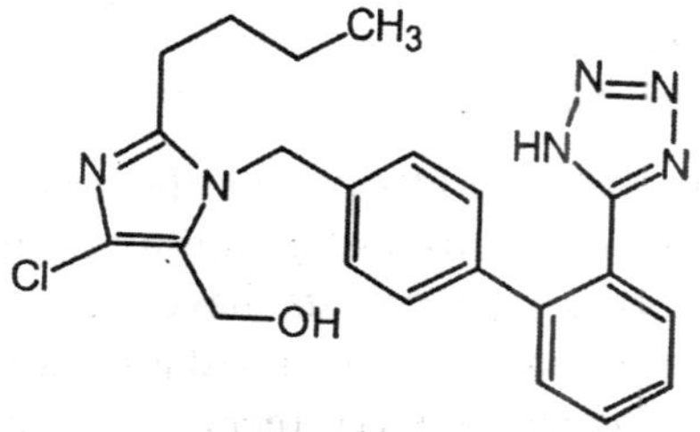

Losartan

CNS Arena

Totally synthetic drugs have been of the utmost importance in the CNS arena since the accidental discovery of the powerful analgetic properties of meperidine in 1937 and continuing through the synthesis of the currently widely used selective serotonin reuptake inhibitors as antidepressants. Along the way, the treatment of mental illness has been revolutionised through the discovery of CNS drugs such as the phenothiazines (chlorpromazine) and the butyrophenones (haloperidol) for psychotic disorders, the tricyclic antidepressants (imipramine) for depression, and the benzothiadiazines (diazepam) for anxiety. Of the numerous synthetic CNS drugs that are now available in medicine, just two of the newer ones will be briefly discussed for illustrative purposes.

Fluoxetine (Prozac™) was the first of the specific 5-HT reuptake inhibitors, also known as the selective serotonin reuptake inhibitors (SSRIs), to be discovered. A number of others have subsequently been introduced. The synthesis and pharmacological activity of fluox etine was first described by Wong and coworkers in 1975. Since its discovery, more than 5000 research papers have appeared on fluoxetine, and in her review of the pharmacology and uses of fluoxetine, Scott notes that the development of the SSRIs is one of the most significant milestones to take place with regard to the progress of psychopharmacology.

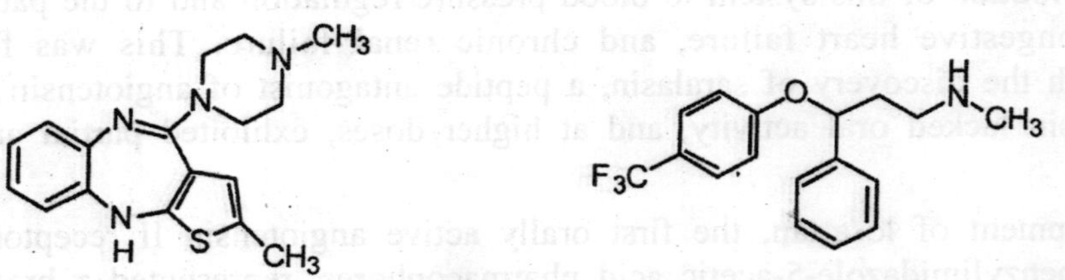

Olanzapine

Fluoxetine

Olanzapine (Zyprexa®) is a recently introduced atypical antipsychotic agent with a high affinity for dopaminergic and serotonergic receptors, and it also has high anticholinergic activity. A key advantage over older agents is that it is less likely to produce extra pyramidal side effects and does not produce granulocytopenia.

Histamine antagonists and proton pump inhibitors

H1 histamine receptor antagonists are used worldwide for the treatment of allergic rhinitis, conjunctivitis, and urticaria. Since the first clinically useful antihistamines were introduced in the 1940s, over 40 first-generation H1 receptor antagonists have been marketed. Adverse CNS effects and poor receptor specificity resulting in marked sedation and anticholinergic effects, have led to the development of second-generation H1 receptor antagonists largely devoid of these adverse effects. Loratidine and fexofenadine represent two of the more common second-generation agents prescribed. Other second-generation drugs of this class include acrivastine, astemizole, azelastine, cetirizine, ebastine, and mizolastine.

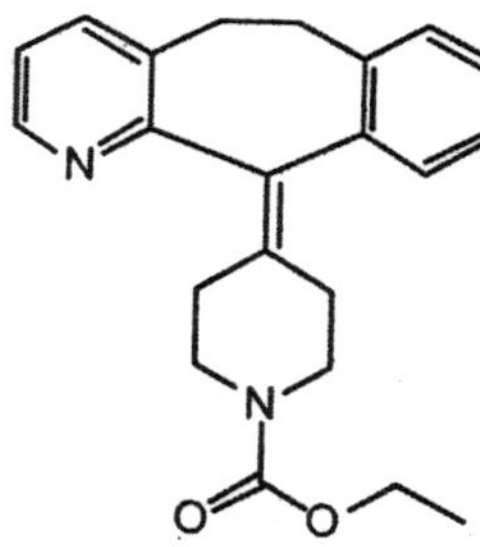

Loratidine

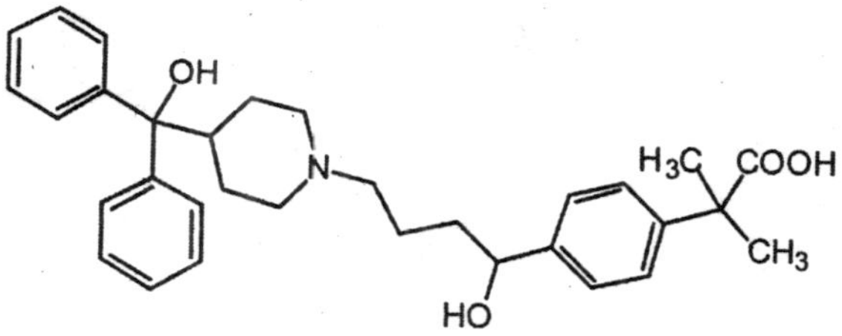

Fexofenadine

The development of cimetidine, the first clinically useful H2 histamine receptor antagonist indicated for the treatment of peptic ulcer disease approved by the U.S. Food and Drug Administration in 1977, clearly defined the role of medicinal chemistry in rational drug design. Cimetidine was constructed using histamine as the lead pharmacophore, and structural optimisation was accomplished by modulation of physicochemical parameters including partition coefficients and ionisation constants. Separation of agonist and antagonist properties initially led to the development of burimamide, metiamide, and finally cimetidine. Additional H2 receptor antagonists introduced later include ranitidine, famotidine, and nizatidine.

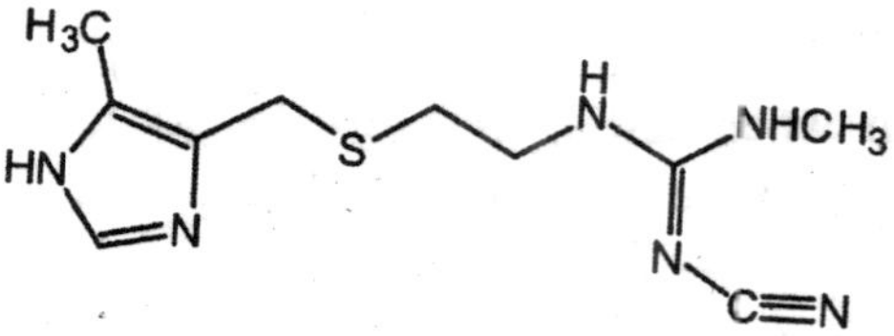

Cimetidine

Proton pump inhibitors represent a class of antisecretory compounds, the substituted benimidazoles, that are devoid of anticholinergic or histamine H2 receptor antagonist proprties, but suppress gastric acid secretion by specific inhibition of the (H^+,K^+)-ATPase enzyme system at the secretory surface of gastric parietal cells.

The proton pump inhibitors have emerged as a therapeutic alternative to H_2-antagonists for the treatment of gastric disorders, including gastric esophageal reflux disease (GERD). Omeprazole and related proton pump inhibitors (lansoprazole, pantoprazole, and rabeprazole) all share a benzimidazole framework, and are considered pro-drugs, since they are stable at neutral pH but rearrange in the presence of acid to form tetracyclic sulphenamides, which covalently inhibit the enzyme.

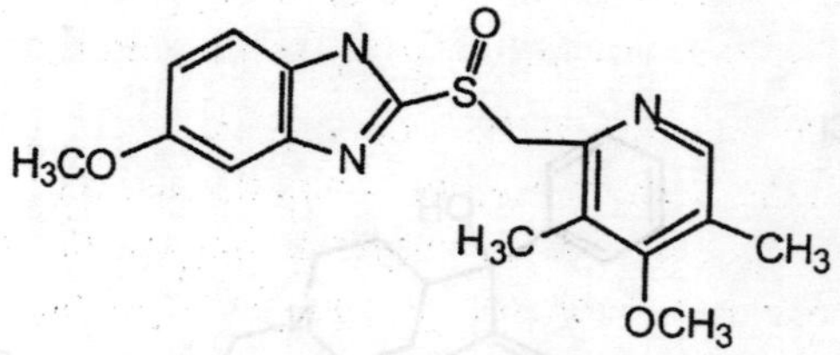

Omeprazole

Herbal Medicine Preparations (HMPs)

Herbal medicine preparations (HMPs), as they are generally known in most countries, are classified as botanical dietary supplements (BDSs) in the United States, where they are not subject to the same safety and efficacy regulations that apply to prescription and over-the-counter drugs. The United States is presently experiencing an unprecedented boom in their use. Between 1990 and 1997, purchases by the general population increased by 380 %. In 1998, the total market was worth USD 3.87 billion, and the herb with the highest annual percentage increase (2801 %) was St. John's wort.

Efficacy

St. John's wort will be used as an example of an HMP for which efficacy and safety have been investigated fairly rigorously. Its main indication is mild to moderate depression. Depression responds extremely well to placebo medications. Thus, it is particularly important to account for this phenomenon when testing the efficacy of St. John's wort for depression. Numerous placebo-controlled trials have been carried out, almost all in Germany where this herbal preparation now outsells Prozac® by about 2:1.

There is much debate about which are the active ingredients of St. John's wort. Even though no final conclusions can be drawn as yet, it is clear that more than one family of active principles is responsible. It seems certain that hypericin and hyperforin both contribute to the antidepressant action. The relative size of these contributions to the total pharmacological action remains uncertain. A systematic review of all trials available by 1996 located 23 placebo-controlled trials.

Collectively, these studies included 1757 patients. A meta-analysis of their results showed that St. John's wort was significantly more efficacious than placebos in treating depression. The chances of benefiting from the herbal remedy were about 2.5 times higher than those for placebo. Eight studies were found where St. John's wort was tested against conventional antidepressants. A meta-analysis of these data suggested that the herbal medicine is equally as effective as imipramine, maprotiline, bromazepam, amitriptyline, or diazepam.

Even though this sounds highly encouraging, several important caveats ought to be pointed out:

— All of these studies were on mild to moderate depressions, and experts are not unanimous about whether these required medication in the first place.

— All of the studies were of relatively short duration (usually 4-6 weeks), and long-term effects remain uncertain.

— There is insufficient information as to relapse rates of patients treated with St. John's wort.

— In the comparative trials, the conventional antidepressants were usually underdosed; thus therapeutic equivalence has not been established beyond reasonable doubt.

— No trials exist that compare St. John's wort against the newer synthetic antidepressants like Prozac®.

Since the publication of this meta-analysis, six further studies have been published. These new data have been recently reviewed. Essentially, the new trials confirm the previous results. Some evidence emerged that St. John's wort is as effective as the newer synthetic antidepressants and that it also works in the more severe types of depression. Yet, so far, this evidence is not fully convincing. A large U.S. study organised by the National Institutes of Health (NIH) is on its way, which is expected to provide a definitive answer.

St. John's wort was chosen here as an example. Other HMPs for which similarly compelling evidence of efficacy (i.e., systematic reviews) has been published include the following:

— ginkgo biloba for cerebral insufficiency and dementias

— saw palmetto for benign prostate hyperplasia

— horse chestnut seed extract for primary varicosis

— peppermint oil for irritable bowel syndrome

Ginseng is a good example of a widely used HMP that is not backed up by convincing evidence of efficacy. It achieved retail sales of USD 96 million in the United States in 1998 as a BDS, placing it third in this category. Considerable confusion has been caused by not distinguishing between Siberian ginseng (Eleutherococcus senticosus), Asian ginseng (Panax quinquefolius), and Japanese ginseng (Panax japonicus). Panax ginseng species have been in traditional use for alleged sedative, hypnotic, demulcent, aphrodisiac, antidepressant, and diuretic activity. Moreover, it is often recommended to improve stamina, concentration, vigilance, and well-being.

The pharmacological activities of the Panax ginseng species range from stimulation of the central nervous system to modulation of the immune defense and anabolic effects. The effects of Siberian ginseng are less well researched but claimed to be similar to those of Panax ginseng. A systematic review of all randomised, placebo-controlled, double-blind trials included 16 studies of all types of ginseng relating to enhancement of physical or mental performance and immunomodulation. Most trials were methodologically poor, with many studies using healthy volunteers rather than patient populations. The results were, in general, contradictory, and none of the above-named indications were supported by evidence from sound clinical trials. A further review of animal and human studies of ginseng as an erogogenic aid to physical performance concluded that there is an absence of compelling research evidence regarding the efficacy of ginseng for this indication.

Safety

Consumers turn to HMPs because they (are made to) believe that they are inherently free of adverse events. Natural is all too often and all too quickly equated with harmless. This notion is misleading at best and dangerous at worst. The argument of the "test of time" is often used: anything that has been used for hundreds of years has to be safe. There are, however, numerous reasons why this argument is a fallacy.

The only HMP that has been systematically reviewed in relation to adverse drug reactions (ADRs) is St. John's wort. Searches of four computerised literature databases were performed for records of ADRs. Manufacturers of hypericum products, the international drug-monitoring center of the World Health Organisation (WHO), and the national drug safety monitoring bodies of Germany and the United Kingdom were also contacted for information. Information on ADRs originated from case reports, clinical trials, post-marketing surveillance, and drug-monitoring studies.

Collectively, these data suggest that hypericum is well tolerated, with an incidence of adverse reactions similar to that of placebo. The most common adverse effects are gastrointestinal symptoms, dizziness/confusion, and tiredness/sedation. A potential serious ADR is photosensitivity, but this appears to occur extremely rarely. This review concluded that hypericum has an encouraging safety profile. However, as most of the current data originate from short-term investigations, more long-term studies would be desirable.

An analysis was carried out to compare the nature and frequency of ADRs of St. John's wort to a range of synthetic antidepressants. This confirmed the safety of this HMP and suggested that ADRs of St. John's wort have an incidence rate that is 50 % or less than that associated with conventional antidepressants. Similarly positive, albeit less thoroughly researched, evidence exists for the other above-mentioned HMPs.

A number of relatively serious adverse effects of Panax ginseng have been reported, ranging from insomnia, diarrhea, vaginal bleeding, and mastalgia to severe headache, schizophrenia, and Stevens-Johnson syndrome. The incidence of these adverse effects seems to be low. Also, a recent study suggested that an oral dose of 3 g of American ginseng attenuated the postprandial glycemic response to a 25 g oral glucose challenge. The effect was notable both in nondiabetic subjects and in individuals with Type 2 diabetes. The authors therefore caution that American ginseng could cause hypoglycemia in nondiabetic subjects.

Yet, other HMPs are clearly burdened with serious ADRs. Examples from this category include: aconite, aristolochia, broom, chaparall, comfrey, germander, liquorice root, and pennyroyal. Plants belonging to the Aristolochia species are rich in aristolochic acids and aristolactams. Aristolochic acids I and II have been demonstrated to have mutagenic potential in several test systems.

A 49-year-old woman was admitted to hospital with clinical signs of hepatitis where this diagnosis was confirmed. The usual causes for hepatitis were ruled out. The medical history revealed that she had recently started to use a Chinese herbal tea to treat her eczema. Examination of the herbal mixture showed that it contained Aristolochia debilis root and seven other medicinal plants. Like Aristolochia fangchi, Aristolochia debilis contains aristolochic acid and is therefore the likely cause of the toxic hepatitis.

In Belgium, nephropathy in about 100 individuals has been attributed to an herbal slimming preparation that included the Chinese herbs Stephania tetranda and Magnolia officinalis. In the preparation, the root of Stephania tetranda (Chinese name "Fangji") was, in all probability, substituted for or contaminated with the root of Aristolochia fangchi (Chinese name "Guang fangji"). The nephropathy was characterised by extensive interstitial fibrosis with atrophy and loss of the tubules. Several patients showed evidence of urothelial malignancy. The preparation was also distributed in France, and two cases have been reported from Toulouse and one possible case from Nice. Contamination is an important safety issue with HMPs. This is especially relevant for Asian HMPs, which have been shown to be contaminated with synthetic drugs and/or heavy metals with an alarming degree of regularity, e.g., ref.. These reports stress the importance of proper quality control of these preparations.

Another issue is herb/drug interactions. This area is grossly under-researched, yet it is important, particularly since we know that users of HMPs tend to belong to that fraction of society that would employ them alongside prescribed drugs. Judging from what we know to date, it is certain that such interactions exist, e.g., ref.. This is perhaps the most urgent issue for future research. For example, important herb/drug interactions are possible if St. John's wort is administered concomitantly with other medications. It acts as a hepatic enzyme inducer via an interaction of the cytochrome P450 system. In addition, it probably activates P-glycoprotein, which further increases the elimination of synthetic drugs. Through these mechanisms, it can decrease the plasma level of a range of prescribed drugs (e.g., anticoagulants, oral contraceptives, and antiviral agents), which can have clinically serious consequences.

Vitamins and Nutrients

Because of the enormous increase of the world population, in particular in the developing countries, the problem of providing people with sufficient food of balanced composition presents a major challenge. Besides the major constituents, fats, carbohydrates, and proteins, other food components such as vitamins, carotenoids, essential fatty acids, dietary fibers, and others play an important role. On the other hand, serious health problems arise nowadays among populations of industrialised countries due to excessive consumption of food with a high-fat and/or sugar content.

The modern trend toward new types of food (health food, fast food, low-calorie food, etc.) in affluent societies requires special methods of processing, supplementation with essential nutrients, and storage to assure good quality and the consumer's acceptance. For example, many food products have added vitamins to increase nutritive value, protect from oxidative degradation, and serve as colorants.

Vitamins

Vitamins are essential nutritional factors that are contained in food as small organic compounds and are required for the normal growth, maintenance, and functioning of the human body. Historically, they have been divided into water-soluble (B_1, B_2, B_6, B_{12}, biotin, C, folic acid, nicotinamide, pantothenic acid) and fat-soluble (A, D, E, K_1) vitamins. The water-soluble vitamins except vitamin C are coenzymes or precursors of coenzymes that catalyse, in the presence of specific enzymes, many biochemical reactions in the metabolism of carbohydrates, proteins, lipids, and nucleic acids.

Others do not function as coenzymes, but are involved in the visual process, growth, and cell differentiation (vitamin A and the derivative retinoic acid, ß-carotene as provitamin A), intestinal absorption of calcium and bone mineralisation (vitamin D). They also serve as antioxidants by scavenging free radicals to protect the organism from oxidative damage (vitamin E). Water-soluble vitamins are poorly stored, and high intake levels lead to excretion in urine, while fat-soluble vitamins can be stored in the body. Therefore, hypervitaminosis, mainly of vitamins A and D, can lead to toxic manifestations caused by excessive intake.

Throughout the history of mankind, deficiencies of nutritional factors-especially vitamins (hypovitaminosis)-and hence disturbance of many metabolic functions, have been a major cause of diseases (e.g., scurvy, pellagra, beriberi, pernicious anemia, blindness, etc.) and death. In fact, vitamins were originally recognised by signs of deficiency. Administration of vitamins reverses the majority of the clinical symptoms. In humans, levels of recommended dietary allowances (RDAs) have been established based on the estimated needs of healthy people. These levels are thought to prevent signs of clinical deficiency, however, they may not be adequate to cover the increased vitamin requirements in humans that smoke, suffer from diseases, or have metabolic or genetic disturbances.

Vitamin C

There have been many clinical studies on the effect of vitamin C on the common cold and the reported results have shown wide variation, most likely due to the different protocols used. However, a decrease in the duration and severity of cold symptoms has been consistently observed for large doses of vitamin C (1-3 g/d, USA RDA for adults is 60 mg/day). Vitamin C is the most efficient antioxidant in human blood. Since the concentration of vitamin C decreases in plasma and leukocytes in episodes such as infections, vitamin C acts as an antioxidant by inactivating excessive amounts of radicals formed during the oxidative burst of phagocytes. These radicals destroy viruses and bacteria in the host, but also induce inflammatory processes that are reduced by vitamin C. The importance of vitamin C as part of the defense system is also shown by the very low concentrations of vitamin C in the plasma of critically ill patients.

Vitamin E

There is experimental evidence that the oxidative modification of low-density lipoproteins (LDLs) is a key process in initiating atherosclerosis. Lipid-laden macrophages (foam cells) containing oxidised LDLs are found in early lesions called fatty streaks. These findings led to the formulation of the antioxidant hypothesis of atherosclerosis. LDL particles carry natural antioxidants (vitamin E, β-carotene, lycopene, ubiquinol), which can inactivate free radicals and hence protect LDLs from oxidative damage.

Epidemiological studies have shown a correlation between high levels of vitamin E in plasma and a lower risk of coronary heart disease. In randomised, placebo-controlled trials of a-tocopherol (vitamin E) in relation to cardiovascular mortality, different results have been reported. In the Alpha-Tocopherol Beta Carotene study, no effect was seen.

The dose of 50 mg of vitamin E per day may have been too low to give treatment effects. However, in the GISSI-Prevenzione trial, using higher doses of vitamin E (300 mg/d), the mortality group was similar to the placebo group. In contrast, the rate of nonfatal myocardial infarctions was significantly reduced in the CHAOS study using 400 and 800 IU of vitamin E per day. Vitamin E is nontoxic and in animal experiments has been shown to be neither mutagenic, teratogenic, nor carcinogenic.

B-vitamins

In the early 1960s, a new, rare disease, homocystinuria, was found to be associated with an inborn error of metabolism due to the absence or reduced activity of the enzyme cystathionine ß-synthase. Elevated levels of homocysteine are found in the plasma and urine of these patients. Characteristic features of patients with homocystinuria are premature vascular disease and death. Homocysteine is formed as an intermediate in the transsulfuration pathway where methionine is converted to cysteine. B-vitamins are involved in many of these enzymatic reactions, and a low intake of these vitamins is correlated with elevated homocysteine levels, especially in the elderly. From these observations, it was hypothesised that homocysteine might be involved in the pathogenesis of arteriosclerosis.

The analysis of many studies on homocysteine and cardiovascular diseases led to the conclusion that a moderate level of homocysteine in blood is an independent, graded risk factor for atherosclerotic disease in coronary, cerebral, and peripheral vessels. Results from the recently published European Concerted Action Project show a dose-response effect between the increase of total homocysteine and the risk of coronary disease. In addition, an inverse relationship could be shown between levels of homocysteine in plasma and mortality of patients with coronary artery disease.

Production of Vitamins

Vitamins are produced on an industrial scale either by chemical synthesis, fermentation, or extraction from natural sources. There has been public concern that vitamins produced either by chemical synthesis or fermentation may be different from vitamins present in fruit or vegetables. However, the majority of industrially produced vitamins are chemically identical, "nature-identical", with the corresponding substances isolated from plants. Generally, the industrially produced vitamins are either nature-identical compounds, derivatives that are transformed into nature-identical forms after hydrolysis in tissues, or they consist of several stereoisomers.

A typical example of a nature-identical vitamin is vitamin C with the official designation L-ascorbic acid, which is produced worldwide on a scale approximating 60 000 tons per year. It is completely identical with the compound isolated from natural sources such as citrus fruits and is widely used in the food and pharmaceutical industries. Vitamin C has two chiral centers at C-4 and C-5. The stereoisomer erythorbic acid (D-araboascorbic acid or D-isoascorbic acid) has a negligible vitamin C activity but similar antioxidant properties to L-ascorbic acid. Erythorbic acid does not occur in nature and is nontoxic. Worldwide, about 5500 tons of erythorbic acid/sodium erythorbate are produced annually. It is used as a substitute for vitamin C as an antioxidant, for instance, to inhibit the formation of nitrosamine in cured meat.

The stability of some vitamins is unsatisfactory when stored over extended periods of time. However, a dramatic increase of stability can be achieved by transforming certain vitamins into their appropriate derivatives. These modifications protect the active molecules from degradation, e.g., by atmospheric oxygen and moisture. Examples of commercially available synthetic derivatives are alltrans-retinyl acetate or RRR-a-tocopheryl acetate. When ingested, these compounds are hydrolysed by enzymes such as pancreatic lipase and retinyl ester hydrolase in the gastrointestinal tract producing nature-identical compounds.

An intense discussion concerning natural vs. synthetic vitamins has been focused on a-tocopherol, which is the most active and most important membcr of the vitamin E group. Vitamin E is a general term employed for tocopherols and the structurally related tocotrienols. They are highly lipophilic and protect membranes and lipoproteins from oxidative damage. a-Tocopherol is used in large quantities in humans for medical purposes, in animal husbandry, and to protect food from oxidative degradation.

The homologs, β-, γ-, and δ-tocopherol, having a lower degree of methylation of the chromanol ring system, are less biopotent. These mixed tocopherols occur in substantial quantities in vegetable oils such as soybean oil or palm oil distillates and can be isolated from the residues, which are available on a large scale from refining processes. From these β-, γ-, and δ-tocopherol homologs, nature-identical RRR-a-tocopherol can be obtained by chemical transformation.

The naturally occurring α-tocopherol (RRR-α-tocopherol or d-α-tocopherol) has three chiral centers. The chemically synthesised all-rac-a-tocopherol (d,l-α-tocopherol) is racemic at all three chiral centers and is, therefore, a mixture of eight stereoisomers. The usual synthetic process is condensation of racemic isophytol with 2,3,6-trimethylhydroquinone and is carried out on a scale of about 20 000 tons per year worldwide.

Using the rat gestation-resorption test, the activity ratio of the biopotencies of all-rac-α-tocopherol and RRR-α-tocopherol amounted to 1:1.36. All eight stereoisomers have been synthesised and tested for their biological activity. The RRR form is the most active (100 %), and the SSR form is the least active (21 %). It appears that the natural form, RRR-α-tocopherol, is preferentially bound and retained in the human body, whereas the synthetic vitamin E, all-rac-a-tocopherol, is metabolised at a higher rate, and the metabolites are more rapidly excreted in the urine.

Carotenoids

The term carotenoid includes both pure hydrocarbons and oxygenated derivatives. Carotenoids are synthesised by photosynthetic bacteria, yeasts, and plants, where they function as accessory pigments in photosynthesis and photoprotection. Because of the enormous variety of chemical structures, they are the most important group of pigments in nature. Most of the naturally occurring carotenoids exist as alltrans isomers.

Due to their conjugated polyene structures, carotenoids can absorb light and quench various energetically excited molecules and inactivate free radicals, particularly in lipid peroxidation by inhibiting the propagation steps. ß-Carotene can be enzymatically transformed by reduction to retinol (vitamin A) or oxidised to retinoic acid, which is important for growth and cell differentiation. Carotenoids capable of yielding vitamin A are termed provitamin A carotenoids. This conversion of ß-carotene to vitamin A is regulated by the vitamin A status of humans, and, therefore, consumption of large doses of ß-carotene does not cause vitamin A hypervitaminosis and toxicity.

A very interesting development with far-reaching health consequences is the genetic manipulation of staple foods such as rice to express ß-carotene. Insufficient supply of either vitamin A or ß-carotene results in eye diseases such as night-blindness, xerophthalmia, and eventually blindness. In addition, diarrhea, respiratory diseases, and viral infections (measles) have been observed in vitamin A-deficient subjects. Particularly in Southeast Asia, where rice represents the main staple food, this project could have enormous impact on the health of the population.

At present, partial and total chemical synthesis are the main manufacturing methods to synthesise the many carotenoids. To produce these on an industrial scale, sophisticated technical syntheses have been worked out yielding nature-identical products. Carotenoids are used for the coloration of various food products such as margarine to achieve the level of pigmentation demanded by consumers. Naturally occurring carotenoids can be extracted from various sources, but the high production cost of natural ß-carotene, for example, limits its use as a colorant to health foods and related products.

Although many in vitro and in vivo studies claim beneficial health effects of carotenoids (e.g., on ischemic heart disease, stroke, immunomodulation, cancer, cataracts, macular degeneration), the biological functions of carotenoids in humans are not yet fully understood. ß-Carotene and/or canthaxanthin have been used successfully to treat certain genetically inherited diseases in humans such as erythropoietic protoporphyria. No serious toxicological problems were found in these patients after long-term use of high doses of ß-carotene (180 mg/d).

The carotenoids lutein and zeaxanthin appear to be important in the prevention of age-related ocular diseases such as macular degeneration and senile cataracts. They may absorb blue light, the most energetic part of sunlight, thus protecting photoreceptors and the retinal pigment epithelium and function as antioxidants limiting oxidative damage. Several commercial products containing lutein are available as supplements on the American market to lower the risk of macular degeneration.

Epidemiological observations have shown an inverse relationship between intake of green/ yellow vegetables containing ß-carotene and the incidence of certain types of cancers. However, in two randomised, placebo-controlled intervention trials, ß-carotene supplements unexpectedly caused a higher incidence of lung cancer in smokers and/or asbestosexposed workers. In contrast, the results of a study involving a large group of male U.S. physicians showed no effect of ß-carotene on cancer or cardiovascular diseases. As a consequence, the use of ß-carotene to limit these diseases remains questionable.

Fats

The excessive energy consumption (fat and sugar) in the affluent societies of Europe and North America and, hence, the increasing incidence of overweight and obesity causes many health problems. Chronic diseases related to obesity are diabetes mellitus, arthritis, hypertension, some types of cancer, and coronary heart disease with increased mortality. Furthermore, childhood obesity is the beginning of chronic diseases that become apparent later in the adult population. Among the many risk factors of coronary heart disease and stroke identified so far, is a high intake of saturated fatty acids (animal fats) which raises the level of blood lipids, low-density lipoproteins, and triacylglycerides. This is associated with atherosclerosis (formation of fatty streaks and fibrous plaques in blood vessels supplying the heart and brain) and increased prothrombotic tendency for platelet aggregation (thrombus formation).

The current Western diet contains high amounts of total fat, trans fatty acids and the *n*-6 and *n*-3 essential fatty acids in an unbalanced composition. As recommended by health authorities, dietary intake of total fat should not exceed 30 % of the total daily calorie intake, but Americans, for example, obtain about 35-40 % of their energy from fat. The energy per gram of dietary fat is 9 kcal. Although recommended daily allowance (RDA) values for *n*-3 and *n*-6 polyunsaturated fatty acids (PUFAs) have not yet been established, the advised intake of PUFAs should be below 8 % with a ratio of *n*-6 to *n*-3 PUFAs between 5-3:1. Since foods that combine fat and sugar are universally preferred by humans of all ages, it seems to be necessary either to reduce, balance the components of, or partially substitute dietary fat intake. In response to consumer demand and the perceived health benefits of lower fat consumption, products containing naturally occurring PUFAs or fat replacers are being developed and marketed.

Generally, fats (lipids, triglycerides) are composed of one molecule of glycerol and three chains of either saturated, monounsaturated, or polyunsaturated fatty acids differing in chain length. PUFAs are grouped into classes depending on the position of the first double bond from the methyl end of the molecule. Accordingly, they belong to the *n*-3, *n*-6, or *n*-9 class of PUFAs with *n*-3 and *n*-6 as the main classes. Linoleic acid (*n*-6) and a-linolenic acid (*n*-3) are essential fatty acids, which have to be provided by the diet as in the case of vitamins. Others can be enzymatically synthesised in tissues using elongases and desaturases to obtain the desired chain length and number of double bonds.

The main members of the *n*-3 PUFAs, eicosapentanoic acid, and docosahexanoic acid, are derived from a-linolenic acid, while the *n*-6 PUFA, γ-linolenic acid, dihomo-?-linolenic acid, and arachidonic acid, are synthesised from linoleic acid. They are structural components of phospholipids in cell membranes and serve as precursors for prostaglandin, eicosanoid, and leukotriene biosynthesis. A number of studies have shown that ingestion of *n*-3 PUFAs reduces cardiovascular morbidity and mortality. Greenland Eskimos and Japanese consuming high amounts of fish and hence *n*-3 PUFA have an increased incidence of hemorrhagic stroke. However, moderate fish oil intake does not reduce platelet aggregation and therefore does not enhance bleeding tendency.

Generally, animal fats contain mainly saturated fatty acids. Vegetable oils (sunflower, linseed, rapeseed, soybean, corn) are rich in *n*-6 PUFA, while green leafy vegetables (e.g., lettuce, spinach, cabbage) and fish liver oils or fish (e.g., herring, salmon, sardine) are relatively rich in *n*-3 PUFA. A wide range of processed foods such as margarine, fat spreads, salad dressings, yogurt, and milk drinks are based on vegetable oils containing PUFAs or are fortified with a single isolated PUFA. The recommended daily intake of *n*-3 PUFA is about 1.2 g.

Fat Replacers

Generally, fat is important both for the consistency and texture of food and as a carrier and conveyer of flavor because many flavor and aromatic components are lipid-soluble. Fat replacers should, therefore, have similar organoleptic characteristics. These products should be safe and physiologically inert, while maintaining functional (appearance, hydrophobicity, temperature behavior), and sensory properties of common high-fat products. They mainly stimulate sensory properties. Influences on other nutrients in the gastrointestinal tract are generally minimal. However, some materials (cellulose, certain gums) are resistant to digestion and may affect absorption of micronutrients such as minerals, fat-soluble vitamins, and carotenoids.

Based on their chemical type, fat replacers can be grouped in three classes of products: (1) lipid-based compounds (e.g., fatty acid esters of sugars and sugar alcohols such as Olestra®, branched and asymmetric triglycerides, polycarboxylic acid and propoxylated glycerol esters, retrofats, specific naturally occurring lipids); (2) proteins (whey protein concentrate, microparticulated proteins, e.g., Simplesse®); and (3) carbohydrates (modified starches, inulins, water-binding and jelling substances, dextrins, polydextrose, modified glucose polymers, dietary fibers such as gums, pectin carrageenan).

Many of the lipid-based substances are resistant to hydrolysis by pancreatic lipases in the gut and are excreted intact. Therefore, they have no metabolisable energy. Examples are retrofats (inverse fats), which are esters of carbonic acids such as citric acid and long-chain alcohols.

Asymmetric triglycerides such as Caprenin® and Saltrim®, containing medium- and long-chain fatty acids, are fat-like in their functional properties with diminished energy values. They are hydrolysed in the gastrointestinal tract, but only partially absorbed due to their extended chain lengths. The caloric value of Caprenin® is 20 kJ/g (5 kcal/g) compared with 36 kJ/g (9 kcal/g) for common dietary triglycerides. It has GRAS (generally regarded as safe) status and is on the U.S. market.

Other products derived from natural food sources are carbohydrate- and protei*n*-based materials that are partially or fully digested and absorbed. Examples are gums (carrageenan, xanthan, alginates, guar) that are used in cream toppings and candies. Derivatives of starches from potatoes, tapioca, and corn obtained by enzymatic or acid hydrolysis are used in salad dressings, frozen desserts, table spreads, dips, or confections. In addition, products derived from hemicellulose and soluble fibers are used as water-soluble bulking agents. Gelatin and whey products provide stability and improve mouth feel. Pòlydextrose is produced by polymerisation from glucose and has no effect on absorption and metab-olism of essential nutrients. It is used for salad dressings, confections, frostings, and frozen dairy deserts.

Carbohydrate fatty acid polyesters are synthetic fat substitutes that çan be obtained by esterification of fatty acids and mono-, oligo-, or polysaccharides with a wide range of physical and biological properties. Sucrose polyesters lower levels of blood cholesterol in hypercholesterolemic patients and also cause reduction of body weight. The product Olestra® is a sucrose polyester of naturally occurring fatty acids of short to medium chain lengths. Approximately six to eight fatty acids are attached to one molecule of sucrose. Olestra® is neither hydrolysed nor absorbed in the gastrointestinal tract and hence is a noncaloric fat replacer. It is registered in the United States and used for deep-frying and for the production of potato chips and snacks.

Microencapsulated proteins are usually prepared from milk, egg white, or whey proteins. The uniform particle size of 0.1-3 μm results in the smooth texture of fat. Products containing microencapsulated proteins have a creamy texture similar to products with a full fat content. They are used in many applications such as frozen and refrigerated products, salad dressings, mayonnaise, sour cream, and cheese spreads.

Toxic Substances

Toxic substances have the ability to damage living organisms-they are poisons. Everyday household products, such as bleaches, disinfectants, and detergents, contain toxic substances. These products are useful to us, and we have learned to minimise the risks that they present. The better we understand such substances, the more we are able to benefit from them. Toxic substances are potentially dangerous, but many perform useful roles in the modern world. If we are to live safely with them, we need to minimise the risks they present.

The importance of dose must be emphasised when considering toxic substances. The dose makes the poison. Even the most toxic substance known is now used in medicine in carefully controlled doses. Too little of a substance can be just as toxic as too much. Oxygen and water are examples of substances essential to life that can be toxic to humans if we have either too little or too much.

Toxicity is not a function of whether or not a substance is of natural or synthetic origin. The toxins most lethal to humans are mostly natural substances and are often part of a chemical defense strategy to protect against predators. Highly toxic synthetic substances are most likely to be nerve gases or pesticides. A representative list of toxic substances is shown in Table 1.

Table 1. A representative list of toxic substances from the chemistry

Name	*Fatal adult human dose in milligrams*
Botulinum toxin A	0.001
Ricin	0.07
Saxitoxin	1
VX (nerve gas)	10
Hydrogen cyanide	50
Nicotine	60
Arsenic	70
Soman (nerve gas)	350
Strychnine	350
Dieldrin (pesticide)	3500
Caffeine	10 000
Pyrethrins (pesticide)	15 000
Ethanol	386 000

Most Powerful Toxic Substance

The dubious honor of being the most powerful toxic substance goes to a protein produced by the bacterium, Clostridium botulinum. This protein is responsible for fatal food poisoning-botulism-being produced when the bacterium grows in the absence of oxygen in canned or preserved food. There are seven closely related proteins produced by the bacterium, and they are named botulinum toxins A to G.

Botulinum toxin A, the most widely studied, is fatal to humans in doses as low as 0.001 mg. This makes it the most powerful toxic substance known. Like its close relative, tetanus toxin, botulinum toxin is a neurotoxin that prevents the release of neurotransmitters from nerve terminals. Botulinum toxin A is a protein made up of 1285 amino acids, and its structure, which has been determined by X-ray analysis, has given us great molecular understanding of its neurotoxic action.

Despite its extreme toxicity, botulinum toxin A is finding increasing use in medicine for certain conditions that involve involuntary muscle contractions. This is a result of pioneering work of scientists at the University of Wisconsin in the early 1980s. In 1989, the U.S. FD A granted approval for its use in the treatment of the eye disorders, blepharospasm and strabismus, that are characterised by excessive muscle contractions. Very small doses of the toxin are injected into the affected muscles.

The toxin paralyses or weakens the injected muscle by blocking the release of acetylcholine, but leaves other muscles unaffected. There is an extensive literature on the therapeutic uses of botulinum toxin A and numerous Web sites, including Botulinum-Toxic.Net by Eric First at the Boston University School of Medicine. The medical use of botulinum toxin A, a natural substance, illustrates that a good understanding of the chemistry and biology of the substance can make even the most toxic substance known useful to humans.

Most Publicised Toxic Substances

Dioxin is the name given to a family of polychlorinated substances, the most widely studied and the most toxic of which is TCDD, 2,3,7,8-tetrachlorodibenzo-*p*-dioxin.

TCDD

Dioxins are undesirable impurities derived from the combustion of substances that contain chlorine. Forest fires, paper manufacture, municipal waste combustion, motor vehicle emissions, and chemical manufacturing plants all produce dioxins. Dioxins are, thus, natural substances, whose production is increased by human activity.

There is great public interest in dioxins as toxic substances in the food chain and in the environment. Serious concerns about dioxins were first raised in 1970 when they were found in significant concentrations in Agent Orange, used as a defoliant in the Vietnam War and suspected of being responsible for birth defects. When most people think of toxic substances they think of dioxins.

In laboratory animals, dioxins have been shown to cause a variety of abnormalities. The LD50 (lethal dose) for TCDD in rats is 0.022 mg/kg. This would translate to an LD of 1.54 mg for a 70-kg human (making it some 1500 times less potent than botulinum toxin A), but very little human toxicity data is available for TCDD.

Most human data has been obtained from occupational settings where workers have been exposed to chemicals containing TCDD. The signs and symptoms of poisoning of humans with chemicals containing TCDD are similar to those seen in animals. Initially, these were severe skin lesions-chloracne -that provided most cause for concern, but major worries about cancer and birth defects created alarm in the community.

Fears concerning the threat of dioxins to human health are much stronger than are justified solely on the basis of the available scientific data. There continues to be extraordinary publicity in the popular press given to dioxins and their threat to public health. No comprehensive studies have been conducted to determine possible adverse effects to the general population from exposure to dioxins in the environment.

There are numerous Web sites on the dangers of dioxins, for example, Dioxin Alert! documents the dioxin scare in Belgium that resulted in June 1999 in the ban of many Belgium foodstuffs in the European Union and in many Asian countries. As a result, Malaysia ordered the recall of all meat, eggs, dairy products, and infant formula produced in the European Union.

The source of the dioxins was identified as a storage tank of oils and fats in a Belgium processing company that sold the fat for use in animal foodstuffs. The origin of the dioxins in the storage tank is unknown. Worries about dioxin contamination also led to the withdrawal of Coca Cola produced in Belgium. A letter to The Lancet said that the widespread reports of Coca-Cola "poisoning" were in fact symptoms of mass sociogenic illness (MSI)-brought about by society's heightened sense of anxiety over the safety of food products!

Unlike botulinum toxin, dioxins currently have no medically useful properties. The unwanted production of these substances is being reduced in line with the overall "greening" of chemical manufacture.

Endocrine-disrupters

Endocrine disrupters have a chemical similarity to natural hormones and are able to interact with receptors for these hormones and thus to disrupt the natural function of the hormones. These substances occur widely in the environment being derived from both natural and synthetic sources.

Public concern about endocrine disrupters in the environment was stimulated by the publication of a book entitled Our Stolen Future: Are We Threatening Our Fertility, Intelligence and Survival?-a scientific detective story in 1996 by Colborn, Dumanoski, and Myers. This issue began with research in the 1970s that indicated that the insecticide DDT and a group of industrial substances known as PCBs (polychlorinated biphenyls) were reducing the fertility of seals. Much scientific work has been carried out subsequently on the mechanisms of action of these substances, which are clearly associated with their chemical similarities to human estrogens.

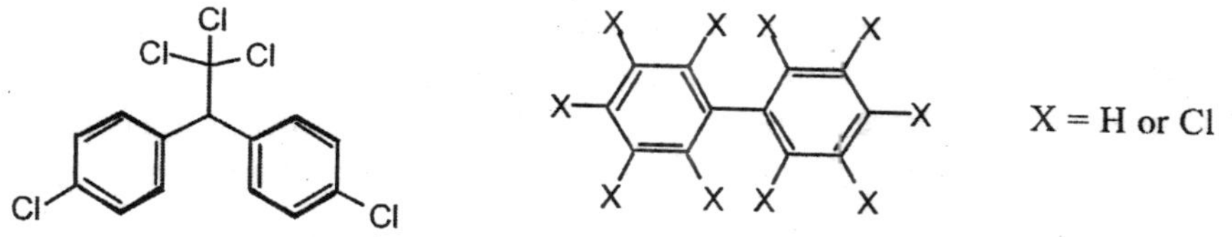

DDT

PCBs

X = H or Cl

It is now apparent that a range of natural and synthetic substances are capable of acting as endocrines in a wide variety of animals. In addition to DDT and PCBs, synthetic substances that can act as endocrine disrupters include phthalate plasticisers used to soften poly(vinyl chloride) (PVC) and synthetic estrogens used in contraceptive pills and in hormone replacement therapy. Natural substances known to interact with hormone receptors include phytoestrogens (mainly isoflavones such as genistein that are widely distributed in plants) and, of course, the hormones excreted by humans and other animals.

Considerations of the risks associated with endocrine disrupters must include assessment of the possible exposure of these substances to humans and animals, that is, the dose of the substance. Clearly, most people consume phytoestrogens in their normal diet. For example, genistein is found in high concentration in soy products, including tofu. There is no evidence to suggest that people consuming substantial quantities of soy products have disrupted endocrine function.

Current studies on endocrine disrupters focus on the development of accurate, sensitive, and selective methods for monitoring these substances in the environment, and a much more detailed understanding of the effects of these substances on biological systems. Assessing the overall toxicity of endocrine disrupters requires more information on environmental levels, dietary intakes, target organ exposures, mechanisms of action, and interactive effects of mixtures.

Assessing Toxicity

We are all familiar with the concept of "risk". Just about everything we do involves risk. Every day we make choices amounting to risk assessment. When dealing with potentially toxic substances, however, risk assessment takes on a more formal role. A variety of government agencies regulate the manufacture, availability, and use of chemical substances in foods, medicines, agricultural and veterinary products, and industrial chemicals. In this context, risk and hazard have different meanings.

A hazard is an intrinsic property of a substance to do harm, while the concept of risk has two elements-the likelihood of an event occurring and the consequence of the event if it does occur. For example, to assess the risk of nicotine in cigarettes, we must consider the toxic substance nicotine as the hazard, the act of smoking cigarettes as the event, and the consequences may include lung cancer and heart disease.

Governments have been urged by organisations, such as Greenpeace, to adopt a precautionary approach to substances that may harm the environment. They are being encouraged to take action to avoid the threat of harm even when firm evidence of cause-effect relationships is unavailable. There is increasing pressure on government agencies to adopt what has become known as the "Precautionary Principle" when assessing substances.

Thus, assessing the likely risk of toxic substances is a complex issue involving data from a variety of scientific disciplines and consideration of both short- and long-term effects on all living organisms. With toxic substances, we must balance any social benefit associated with a particular substance against the risks posed by its use, manufacture, and disposal.

Chapter 2

Action of Synthetic Drugs

To the field of pharmacology, isolation of the pure active compounds is of very great importance, as it opened up the possibility of an accurate study of the effect of dosage and enabled quantitative measurements of the physiological effect of the drug to be made. Such progress was obviously impossible as long as investigators were compelled to deal with a crude drug of unknown strength and composition. By this means, also, secondary by-effects, often unpleasant and harmful, could be got rid of in those cases where they were due to the presence of substances other than the desired active principle.The study of the constitution of the active principles led to attempts at their synthetic production. For a long time, however, synthetic chemistry was not far enough advanced to succeed in the synthesis of such complex substances, and efforts were therefore made to find which portion of the molecule gave rise to the physiological effect in order that simpler analogous compounds might be prepared possessing the characteristic action of the drug. In this way it was found that the physiological action of the drug was in general dependent on its chemical nature, although naturally modified by differences in physical properties, such as solubility, volatility, etc. For instance, certain types of compounds often possess a characteristic action, e.g. the quaternary ammonium compounds have a paralytic action on the motor nerves. On the other hand, in the majority of cases the relation between chemical constitution and physiological action is extremly obscure, and is usually to be found only within very narrow limits, outside which any generalizations that may have been made are apt to break down. Thus a very small change in the chemical constitution of a compound is often accompanied by a complete change in its physiological action, as in the case of cocaine, α-cocaine, and α-eucaine, the first of which has a strong local anaesthetic action, whilst the second has no local anaesthetic action whatever, and the third which resembles a-cocaine in its chemical constitution more than it does cocaine has a marked local anaesthetic action, just like cocaine itself. In this connection mention should be made of the fact that a very considerable difference in physiological action is often shown by stereo-isomerides, as in the case of d- and Z-nicotine, d- and l-hyoscyamine, pilocarpine, and isopilocarpine, etc.

At this point it may be advisable to indicate the relation between the chemical composition and the physical properties of a substance. It is held by some pharmacologists that the physiological action of a drug is conditioned solely by its physical properties, but even if this were always the case, it would not exclude a connection between the physiological action and the chemical constitution. It may be that the effect of some drugs is due to a subtle combination of various physical properties; these cannot, however, be readily ascertained in the majority of cases, but we can endeavour to correlate the chemical constitution which is the foundation of these properties with the physiological action.

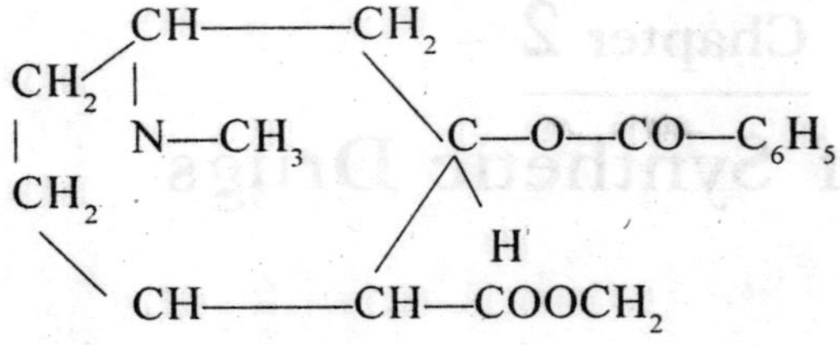

```
     CH———————CH2
CH2/ |          \
|    N—CH3       \C—O—CO—C6H5
CH2              /\
   \            /  H
    CH———————CH—COOCH2
```

Cocaine

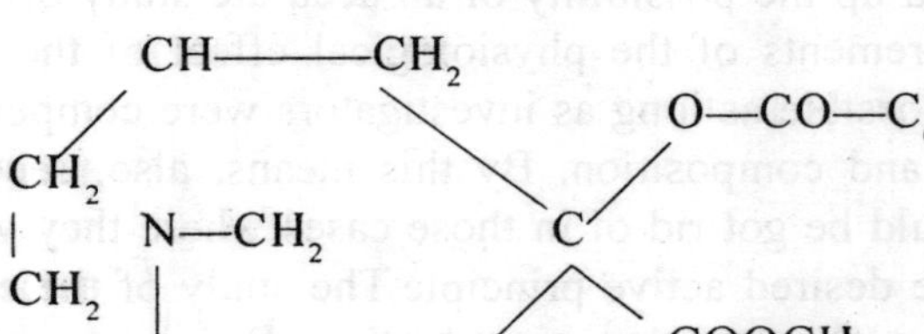

```
      CH———CH2
       |     \   O—CO—C6H5
    CH2|      \ /
     | N—CH2   C
    CH2|      / \
       |     /   COOCH3
      CH———CH2
```

α-cocaine

```
          CH3
          |
     CH2—C———CH2
         |     \  O—CO—C6H5
         N—CH2  C
         |     /  COOCH3
     CH2—C———CH2
          |
          CH3
```

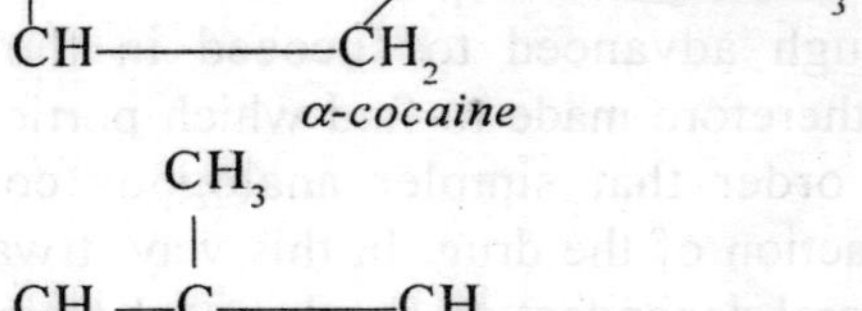

α-eucaine

Another point on which misconception seems to have arisen, is the question of whether a difference in properties between two substances is to be attributed to chemical or purely physical causes. The following quotation from a leading text-book of pharmacology will serve to illustrate this:

> Yellow phosphorus, given by the mouth, is a violent poison, but red phosphorus, which only differs from it in its physical properties is comparatively innocuous. Yet were the two differently constituted chemically, 'the difference in their action would, no doubt, be referred to this fact. Now, according to all modern views, there is a considerable difference in the chemical constitution of the molecules of the two different kinds of phosphorus, a difference which is manifested in the great difference of their reactivity and of their toxic action. There is probably as great a chemical difference between yellow and red phosphorus, as between acetylene, $(CH)_2$, and benzene, $(CH)_6$, but at present we have no means of accurately investigating the molecular complexity of the two solid varieties of phosphorus.

Although the action of some drugs may be conditioned chiefly by their physical properties, yet there are others in which it is undoubtedly due to their chemical character (e.g. organic arsenic derivatives certainly owe their trypanocidal and anti-syphilitic action to the presence of the arsenic). The term chemical character rather than constitution is used deliberately, as in many cases the physiological action of a substance may be traced to some specific chemical property, which is not necessarily closely connected with its constitution. For example, the poisonous character of oxalic acid and its soluble salts could hardly be deduced from its structural formula, but its toxicity is sufficiently explained by the fact that it forms an insoluble calcium salt, a sufficient quantity of calcium in solution being essential for the welfare of the organism.

The higher members of this series of dibasic acids, having more soluble calcium salts, are proportionately less toxic. In other cases, however, some kind of relation has been traced between the presence of certain groupings in the molecule and the appearance of certain more or less characteristic physiological effects. These relations may be used in attempting to prepare new synthetic drugs, but it is impossible to forecast the action of any new substance with certainty, and careful pharmacological experiments must be made before a new compound can be introduced as a drug. Guidance in our efforts to obtain synthetic drugs may also be obtained from a study of the changes undergone by compounds in the organism. These changes are usually in the direction of the transformation of an active and poisonous substance into a less active and less harmful one. Thus, in order to replace a drug by a new product in which harmful by-effects are eliminated, we can often attain the desired object by studying the changes undergone by the former in the organism, and by taking for the starting-point of the new drug the product of the metabolism of the original drug. For example, it was found that aniline is oxidized by the body into para amino-phenol, and this observation led to the introduction into medicine- of a large number of derivatives of para amino-phenol, of which phenacetin is the best-known example.

The theories that have been advanced to explain the action of various drugs are chiefly concerned with carbon compounds, as the latter offer practically the only means of comparing substances possessing definite structural relationships with one another. The influence of the chemical constitution of a substance on its physiological effect can be seen from the fact that definite changes in the chemical constitution of substances belonging to the same class, are usually accompanied by definite changes in the physiological effect, and further that the addition of certain molecular groups to differently acting substances can change them into similarly acting or equally inactive bodies.

As an instance of this point, mention may be made of the discovery of Crum Brown and Fraser, that various alkaloids possessing the most diverse physiological actions, on combination with alkyl halides to form quaternary ammonium derivatives

$$NR_1R_2R_3 + RX = \begin{matrix} R \\ X \end{matrix}\!\!>NR_1R_2R_3$$

where $R_1R_2R_3$ are organic radicles of any complexity, and BX stands for an alkyl halide such as methyl iodide, CH_3I, ethyl bromide, C_2H_5Br, etc. yield substances which in almost every case possess the property of paralysing the motor-nerve endings in the same manner as curare. In this way, by the process of methylation, one can obtain from all tertiary bases ($NR_1R_2R_3$), quaternary ammonium compounds,

$$\begin{matrix} R \\ X \end{matrix} > NR_1R_2R_3$$

which are very, and often disproportionately, poisonous compared with the original bases. Curare itself contains a tertiary base, curine, which is not very poisonous, as well as the far more poisonous ammonium base, curarine. The former on methylation yields the latter, which is 226 times as poisonous as the original substance.

On the other hand, the following examples may be quoted to show that the addition of - particular groups to certain substances may weaken or destroy their action. This is well illustrated by the effect of the entrance of acidic groups into the molecule, thus substances containing a hydroxyl group, on combination with sulphuric acid, lose their toxic properties. For instance, phenol is very poisonous, but phenol-sulphuric acid, in the form of its sodium salt, C_6H_5—O—SO_2—ONa, is harmless. Morphine, $C_{17}H_{17}NO(OH)_2$, is an extremely powerful and poisonous drug, but morphine-sulphuric acid, $C_{17}H_1$7NO(OH). O. SO_2. OH, is practically inert, and can be given in doses of 5 grams without any harm. This fact should be borne in mind in considering the attempts that are often made of introducing the sulphonic acids of various drugs into medicine, as in nearly every case these are bound to be useless. The effect of introducing the carboxylic acid group (CO—OH) is often the same as that of introducing the sulphonic acid group (SO_2—OH). For example, the toxic substance methylamine, NH_2—CH_3, is thus changed into the harmless glycine, NH_2—CH_2—COOH. The mere addition of acid radicles, such as acetyl, CO—CH_3, without actually converting the substance into an acid may often be sufficient; thus acetamide, NH_2—CO—CH_3, is practically harmless, but ammonia, NH_3, is poisonous; acetanilide, C_6H_5 NH. CO. CH_3, is less poisonous than aniline, C_6H_5.NH_2.

An example of an opposite kind is furnished by the effect of the addition of hydrogen to cyclic bases, which is almost invariably to increase their activity, and also their toxic properties. Other examples of this kind will be adduced later, but sufficient have been mentioned to indicate that similar alterations of differently acting compounds often produce similar or identical alterations in the physiological effect.

Ehrlich considers that the selective action of a compound for certain cells depends on the coming together of particular groups in the molecule in some sort of chemical connection with the cell substance. It is only when the compound is held to the tissues (anchored) by these groups, that the whole complex molecule can take effect and exert its characteristic physiological action. If, therefore, the character of these special anchoring groups be altered, the substance can no longer exert its action on these particular cells, but it may happen that the alteration of the anchoring group causes the substance to become anchored to a different set of cells, and so to produce a different physiological effect. Examples of this type will be quoted subsequently.

These views may be more readily understood by analogy with Witt's theory of dyeing, according to which the colour of a substance is due to the presence of certain chromophore groups, such as the azo group, —N = N—, while, for the coloured substance to have dyeing properties, it is necessary for another, salt-forming, group to be present, by which it can be held fast to the fibre. A dye, therefore, must contain both a chromophore group and a salt-forming group, and in the same way, a drug is supposed, besides containing an active group, the pharmacophore, corresponding to the chromophore, to contain also an anchoring group, corresponding to the salt-forming group. This analogy may be extended, and the phenomena observed in staining nerve tissues may be compared with the biochemical processes which take place between poisons and the living tissue.

According to Ehrlich, the process of dyeing is similar to that which takes place when a poison is injected into the body. Thus, if a wool fibre be immersed in a dilute solution of picric acid, the colour is withdrawn from the solution and enters the fibre. The fibres of an animal tissue are regarded, in the same way, as withdrawing the dyestuff from the solution, and fixing it, if it is more soluble in them than in the original solution. If the dyestuff is more soluble in another solvent, e.g. alcohol, than in the tissue, then the latter can be again decolorized by shaking it in alcohol.

Now, all dyes capable of dyeing nerve cells lose this property if a sulphonic acid group is introduced into the molecule. The bulk of nerve-dyes can be extracted from their aqueous solutions with ether, but the sulphonic acids cannot owing to their greater solubility in water. It is therefore supposed that there are certain substances in the nerve tissues which act like ether does in the test-tube, so that the strong action of certain poisons on the brain and nervous tissues is due to their being extracted in the same way as they can be extracted with ether. Ehrlich's views have proved to be very stimulating and fruitful, especially in recent years.

A theory of the action of poisons, in which the subject is regarded from a different point of view, is that due to Loew. It states that all substances which are capable of acting on aldehyde or amino groups, even when in dilute solution, must be poisons for living tissue, on which they will exert a substituting action. The greater the reactivity of a substance for aldehyde (CHO) or amino (NH_2) groups, the greater will be its physiological effect and its toxicity. For example, phenylhydrazine, C_6H_5. NH. NH_2, and hydroxylamine, NH_2. OH, which are so reactive to ketone and aldehyde groups, are strong poisons, both to plants and animals. Aniline, C_6H_5. NH_2, which reacts less readily with aldehydes than does phenylhydrazine, is less poisonous; and similarly ammonia, NH_3, is less poisonous than hydrazine, H_2N NH_2. Substances containing tertiary combined nitrogen are usually less toxic than the corresponding substance where the nitrogen is present in the more reactive form of a secondary imino (NH) group. This is exemplified by the fact that if the hydrogen of the NH group in many alkaloids is replaced by a methyl group, the resulting tertiary base is far less poisonous than the original alkaloid. In the same way, if one of the hydrogen atoms of the NH_2 group in aniline is replaced by an alkyl group, the toxicity is diminished, as the substance reacts less readily with aldehydes. A similar explanation is offered of the fact that piperidine is far more toxic than pyridine, and tetrahydroquinoline far more toxic than quinoline, namely, that the reduced compounds, which contain secondary nitrogen in the place of tertiary, have a greater reactivity with protoplasm. This view is supported by the fact that the toxicity of substances with labile amino groups is increased by the addition of a second amino group, but is lessened when an amino group is changed into an imino, NH, group. Thus, para-phenylene-diamine, NH_2⟨⟩NH_2, is more toxic than aniline. In general, if the chemical character of a poison is made more labile by any change in the character of the molecule, then it becomes more toxic, and vice versa. An example of this is shown by the hydroxybenzenes, in which the increased reactivity due' to the entrance of hydroxyl groups into the benzene ring is accompanied by an increase in the toxicity of the substance. Thus, trihydroxy-benzene, pyrogallol, is more poisonous than catechol., which contains only two hydroxyl H groups, and this in turn is more poisonous than phenol (monohydroxybenzene). The labile substance, ammonium sulphocyanide, is toxic to plants, but the more stable isomer, thio-urea, is not. Substances with carbon atoms united by double bonds are generally more reactive and more toxic than closely related compounds with simple linkage. Neurine is more poisonous than choline. The toxicity of phenols is attributed to their reactivity, especially with aldehydes. This reactivity is diminished by the entrance of acid groups, such as the carboxyl, COOH, and the sulphonic acid group, SO_2—OH, and this diminution in the reactivity is accompanied by a diminution of their toxic power. For example, salicylic acid, is less poisonous than phenol,.In the same way, saccharin, is not at all poisonous, as the presence of the carboxyl and the sulphonic acid groups so much diminishes the reactivity of the imino group.

These views of Loew, which are both simple and very suggestive, can nevertheless only be applied to certain groups of substances which react with aldehyde and amino groups. They offer no explanation of the selective' action of different tissues for various drugs, as they are applied simply to protoplasm, and the protoplasm of any and every tissue has labile aldehyde and amino groups, which, according to this view, should react with the drug. Nevertheless, the theory is of some value in co-ordinating the action of the various members of certain classes of compounds.

In general it is supposed that there are two kinds of groups in the molecule of an active substance, necessary for the action to appear, but in some cases the active group and the anchoring group may be one and the same. If the anchoring group is altered or removed, then a different physiological effect from the original may become apparent if there is present a second possible anchoring group, which can now manifest itself more strongly. As these other anchoring groups may effect different tissues from those affected by the original anchoring groups, so a different physiological effect may be produced, or one of the special physiological effects of the substance may become enhanced. In the first place, the most chemically reactive anchoring group dominates the situation.

For example, morphine has a strong hypnotic effect, and the anchoring group is probably one of the hydroxyl groups.If these are combined with sulphuric acid, then the morphine cannot become anchored to the nerve tissues l of the central nervous system, and the resulting substance has no effect at all. If the hydroxyl is only altered by the entrance of an organic radicle, with the formation of substances such as methyl, ethyl, or acetyl derivatives, then the hypnotic effect is thrust into the background, while the action on the respiratory centres, produced by morphine to a slight extent, becomes much enhanced and dominates the physiological effect (codeine, heroin, etc.).

In those cases where the presence of an acid group prevents the substance from acting on certain tissues in spite of the presence of an anchoring group, the esterification of the acid group causes the physiological action to appear. To illustrate this, the esterifying group must of itself play no part in the physiological effect, which must reside solely in the original substance, and only be hindered from appearing by the presence of the acid group. For example, arecaidine has no action at all on animals, but arecoline, its methyl ester, is poisonous, and has physiological effects similar to those of pilocarpine and muscarine. The ethyl ester acts similarly to the methyl ester.

A precisely similar example is furnished by cocaine, which is the methyl ester of benzoyl-ecgonine. Benzoyl-ecgonine itself is far less active, and is twenty times less toxic than cocaine, owing to the presence of the carboxyl group. It is only when this group is esterified, that the typical action of cocaine appears, and it is immaterial by which alcohol the ester formation is accomplished. In every case the typical action of cocaine appears when the carboxyl group is masked, and therefore it is not the alkyl group which is active, but its presence simply serves to destroy the effect of the group which is inhibiting the characteristic action of the drug. When this group is masked, then the typical anchoring group can take effect. Thus it may possibly be that for cocaine to act as a local anaesthetic, it must be anchored to the tissues by the $>N-CH_3$ group, or at any rate by some portion of the molecule other than the carboxyl group, but that if a free carboxyl group be present, it becomes anchored by this instead, and so fails to produce its characteristic action.

One can thus alter inactive compounds into active ones, or vice versa, or can alter the nature of the action of some compounds by means of a chemical change which alters the group. In this way, if the most prominent property of a drug becomes suspended or diminished, a comparatively masked property becomes developed.

Closely connected with the relation between the constitution and the physiological action of a drug is the second, question of whether there is any connection between the action and the chemical changes which the substance undergoes in the organism. Robert considers that the strength of the action of a drug is in no way proportional to the amount of the chemical change which it undergoes in the organism. For instance, very active substances, like atropine and strychnine, pass through the body practically unchanged, but an inactive body like tyrosine is completely oxidized by the organism into carbon dioxide and water. Conversely in the case of many compounds, it can be seen that when they have an effect on the body, they have undergone a chemical change. This is also the case with many physiologically active organic compounds, which often show a chemical alteration in the organism. It would therefore seem as if there were no relation between the changes undergone by a drug and its physiological action, but nevertheless some facts can be cited against the view that there is absolutely no connection to be traced.

In discussing the relation between chemical constitution and physiological action, that wide generalizations cannot be drawn, and that it is only in the case of closely related compounds that relationships can be traced. In the same way, with regard to the relation between chemical change and physiological action, a connection can be found in certain cases of closely related compounds. In many cases physiologically active compounds show a degradation of the molecule, and these same substances, if they are made so resistant that they no longer suffer any alteration in the organism, then become inactive. For example, xanthine has no tonic action on the heart muscle, but theobromine (dimethylxanthine) has a slight tonic action, and caffeine has a still more marked tonic action, from which it will be seen that the addition of methyl groups to the nitrogen of xanthine adds a special action on the heart to the ordinary physiological action of xanthine. Experiments, made to ascertain the fate of these compounds in the organism, show that the products of metabolism found in the urine after the administration of caffeine and theobromine contain xanthine bases, poorer in methyl groups than the substances originally administered.

```
HN———CO                 HN———CO
|     |    H             |     |    CH3
|     |                  |     |   /
OC    C—N                OC    C—N
|     ||  \              |     ||  \
|     ||   CH            |     ||   CH
|     ||  //             |     ||  //
HN    C—N                CH3—N————C—N
```

Xanthine *Theobromine*

```
CH3—N————————CO
    |         |
    |         |    CH3
    |         |   /
    OC        C—N
    |         ||  \
    |         ||   CH
    |         ||  //
CH3—N         C—N
```

Caffeine

This indicates that the methyl groups have been split off in the organism. In experiments on dogs, caffeine gives first theophylline and then 3-mono-methyl-xanthine, smaller quantities of the two other dimethyl compounds, theobromine and para-xanthine, being also formed. In man it is degraded into theophylline. In both cases, therefore, we have a splitting off of some of the methyl groups, the groups which appear to be responsible for the tonic action on the heart. This case indicates that there is sometimes a relationship between the physiological action and the changes undergone by the substance in the organism. Another well-known instance is afforded by the sulphone derivatives, the hynotic action of which has been shown to be connected with the presence of ethyl groups in the molecule. The methyl derivatives are inert, and pass through the organism unchanged; the ethyl derivatives produce sleep, and are almost completely decomposed by the organism.

In the case of the alkaloids, it is very difficult in most cases to obtain any knowledge of the mechanism of their action, and generally they seem to pass out of the organism for the most part unchanged. Hence it has been assumed that the action of some of these is catalytic, an assumption that explains nothing. Another view is that the small portion of the alkaloid that cannot be recovered from the body has undergone a chemical change to which its action is due. Compared with food-stuffs, most drugs are destroyed by the organism with difficulty, and they owe their, activity to this property, but if they are absolutely resistant they are quite inactive. Those substances which have a specific action must be fairly resistant, otherwise they would react with all protoplasm. In the synthesis of drugs we make use of this property by conferring an artificial resistance on the synthetic substance, so as to prevent it from reacting too suddenly, and also to prevent it from reacting with all tissues, so that it does not show undesirable by-effects. For this reason, all substances which can react with every kind of tissue cannot be used as drugs, unless the reactivity is artificially diminished, or the dose made very small, in which case they can be made useful (e.g. formaldehyde, hydrocyanic acid).

In some substances the physiological effect seems to be due more to the stereo-chemical configuration than to the chemical constitution. One general instance will be a sunicient support for this statement. It has already been pointed out that all ammonium bases have a paralytic action on the motor nerves, and that this action is quite independent of the structure of the rest of the molecule. Even the presence of nitrogen is not necessary, as a similar effect is shown in those bases where the nitrogen is replaced by phosphorus, arsenic, or antimony, and it therefore seems that the effect is due to the configuration of the ammonium, phosphonium, or arsonium group, this being a three-dimensional (space) one, while that of the original amines almost certainly lies in one plane. This view is strongly supported by the fact that dimethyl sulphide, $(CH_3)_2S$, and diethyl sulphide, where the sulphur is divalent, and where the configuration must lie in one plane, are practically inert, but that trimethyl sulphonium hydroxide, $(CH_3)_3SOH$, and trimethyl sulphonium iodide resemble the ammonium bases in having a curare-like action on the motor nerves. In these compounds, where the sulphur is tetravalent, its valencies are distributed in three dimensions, as shown by the fact that optically active sulphur compounds of this type have been prepared.

Finally, stress must again be laid on the incompleteness and limited scope of all theories connecting chemical constitution and physiological action. No single one of them can explain all the facts, but most of them seem to contain some element of truth and the different theories should not be regarded as mutually exclusive or contradictory, but rather as representing different aspects of a very complex problem.

Effect of Elements and Radicles

In this section, an outline is given of such generalizations as can be drawn concerning the effect of various elements and radicles on the physiological action of chemical compounds. These furnish merely a rough guide to what may be expected when a new element or group is introduced into the molecule, and in every case the result must be tested by experiment. Inorganic Elements. In 1839 Blake noticed that the action of salts, injected into the blood, depended only on the electro-positive half, and hardly at all on the electro-negative. This is analogous to the action of most esters, which generally resembles that of the alcohol from which they are derived, the explanation in both cases being that acids are usually physiologically inert. Later on it was shown that in any given group of isomorphous substances the action is similar, and is usually increased with increase in atomic weight. This was shown to be the case with Li, Na, Eb, Cs, Ag, Tl; and with Mg, Mn, Co, Ni, Cu, Zn, Cd, and with Ca, Sr, Ba. The only exceptions found were in the case of potassium and ammonium, which differ from the other isomorphous substances of the group. But these salts are also exceptions to Mitscherlich's law that isomorphous substances have similar spectra.

It was therefore supposed that physiological action depends on intramolecular vibrations in the same way as the spectra depend on these. The statement made above that isomorphous elements have a similar physiological effect, which becomes stronger with increase in atomic weight, only holds good in the case of electropositive elements; in the case of negative elements, such as the halogens, there is no relation between physiological effect and atomic weight. Elements forming two series of salts show different effects, according to which class the salt belongs, ferric salts, for instance, differing considerably from ferrous. In many salts the action appears to be due to the ions into which the salt dissociates; for example, potassium ferrocyanide is excreted for the most part unchanged, and has neither the action of a ferrous salt nor of a cyanide, and similarly sodium platinocyanide, which is found unchanged in the urine, is almost free from poisonous effects, thus differing both from the cyanides and the platinum salts.

The effect of ionization is strikingly indicated by the mercury salts. Mercuric chloride, $HgCl_2$, is ionized and extremely poisonous, but the cyanide, $Hg(CN)_2$, though soluble, is almost non-ionized and is far less poisonous. The insoluble and practically non-ionized mercurous chloride (calomel), HgCl, is comparatively non-poisonous.

Phosphonium, arsonium, and stibonium bases do not show the action of the other phosphorus, arsenic, and antimony compounds. On the contrary, these compounds resemble the substituted ammonium bases, such as those present in curare, in having a paralytic action on the motor nerves. In these cases we do not get the characteristic action of the poisonous elements themselves, but rather one resembling the analogous compounds of the indifferent element nitrogen.

Radicles and Elements in Organic Compounds

Action of Hydrocarbons and Effect of Alkyl Groups

According to Schmiedeberg, the action of aliphatic compounds is governed by the following rules;

1. The physiological activity of substances (especially aliphatic) depends chiefly on physical properties. The readiness with which a substance is absorbed is a very important point, as obviously a substance that is not absorbed can have no action on the system. Volatility and solubility in water are of great importance, a fact which is illustrated by the paraffin

series, the lower and more volatile members of which show the characteristic narcotic effect of the hydrocarbon groups, while the insoluble non-volatile higher paraffins are without any action at all.

The following rules show the effect of substituent alkyl groups;

1. Very poisonous radicles, on substitution by simple alkyl groups, lose the intensity of the original character of the group. For example, by substituting alkyl radicles for the hydrogen of HCN, the -nitrites, RCN, and isonitriles, R—N ≡ C, are obtained, and only become poisonous when HCN is split off in the organism. Also cacodyl oxide, (CH_3) 2As—O—$As(CH_3)_2$, where the oxygen of arsenious oxide is replaced by two methyl groups, does not show the characteristic arsenic effect until after it has begun to decompose in the body.
2. On the other hand, the effect of the alkyl groups can be lessened or altogether lost by combination with other atoms or groups. For instance, the amines of the fatty series behave like ammonia and have no narcotic action. Nevertheless, the first rule holds in this case as well, as these amines are less toxic than ammonia.
3. When a compound is formed by the union of two groups through an oxygen atom, then the physiological effect depends upon the nature of the two components, each of which acts independently of the other. In those cases where both parts of the compound are similar or equivalent alkyl groups, as in the case of the simple and mixed ethers, then the action of the whole compound is a simple one, and these substances resemble the corresponding alcohols in their physiological effect. To this class of substances should be added those esters the acids of which yield neutral salts without any specific physiological action. For this reason, acetic ester and its homologues are classed with the alcohols. If, on the contrary, the acid has a specific action of its own, then this becomes apparent in the ester, and exerts a modifying influence on the physiological effect of the alkyl group.

These rules of Schmiedeberg are of value, not only in summarizing the effect of alkyl groups, but, as can be seen above, are also useful as applied to other groups of substances, especially the ethers and the alcohols, the effect of which is due to the alkyl groups contained in them. The hydrocarbons of the methane series are, as would be expected from their chemical nature, less active physiologically than those of the ethylene, acetylene, or benzene series. The lower members, when inhaled, produce anaesthesia and sleep, and in larger quantities asphyxiation. The toxicity and the strength of the narcotic action increase as the series is ascended, but on further ascending the series, the action diminishes owing to their diminished volatility and solubility, so that the higher members are quite inert. The members of the ethylene series have a similar narcotic action; amylene, C_5H_{10}, resembles chloroform in its narcotic properties.

The benzene hydrocarbons have a paralysing action on the motor nerves, and a more noteworthy action on the brain and cord, causing lethargy and somnolence. Bromobenzene and chlorobenzene act in the same way as benzene itself. Naphthalene, $C_{10}H_8$, is less toxic than benzene, but slows the respiration and, in fever, lowers the temperature. It also has the property of decreasing nitrogen metabolism. Diphenyl is physiologically inert, while thiophene, furfurane, and pyrrol resemble benzene to a certain extent.

Effect of Alkyl Groups

Some of the effects of introducing alkyl groups into a compound have already been mentioned in connection with Schmiedeberg's rules, but there are many different effects to which attention may be given. The convulsive properties of ammonia are diminished by the entrance of methyl groups, trimethylamine being free from these effects. In aniline, replacement of the hydrogen of the amino group causes diminution of the convulsive properties as with ammonia, but replacement of the hydrogen of the nucleus by methyl groups causes an increase in these effects. The marked influence of the entrance of methyl groups into the xanthine molecule has already been pointed out. The foregoing examples show that the addition of a methyl group to a nitrogen atom can produce very diverse effects, and a similar variety can be found in the effect of the methylation of a hydroxyl group.

The alkyl ethers of the type R.O.R, where R represents an aliphatic hydrocarbon radicle, are distinguished by their resistance towards oxidation, and physiologically they show a marked hypnotic action, as in the case of ordinary ethyl ether, $C_2H_5—C_2H_5$. This hypnotic action is also shown by many other compounds containing an ethoxy group, $O—C_2H_5$, such as ethoxy-caffeine. In many cases, replacement of the hydrogen of a hydroxyl group by a methyl group diminishes the physiological activity. For instance, catechol, is more poisonous than guaiacol, and this in turn is more poisonous than veratrole, and ortho-methoxy-benzoic acid, and anisic acid are less active than salicylic acid. On the other hand, in some cases the activity of a compound is increased by the methylation of a hydroxyl group. For instance, dimethylresorcinol is extremely poisonous, far more so than resorcinol.

OCH_3 / OCH_3 (benzene ring) OH / OH (benzene ring)

In some cases, the methylation of a hydroxyl, and even more often of a carboxyl group, may cause a very marked change in the physiological action of a compound, owing to the methylation producing a new anchoring group for the molecule. It has already been pointed out how the entrance of an alkyl group into certain acids often causes the full appearance of certain previously masked properties, as in the case of cocaine, arecoline, etc. Possibly a similar explanation may be given of the antipyretic action of phenyl-dimethyl-pyrazolone (antipyrine), phenyl-methyl-pyrazolone being inert;

```
CH3—C═══CH          CH3—C═══CH
    |   |               |   |
CH3—N   CO           H—N   CO
     \ /                 \ /
    N—C6H5              N—C6H5
```

Phenyl-dimethyl-pyrazolone (antipyrine) Phenyl-methyl-pyrazolone.

Effect of Hydroxyl Groups

The entrance of a hydroxyl group into aliphatic compounds usually produces a weakening of their physiological activity, and this weakening effect is roughly proportional to the number of the hydroxyl groups. For example, the narcotic and poisonous alcohols give rise to the inactive glycols, glycerol, mannitol, etc., and from the very active aldehydes are obtained the less active aldols, such as

$$CH_3\text{—}CH\begin{cases} OH \\ CH_2 \cdot CHO_3 \end{cases}$$

and by the entry of more hydroxyl groups, the totally inactive aldoses (glucose, etc., CH_2. OH(CH. OH)$_4$ .CHO). A similar effect is often produced in many other compounds, for example, caffeine, the physiological effect being lost in hydroxy-caffeine. In the aromatic compounds, the entrance of a hydroxyl group usually causes an increase in both the physiological effect and the toxicity. The entrance of a hydroxyl group in benzene itself causes a great increase in the toxicity, together with the appearance of the strong antiseptic properties for which phenol is well known, In the case of a. more inert aromatic substance 22 SYNTHETIC DRUGS such as benzoic acid, the entrance of a hydroxyl group is again accompanied by an increase in physiological activity, orthohydroxy-benzoic acid (salicylic acid) having well-marked antiseptic properties as well as an apparently specific action in rheumatism.

The large number of facts known as to the action of compounds containing a hydroxyl group do not lend themselves to the view that this group has an action of its own, but they point rather to the fact that it very often performs the function of an anchoring group. The modification of such groups by esterification or alkylation neutralizes or alters the effect of these groups in anchoring the subtance to a particular tissue. If an acyl group enters the hydroxyl group (esterification), then various different effects may be produced. In those cases where the ester is hydrolyzed in the organism, its action is due to that of the original hydroxylic substance, together with that (if any) of the sodium salt of the acid. As the greater number of acids are inert, the action of most esters is due to a delayed action of the alcohol from which they are derived. In some cases, the esters do not appear to be hydrolyzed by the organism, and they then do not behave in the fashion indicated by the foregoing statement. For example, triacetyl glycerol (triacetin)

$$\begin{array}{l} CH_2\text{—}O\text{—}CO\text{—}CH_3 \\ | \\ CH\text{—}O\text{—}CO\text{—}CH_3 \\ | \\ CH_2\text{—}O\text{—}CO\text{—}CH_3 \end{array}$$

does not show the action of glycerol or of sodium acetate, both of which are practically inert, but possesses a specific action on the nervous system, and is poisonous. Glyceryl ether

$$\begin{array}{ccccc} CH_2 & — & CH & — & CH_2 \\ | & & | & & | \\ O & & O & & O \\ | & & | & & | \\ CH_2 & & CH & & CH_2 \end{array}$$

is also a hypnotic, this being an example of the fact that alkyl compounds of this kind, obtained by the addition of an alkyl residue to a hydroxyl group, often show hypnotic properties; for example, ethoxy-caffeine. In the case of the ethers, such as an ordinary ethyl ether, the narcotic properties are strictly in accordance with the close resemblance between the chemical properties of the ethers and the parent hydrocarbons themselves, the narcotic action being simply that of the alkyl groups, as stated by Schmiedeberg in the third of his rules.

Other conditions being equal, primary alcohols are less active than secondary, and these are in turn less active than tertiary. In homologous series, those members with long side chains are in general more active, a rule which also holds for benzene derivatives with side chains. The chief action of ethyl alcohol is on the nervous system, the narcotic effect being preceded by a loss of control of the higher centres, an effect which has led to the mistaken belief that alcohol in small quantities acts as a stimulant. Although the subject is somewhat controversial, the balance of the pharmacological evidence suggests that there is no justification for the medicinal use of alcohol.

Effect of Halogen in Organic Compounds

The most important effect of the entrance of chlorine into the molecule of aliphatic compounds is an increase in their narcotic action, but this useful property is accompanied by an increase in their depressant action on the heart and blood-vessels. Narcotic action and lowering of blood-pressure appear to be general properties of chlorine compounds, and an illustration of the fact that narcotic action, and also the toxicity of chlorine compounds depends upon the amount of chlorine in the substance, is furnished by the chlorine derivatives of glycerine. Glycerol itself is inert, but the chlorhydrins have a narcotic action, and dilate the blood-vessels, this effect being the greatest in the case of trichlorhydrin, $CH_2Cl—CHCl—CH_2Cl$, and least in the case of monochlorhydrin, $CH_2Cl—CH(OH)—CH_2OH$. This fact is also illustrated by methyl chloride, CH_3Cl, methylene dichloride, CH_2Cl_2, chloroform, $CHCl_3$, and carbon, tetrachloride, CCl_4, a series in which increase in the amount of chlorine is accompanied by increased narcotic action and increased toxicity. But in the case of the chlorinated fatty acids, the toxicity decreases with increase of chlorine, trichloracetic acid being practically non-toxic, while monochloracetic acid is strongly poisonous. The chlorinated fatty acids are also anomalous in another respect, as the narcotic action of the sodium salts of the fatty acids increases with rise of molecular weight from acetic acid to valerianic acid, while in the case of the corresponding chlorinated acids it diminishes with increase of molecular weight.

In chloro-caffeine, there is an antagonism between the tonic effect of caffeine.on the heart and the depressant action of the chlorine, and it is found that the tonic effect of this substance is less than that of caffeine, but the diuretic effect and the stimulation of the brain due to caffeine are not affected. The entrance of halogen into the benzene nucleus produces only a slight change in the physiological properties. None of these derivatives have an anaesthetic or hypnotic action, but there is usually an increase in antiseptic power. In general, there is a close resemblance between bromine derivatives and those of chlorine, both in the aliphatic and aromatic series. Organic iodine compounds differ from those of the other halogens in respect of their increased antiseptic power and their diminished hypnotic properties. The toxicity of the iodine compounds markedly exceeds that of the analogous chlorine and bromine compounds, but this has not seriously hampered their use as antiseptics, and a very large number of iodine compounds have been produced for this purpose.

Effect of Nitro and Nitroso Groups

The entrance of a nitro (NO_2) or nitroso (NO) group in general causes a marked increase of toxicity, independently of whether it replaces a hydrogen atom attached to carbon or one in a hydroxyl group. The aliphatic nitrites give rise to dilatation of the bloodvessels, and are therefore used to lower the blood-pressure. The strength of this effect diminishes in descending the series from amyl nitrite to methyl nitrite. All the nitrites act in this manner, the secondary and tertiary being stronger in their action than the primary, probably owing to the fact that they are more readily hydrolyzed to alcohol and nitrite. A similar action is shown by the esters of nitric acid, nitro-glycerin, $CH_2(ONO_2)$ $CH(ONO_2)$ $CH_2(O.\ NO_2)$, and erythrol tetranitrate being largely used in medicine to dilate the bloodvessels.

Aliphatic nitro-compounds, such as nitro-methane, which are isomeric with the alkyl nitrites

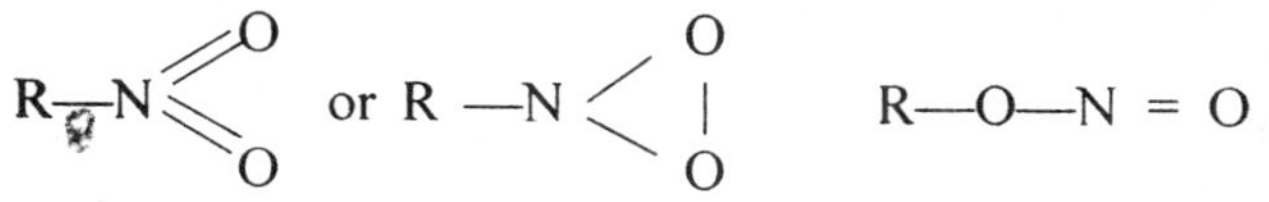

$$R{-}N{\genfrac{}{}{0pt}{}{\nearrow O}{\searrow O}} \quad \text{or } R-N\genfrac{}{}{0pt}{}{/ O}{\backslash O}\Big| \qquad R{-}O{-}N=O$$

Nitro-compound *Nitrite*

differ from them in their physiological action, being poisonous, but without the property of dilating the blood-vessels. The entrance of a nitro group into aromatic compounds usually increases the toxicity, nitro-benzene, nitro-naphthol, and nitro-thiophene, for example, all being more poisonous than the substances from which they are directly derived. Paranitro toluene, is not very poisonous, because it is oxidized to para-nitro-benzoic acid, which is then eliminated as para-nitro-hippuric acid, Tri-nitrotoluene, is distinctly poisonous. As is usually the case, the entrance of negative groups diminishes the toxicity, the aromatic nitroaldehydes being non-poisonous, as they are readily oxidized to the inert nitro-acids.

Effect of Basic Nitrogen Groups

The entrance of basic nitrogen radicles into aliphatic or aromatic compounds', or the presence of nitrogen in cyclic bases, can produce very important. The workers employed in explosive factories often suffer from headache caused by the vapour of nitro-glycerin pharmacological effects.

The convulsive action of ammonia and its disappearance on the entrance of alkyl groups has already been referred to. Keplacement of the hydrogen of ammonia by acid groups also diminishes the activity of the substance, and produces inert compounds which are found unchanged in the urine.

For example;

Carbamic acid, NH_2—CO—OH, is poisonous, probably on account of its unstable character, but its ester (urethane), NH_2—CO—OH—OC_2H_5, is more stable and has a hypnotic action. Hydrazine, NH_2—NH_2, is far more toxic than ammonia, but the tetra- and penta-methylene diamines, $NH_2[CH_9]_4NH_2$ and $NH_2[CH_2]_5NH_2$, are quite non-toxic. Hydroxylamine, NH_2—OH, is very toxic on account of its reactivity with aldehydes (Loew's theory), and hydrazoic acid and its sodium salt are very toxic. The action of oximes resembles that of the aldehyde from which they are derived together with that of a nitrite, the =NOH group probably being oxidized to a nitrite. Acetoxime lowers the blood-pressure and has a narcotic action.

Guanidine is toxic, owing to the presence of the labile imino (NH) group, and cyanamide, CN NH2, has a toxic action similar to that of guanidine. The salts of the platinum-ammonium bases of the type, (NH_4) $2PtCl_6$, resemble the other ammonium bases in having a curare-like action. Very great interest is attached to the experiments that have been made on the effect of the entrance of an amino group into the benzene nucleus, as they form the basis of a large number of antipyretics and analgesics.

Amino-benzene (aniline) resembles ammonia in many respects as regards its physiological action, but it also resembles benzene in some of its properties. It causes convulsions, but also paralysis of muscles and nerves, and if one of the hydrogen atoms of the amino group be replaced by an alkyl group, the convulsions no longer appear, but only the paralysing effect remains. If one of the hydrogen atoms in the nucleus of the aniline molecule be replaced by a simple atom, such as bromine, the convulsive effect is retained, and if it is replaced by an alkyl group the effect is increased; but if a complex group, especially an acid group, enters the nucleus, the effect is lost, as for example in amino-NH_2 benzene-sulphonic acid. Another property of all these derivatives, such as aniline, is that they have a strong toxic action on the blood, forming methaemoglobin. The entrance of a second amino group into the benzene nucleus causes a great increase in the toxicity, all three phenylenediamines, C_6H_4 $(NH_2)_2$, being extremely poisonous.

While all the aromatic derivatives of ammonia and hydrazine possess the property of lowering the temperature of the body, alicyclic-tetrahydro-β-naphthylamine,

CH CH_2
HC CH—NH_2
C
HC C CH_2
CH CH_2

produces a marked rise in the body temperature, and an increase in the albumin-metabolism. Substances containing tertiary nitrogen are often only slightly toxic, and are frequently without acy effect at all. In many cases, if the tertiary nitrogen is changed into secondary by reduction, powerfully active substances are obtained.

Effect of the cyanogen radicle

Hydrocyanic acid, HCN, is well known as an exceptionally strong poison, and this fact is probably connected with its great chemical reactivity. The action of cyanogen is similar, but only about one-fifth as powerful. In general, the isocyanides (isonitriles) cause paralysis of the respiratory centre, and the true cyanides (nitriles) produce coma. In this respect, therefore, the behaviour of hydrocyanic acid resembles that of an isocyanide (KNC) rather than that of a true cyanide (ECN). Neither the nitriles nor the isonitriles, however, show the intensely poisonous action of HCN, this only becoming apparent when HCN is again liberated by the organism. The lower members of the series of fatty nitriles, CH_3—ON and C_2H_5—ON, are. less poisonous than the higher members. Cyanacetic acid, CN—CH_2—COOH, is practically non-toxic. Cyanogen chloride, CN—Cl, is very poisonous, as it readily yields HCN. Potassium sulphocyanide, KCNS, is weakly poisonous for warm-blooded animals, but sodium nitroprusside, $Na_2Fe(CN)_5NO$, causes death with the appearance of prussic acid poisoning. In sodium ferrocyanide, $Na_4Fe(CN)_6$, neither the iron nor the CN group has any physiological action, and sodium platinicyanide shows no poisonous effect, though the ordinary platinum salts are very poisonous.

Effect of Aldehyde Groups

The physiological action of aldehydes appears to be closely related to their chemical reactivity. Formaldehyde, H. CHO, is very reactive, and has a strong irritant action on the mucous membranes, together with powerful antiseptic properties and a hardening action on the tissues. Acetaldehyde, CH_3. CHO, shows the action of the aldehyde group combined with that of the methyl group, as it produces an excitation followed by anaesthesia. The action of its polymeride, paraldehyde, $(C_2H_40)_3$, is stronger and more continued, and that of the higher polymeride, metaldehyde, $(C_2H_{40})_n$, is more toxic. By the entrance of hydroxyl groups into the aldehyde molecule, and also by the condensation of these substances to form aldols, the reactivity of these bodies is appreciably depressed, and so also is their physiological action. The sugars, for example, are physiologically inert.

Effect of Ketones

The ketones in general possess pharmacological properties similar to those which characterize the corresponding alcohols, i.e. they have a narcotic action. In the case of the aliphatic ketones this is fairly well marked on account of the alkyl groups, and a hypnotic action is also shown by the mixed ketones, such as acetophenone, C_6H_5—CO—CH_3 (hypnone). A large number of hypnotic substances of a more complex constitution contain a ketonic group, usually together with an ethyl group.

Effect of Acid Groups

It has already been pointed out how the entrance of acid groups into the molecule causes a marked decrease or a total cessation of the physiological action. Phenol, C_6H_5—OH, is poisonous, but phenyl-sulphuric acid, C_6H_5—SO_2—OH, is harmless. Morphine has a very powerful physiological action and is very poisonous, but morphine- sulphuric acid is quite inactive. In both these cases the diminution of the physiological effect is accompanied by the disappearance of a free hydroxyl group, the hydrogen of which is replaced by the SO_2—OH group. It might therefore be thought that this change is due to the removal of the anchoring group, but the entrance of acid groups has the same effect in many substances where it produces no change in the anchoring or active group. For example, substances containing a nitro group are strongly poisonous, but the entrance of acid groups lowers or destroys the toxicity without altering the nitro group. For example, Martius yellow (dinitro-naphthol) is markedly toxic, but its sulphonic acid (naphthol yellow S) is harmless. Nitrobenzene, is poisonous, but nitro-benzoic acid, is harmless. It does not matter whether the sulphonic acid group is united to carbon or to OH oxygen.

The entrance of carboxyl (COOH) groups into the aromatic nucleus is of great importance from the point of view of the synthesis of drugs, as it generally lowers the toxicity so very much. Benzene itself can be tolerated in doses of eight grams per day, but from twelve to sixteen grams of benzoic acid per day can be eliminated by the organism as hippuric acid, and still larger quantities can be administered without toxic effects, the excess being eliminated unchanged. Aniline, which is more toxic than benzene itself, is rendered practically harmless by the entrance of a carboxyl group, meta-amino-benzoic acid, being well tolerated by the organism. On the other hand, physiological properties which have been lost by the entrance of acid groups can be restored if these groups are esterified. For example, tyrosine, is not poisonous, but the hydrochloride of its ethyl ester has been shown to be strongly poisonous when administered to dogs. The addition of acid radicles to active basic substances is of special importance for the preparation of synthetic drugs. This is especially the case with regard to the acetylation of the amino group. By this means, the basicity is weakened and the action of the substance retarded, as it is only after hydrolysis that the active basic portion of the compound becomes free to exhibit its physiological effect. The acid group is usually physiologically inert, and the choice of the particular acid group to be used to combine with the amino group is governed chiefly by the physical properties of the compound thus formed. Lactyl derivatives are, as a rule, more soluble than acetyl, and these in their turn are more soluble than benzoyl, while salicyl derivatives are usually almost insoluble.

Now, in most cases the speed of hydrolysis depends chiefly on the solubility, and hence the rapidity with which a drug of this type acts depends on the acid group, but the nature of the action is usually not affected thereby. Usually, acetyl derivatives are the most convenient, not only because they are the cheapest to prepare, but also because the lactyl derivative is sometimes too rapidly hydrolyzed, whilst the benzoyl derivatives are generally hydrolyzed so slowly that they are excreted largely unchanged, and therefore without having exerted the desired effect. The only one of the previously mentioned acids that has a marked physiological action of its own is salicylic acid, but the salicyl derivatives of basic compounds are so insoluble that they usually escape from the organism almost entirely unchanged, and therefore are therapeutically useless. The presence of the benzoyl group is of great importance in a large number of substances, especially in some of the alkaloids. Ecgonine methyl-ester is without any noteworthy action, but its benzoyl derivative, cocaine, has a very important physiological action, producing local anaesthesia and other powerful effects.

Effect of Unsaturated Linkages

Unsaturated compounds are usually far more toxic than the corresponding saturated ones, a fact which is in accordance with their greater chemical reactivity. For example, propyl alcohol, CH_3 CH_2—CH_2—OH, is a narcotic, and causes intoxication, but is not really poisonous, whereas allyl alcohol, CH_2=CH—CH_2—OH, has strong poisonous properties, but is without the narcotic action which is characteristic of the saturated alcohols. Indeed, the Unsaturated alcohols are distinguished by their highly poisonous character.

Safrole,

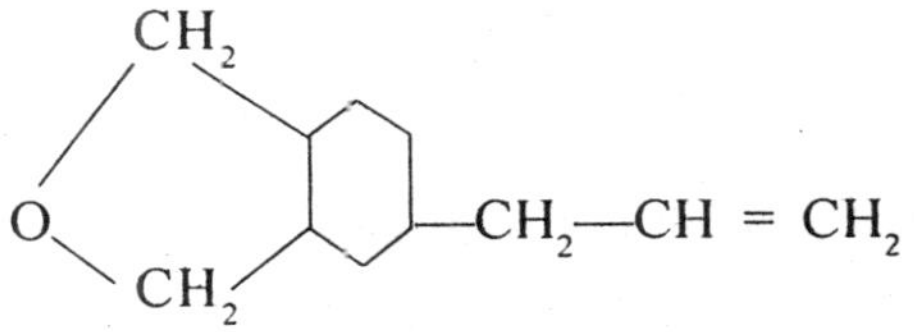

```
        CH2
       /    \
      /      [ring]—CH2—CH = CH2
     O      /
      \    /
       CH2
```

is far more poisonous than any other ethereal oil that has been experimented upon, and the isomeric compound, iso-safrole,

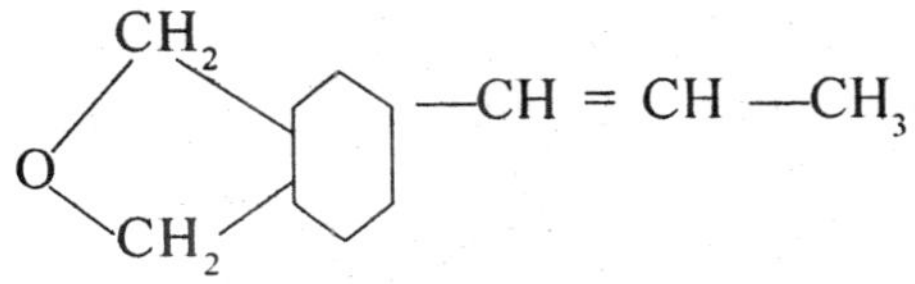

```
      CH2
     /    \
    /      [ring]—CH = CH —CH3
   O      /
    \    /
     CH2
```

is also poisonous, but not so strongly as safrole. Menthone is far

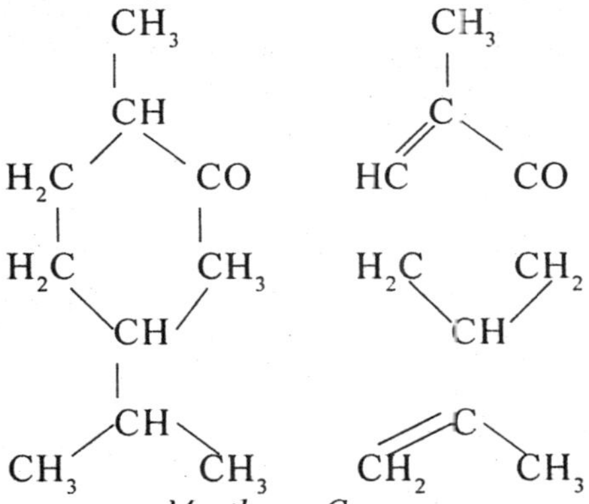

```
         CH3                    CH3
          |                      |
          CH                     C
        /    \                 //  \
    H2C        CO           HC       CO
     |          |
    H2C        CH3          H2C      CH2
       \      /                \    /
         CH                      CH
          |
          CH                      C
        /    \                 //  \
    CH3        CH3          CH2      CH3
```

Menthone *Carvone*

less toxic than carvone, which differs from it only in containing two unsaturated linkages, and acrolein, CH_2=CH—CHO, and crotonic aldehyde, CH_3—CH=CH—CHO, are far more toxic than the corresponding saturated compounds. The influence of increasing unsaturation is well illustrated by the following series of compounds;

CH_3 CH_2—CH_2—OH
CH_3—N
CH_3 OH

Chlorine (slighly toxic)

CH_3 CH=CH_2
CH_2— N
CH_3 OH

Neurine (very toxic)

CH_3 CH≡CH$_2$
CH_2— N
CH_3 OH

Farm ore toxic than Neurine

Allyl-tri-methyl-ammonium hydroxide, a homologue of neurine, is exceptional in being only slightly poisonous. When other elements are compared with carbon, a similar effect is often noticed, namely, that compounds of a lower degree of saturation are frequently more toxic than similar more highly saturated ones. For example, compounds of trivalent arsenic are usually more toxic than those of quinquevalent arsenic. The very poisonous substance known as mustard gas, β–β'-dichlorethyl sulphide, (Cl. CH_2. $CH_2)_2S$, on oxidation to the more highly saturated sulphone, (ClCH$_2$. CH$_2$) 2SO$_2$, is converted into a relatively harmless and non-toxic substance. Numerous other examples of similar kind are known.

Effect of Molecular Weight, Isomerism, etc.

Polymerides often show a different action from that of the original substance, but up to the present no regularities have been observed. The effect of increasing molecular weight in the series of the paraffins and the alcohols has already been referred to. In the substituted ureas, the narcotic effect increases with the number of carbon atoms in the branched side chain in the same way as it does in the alcohols, and a similar effect is found in the case of the pinacones.

The fatty acids are generally innocuous. Oxalic acid is poisonous, but the toxicity rapidly diminishes as the series is ascended. In the case of the homologues of pyridine, the toxicity increases very rapidly with increase of molecular weight. Pyridine, C_5H_5N, is the weakest in its action, and the intensity of toxic effect increases as the series is ascended through picoline, lutidine, and collidine, to parvoline, $C_5HN(CH_3)_4$, which is eight times as strong in its action as pyridine.

There is often a surprisingly great difference in the activity of stereo-isomerides. Isopilocarpine is probably stereo-isomeric with pilocarpine itself, but it is far weaker in its physiological action. Maleic acid is toxic for dogs, but its stereo-isomeride, fumaric acid is said to be harmless.

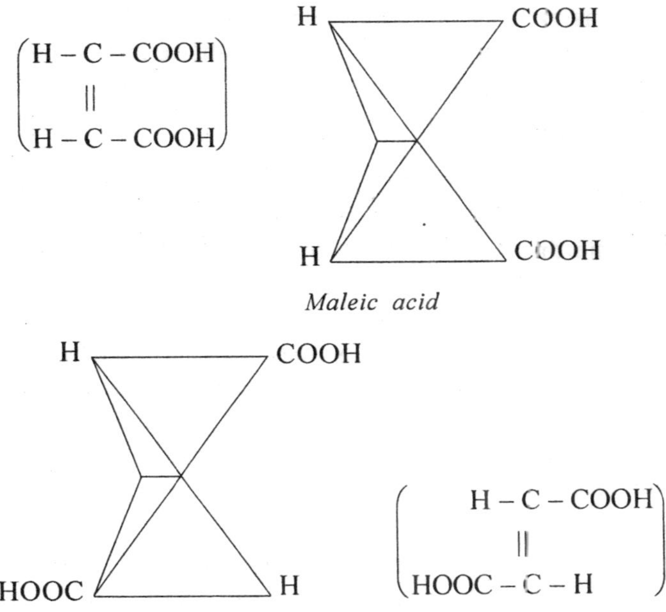

Fumaric acid

The difference may, however, be due to the different degree of ionization, maleic acid being the more highly ionized of the two.

In the case of stereo-isomerides, which are optically active, marked differences in the physiological action are very often encountered. At this point it will suffice to mention briefly a few examples. Atropine (racemic hyoscyamine) differs in some respects from Zoew-hyoscyamine, and Zaew-nicotine is twice as poisonous as the dextro variety. One of the most striking examples of this type is that of adrenaline, the natural laevo form being about eleven or twelve times as active as the dextro. Optical isomerides sometimes show differences in taste or smell, d&ztfro-asparagine, for example, being sweet, and ʻaew-asparagine tasteless.

Differences are often met with in the physiological action of ortho, meta, and para compounds, and in some types of compounds regularities can be traced. It has been suggested that para compounds are.more poisonous than ortho, 5 and this is often the case, but in many compounds, the reverse is true, ortho nitro-benzaldehyde, for instance, being more toxic than para. In fact, although many differences have been noted between the physiological effects of isomeric benzene derivatives, no general regularities have been traced between the ortho, meta, and para compounds, and between the different tri-substituted derivatives. In this connection it is interesting to note that saccharin, which is an ortho compound, is five hundred times sweeter than sugar, while the corresponding para compound is tasteless. Some ingenious theories have been advanced to account for the sweet taste of certain compounds, but they are none of them very satisfactory and call for no detailed account.

Chemistry of Drug Metabolism

The metabolism of drugs, their biochemical modification or degradation, usually through specialized enzymatic systems is konw as drug metabolism. The process of drug metabolism often converts lipophilic chemical compounds into more readily excreted polar products. Its rate is an important determinant of the duration and intensity of the pharmacological action of drugs. The chemical processes taking place in the organism consist of hydrolytic cleavages in the alimentary canal, and more profound changes of oxidation, and sometimes reduction, in the tissues or blood. In the case of the changes taking place in the alimentary canal, it is found that the saliva acts on but few drugs, owing to the short time that they remain in contact with it, and the fact that only one enzyme diastase is present. The contrary is the case in the stomach, in which many drugs can be absorbed, and where unpleasant by-effects are often manifested. For this reason, a great proportion of the work in connection with the synthesis of new drugs consist in modifying previously existing compounds so that they are rendered incapable of being absorbed or of exerting any action in the stomach. The gastric juices contain hydrochloric and other acids, and also an enzyme, pepsin, but it is to their acid character that their action on drugs is mostly due. Salts of organic acids are generally decomposed into the free acid and a chloride of the base, but esters and similar compounds are, in the great majority of cases, undecomposed by the gastric contents.

In the small intestine, substances enter an alkaline medium and come into contact with the pancreatic enzyme, trypsin. The latter has a marked hydrolyzing action on esters, anilides, and similar bodies, and it is only those substances which are hydrolyzed with great difficulty by all ordinary reagents that escape hydrolysis in the intestine. After saponification in the intestine, the components are able to exert their specific action, and advantage is taken of this fact in preparing derivatives the components of which would exert unpleasant by-effects on the stomach, but which remain undecomposed in that organ, and are then hydrolyzed in the intestine, enabling the components to exert their desired effect. For example, salicylic acid and its salts often give rise to unpleasant symptoms in the stomach, but acetyl-salicylic acid is comparatively inert and passes through the stomach unchanged. In the intestine it is hydrolyzed into sodium salicylate, which can then exert its useful action, and sodium acetate, which is inert.

Most aliphatic substances are oxidized to carbon dioxide, water, and urea, but there are numerous exceptions. Many substances are oxidized to acids, but aldehydes are never formed by oxidation in the body. On the contrary, aldehydes are often reduced to the corresponding alcohol, chloral, CCl_3 CHO, for example, being reduced to trichlorethyl alcohol, CCl_3. CH_2. OH. Many substances containing methyl groups are oxidized with difficulty; isopropyl alcohol is said to be partly oxidized to acetone and partly excreted unchanged. Acetone itself is oxidized with difficulty, methyl-ethyl-ketone more readily, while diethyl-ketone is almost completely oxidized. Primary and secondary alcohols are readily oxidized, but tertiary and all halogen-substituted alcohols are difficultly oxidized. Similarly, fatty acids are completely oxidized to carbon dioxide and water, but the chloro-substituted acids are not at all easily oxidized.

Aromatic compounds are not so readily oxidized by the organism as the aliphatic compounds. In all but a few exceptional cases, the aromatic nucleus remains unchanged, the process of oxidation being confined to the side chains. In those substances which contain a side chain of three carbon atoms, the middle one of which bears an amino group, the substance is completely oxidized. For example phenylalanine, tyrosine, etc.

In dogs also, phthalic acid and phthalimide are completely oxidized. Many aromatic compounds are oxidized in the organism by the entrance of a hydroxyl group in the para position to a previously present substituent group, but if the para position is already occupied, no hydroxylation takes place in the animal body. For example, aniline is oxidized to para-aminophenol HO⟨ ⟩NH_2. Ortho compounds are far more readily oxidized than para or meta. Aldehyde groups are oxidized to carboxyl groups, and in general a substance is usually oxidized to a carboxylic acid if this process takes place at all readily. Toluene, for example, gives benzoic acid, but benzene, on the other hand, is oxidized to phenol.

Reduction of a substance in the organism sometimes takes place, although it is a much more unusual process than oxidation. An interesting example is furnished by chloral, CCl_3. CHO, which is reduced by the organism to trichlorethyl alcohol, $CCl_3\ CH_2\ OH$, a reduction which can only be carried out with extreme difficulty in the laboratory. Quinone is reduced to hydroquinone. Keduction of a nitro group to an amino group is exceptional, nitrobenzene, for instance, never being reduced to aniline, but this type of action sometimes does occur. For example, both meta- and para-nitro-benzaldehydes undergo reduction of the nitro group, accompanied by oxidation of the aldehyde group, the product finally formed by the organism being the acetyl derivative of the corresponding amino acid

$$C_6H_4\begin{cases}NO_2\\CHO\end{cases} \rightarrow C_6H_4\begin{cases}NH_2\\COOH\end{cases} \rightarrow C_6H_4\begin{cases}NH.CO.CH_3\\COOH\end{cases}$$

Another interesting example of this type of change is furnished by ortho-nitrophenyl-propiolic acid, which is transformed by the organism into indoxyl, this being excreted as potassiumindoxyl-sulphate

$$C_6H_4\begin{cases}NO_2\\C{\equiv}C{-}COOH\end{cases} \rightarrow C_6H_4\langle{}^{NH}_{C(OH)}\rangle C{-}COOH \rightarrow C_6H_4\langle{}^{NH}_{C(OH)}\rangle CH$$

Picric acid is partially reduced! to dinitro-amino-phenol

$$C_6H_2(NO_2)_3OH \rightarrow C_6H_2(NO_2)_2NH_2.\ OH_4$$

The final stage in the transformation of drugs into inert, substances consists in the formation of acidic substances from the products of oxidation, reduction, etc., by means of the previously mentioned syntheses with sulphuric acid, glycuronic acid, etc. Other synthetic processes sometimes met with are the formation of urea derivatives and sulphocyanides, and more rarely, the introduction of acetyl or methyl groups, and the production of cystine derivatives. Usually no single synthetic process takes place to the exclusion of all others, but, on the contrary, a substance is usually excreted in more than one way, for example, as a glycuronic acid derivative, and also as a sulphonic ester.

Sulphonic Esters

The sulphuric acid required for these syntheses may be formed by the oxidation of albuminous bodies containing sulphur. It has already been pointed out that phenolic substances are often formed by the oxidation of aromatic compounds, and most of these phenolic compounds are excreted primarily combined with sulphuric acid as alkali salts, and secondarily combined with glycuronic acid. Phenol itself is found in the urine as the sodium salt of phenylsulphuric acid, C_6H_5. O. SO_2. ONa, a perfectly non-toxic substance.

Similar syntheses take place with other hydroxyl derivatives, but if these themselves are non-toxic, they are excreted unchanged without undergoing a synthesis with sulphuric or glycuronic acids. For example, homogentisinic acid, (CH_2 COOH, which is non-toxic, is eliminated unchanged, but the corresponding gentisinic acid, which is toxic, is eliminated partly as the non-toxic sulphuric acid derivative. The entrance of acid groups into phenols, with the corresponding loss of toxicity, destroys the property of combining with sulphuric or glycuronic acids, as in the case of salicylic acid, which is eliminated in combination with ammo-acetic acid in the same way as benzoic acid (i.e. it forms a compound analogous to hippuric acid).' Loss of acid character causes this property of combining with sulphuric acid to rappear and methyl salicylate and salicyl amide are found in the urine as sulphuric acid derivatives.

Introduction of more hydroxyl groups also causes this property to reappear, as in the case of gentisinic acid, and also protocatechuic and vanillic acids ; but veratric acid, which contains no free hydroxyl, does not unite with sulphuric acid. Aromatic ketones are usually oxidized to acids, but if they contain a hydroxyl group the formation of esters with sulphuric or glycuronic acids takes place, to the exclusion of the oxidation of the ketone to the corresponding acid. For example, acetophenone, C_6H_6. CO. CH_3, is oxidized to benzoic acid, but paeonol, gallacetophenone, and resacetophenone are found in the urine as sulphuric and glycuronic acid derivatives.

HO, HO, COOH — *Protocatechuic acid*

CH_3O, HO, COOH — *Vanillic acid*

CH_3O, CH_3O, COOH — *Veratric acid*

OH, CH_3O, CO.CH_3 — *Paeonol*

HO, OH, OH, CO.CH_3 — *Gallacetophenone*

HO, OH, CO.CH_3 — *Eesacetophenone*

Glycuronic Acid Derivatives

This acid, which has the constitution CHO (CH. OH)$_4$ COOH, is of great interest, as a very large number of poisonous substances are found in the urine as derivatives of it. It seems likely that in the first instance combination takes place between the drug and glucose, the CH_2OH group of which is then oxidized to COOH, with the formation of glycuronic acid. In the case of aliphatic compounds, the resulting compounds are formed with elimination of water, and are probably analogous in structure to the simple glucosides.

Derivatives of Amino-acetic Acid

Combination between glycine, $NH_2.CH_2.COOH$ and benzoic acid takes place in the kidney with formation of hippuric acid. This sysnthesis is of great importance as so many compunds are oxidised in the body to benzoic acid and it is also typica of a large number of precisely similar syntheses undergone by other carboxylic acids, such as salicylic acid, para-hydroxybenzoic acid, chloro-, nitro-, and bromo-benzoic acids, anisic acid, naphthoic acid, and many others.

Other synthetic processes that may be mentioned are the formation of the less toxic sulphocyanides from the toxic nitriles, the transformation of pyridine into' methyl pyridyl ammonium hydroxide, and the introduction of an acetyl group into a compound, as in the previously mentioned case of meta-nitro-benzaldehyde, which is transformed into meta-acetyl-amino-benzoic acid.

Chapter 3

Narcotics and Anaesthetics

The discovery of many synthetic substances having a powerful narcotic action, and almost free from the dangers and other drawbacks of preparations containing morphine, is one of the great triumphs of the application of synthetic chemistry to pharmacology. Still more important are those substances which are used as general anaesthetics. These compounds do not differ fundamentally from the other narcotics in their physiological action or chemical constitution, but they are usually volatile substances which are administered by inhalation, so that their effect can be rapidly produced, and the duration easily regulated. On the other hand, for use as a narcotic it is more convenient to employ non-volatile substances capable of offering resistance to oxidation by the organism, and which are therefore slower and more prolonged in their action.

The aliphatic hydrocarbons possess narcotic properties, and these are increased by the introduction of a hydroxyl group to form alcohols. The introduction of more hydroxyl groups, as in glycerol, causes the narcotic action to disappear, the hydroxyl merely playing the part of an anchoring group.

On the other hand, the narcotic action of many different substances is associated with the presence of alkyl groups, especially ethyl groups. It therefore appears that in the alcohols, the alkyl group, and not the hydroxyl group, is the active portion of the molecule. Emplacement of hydrogen atoms in a hydrocarbon by halogen, and especially by chlorine, also greatly increases the narcotic action of the substance, and in, these cases the chlorine actually seems to play an important part in the action of the substance, because, unlike the case of the alcohols, the strength of the action tends to increase with an increase in the number of chlorine atoms in the compound. This narcotic property is characteristic only of the aliphatic halogen compounds ; in the case of the halogen derivatives of benzene it is absent.

The inhalation general anaesthetics comprize, therefore, two groups, those in which the action is associated with the presence of halogen in aliphatic combination, and those in which it is associated only with the presence of alkyl groups. The narcotics likewise' include substances of these types, and also many compounds, the narcotic action of which seems to be connected with the presence of the carbonyl group (—CO—) in the molecule, and in some cases with the presence of a ring system containing basic nitrogen.

The foregoing statements will show that the anaesthetics and narcotics comprize a number of substances, which, from the chemical point of view, have really very little in common. Attempts have therefore been made to find a relation between some of the physical properties of these substances, and as a result, some interesting facts have been brought to light, which tend to show that in many cases there is a close parallelism between the hypnotic action and certain physical properties of the substance. Not only is this the case, but these physical theories of narcosis are superior to any purely chemical theory that can be devised, in that they throw some light on the mode of action of the hypnotic substances.

This is well illustrated by the work of Overton, and the suggestive theory that has been put forward by Hans Meyer.

It has been shown by Overton that substances may be divided into different groups according to the rapidity with which they diffuse into protoplasm, the rate of diffusion as a genera'l rule depending on the solubility of the substances in fat, lecithin, and lipoid substances of that type. If S_f denotes the solubility of a substance in fat, and S_w that of the same substance in water then the ration $\frac{S_f}{S_w}$ is called the distribution coefficient of the substance, and according to Overton, the value of this coefficient determines the velocity of diffusion into the cell protoplasm. Now, it was pointed out by Overton and Meyer that anaesthetics and narcotics are generally substances which diffuse rapidly, and therefore these compounds should possess a high distribution-coefficient. Meyer compared the aliphatic narcotics, and found that the narcotic power of these was roughly proportional to the magnitude of the distribution-coefficient. This fact is sometimes expressed by saying that the strength of the narcotic action of a compound is dependent on its solubility in lipoid substance. This is not strictly true, as it depends not so much on its actual solubility in the lipoid substance as upon the ratio of its solubility in lipoids to that of its solubility in water.

Meyer compared the narcotic power of these substances by finding the smallest concentration which would produce a definite physiological effect, and expressed the values as fractions of the normal solution (one gram mol. litre), calling these the liminal values. These liminal values can be taken as being approximately inversely proportional to the strength of the narcotic action. On comparing them with the distribu tion-coefficient, it is found that they vary in the opposite

$(CH_3)_2\ C(SO_2.CH_3)_2.$ $CH_2\ (SO_2.\ C_2\ H_5)_2.$

direction, being small when the distribution-coefficient is large, and these results therefore indicate that the strength of the narcotic action of these substances is approximately proportional to the distribution-coefficient.

In the preceding table the substances are arranged together in accordance with their chemical nature, but in order to show the close parallelism of the narcotic effect and the distributioncoefficient, it is worth while arranging them in order of magnitude.

Substance.	Distribution coefficient.	Liminal Value.
Trional.	4-46	0-0018
Tetronal	4-04	0-0013
Bubyl-chloral-hydiate	1-59	0-002
Sulphonal	1-11	0-006
Bromal-hydrate	0-66	0-002
Triacetin	0-30	0-010
Diacetin	0-23	0-015
Chloral-hydrate	0-22	0-02
Ethyl-urethane	0-14	0-04
Monacetin	0-06	0-05
Methyl-urethane	0-04	0-4

The substances are here given in order of decreasing distribution-coefficient, and it will be seen that with two slight exceptions, the liminal values of the substances, given in the same order, steadily increase. This result is of great interest, especially as the substances tabulated above represent many widely different chemical types. In the case of the sulphone derivatives, a high distribution-coefficient seems to be connected with the presence of ethyl groups, and it had already been pointed out by Baumann and Kast that the narcotic properties of these compounds depended on the ethyl groups. This will be referred to in greater detail in connection with these compounds.

Not only is the theory of Overton and Meyer well supported by the close parallelism shown above, but numerous subsidiary facts appear to give it additional support. One of these is an observation by Mansfeld that some narcotics have a more powerful action when given to starved animals, the explanation suggested being that in these there is less tissue-fat to absorb some of the narcotic, so that a greater portion of the latter is absorbed by the central-nervous system. No doubt, however, an alternative explanation could be offered.

Although the theory of Overton and Meyer suggests one factor which may be of great importance in determining narcotic action, yet there is no doubt that some part must be played by many other factors such as colloidal state, osmotic permeability, selective action on particular cells, etc. One fact which indicates the incompleteness of the theory is that the peripheral nerves contain a large amount of lipoid substance but are much less affected by the aliphatic narcotics than is the central nervous system.

It has also been pointed out" by Cushny that many aromatic compounds have a high distribution-coefficient, but are without narcotic action. A possible explanation of these facts is suggested by the views of Traube, according to which it is the osmotic permeability of a substance which is of primary importance in determining its narcotic action. In support of this view, it is pointed out that pyridine, nicotine, and antipyrine rapidly penetrate the membranes, although their distribution-coefficients are less than unity. Traube considers that surface tension is the force producing osmosis, and he therefore concludes that surface tension and narcotic power should run parallel. This was found to be the case with a large number of narcotics of varied types. It certainly seems reasonable to suppose that rapid penetration of the cells should be the most essential condition for enabling a substance to exert its effect on the interior of those cells. When the substance has once gained an entrance into the cell, its solubility in the cell lipoids may be an important factor in determining its narcotic power. The theories of Traube and Overton-Meyer are therefore not altogether antagonistic, and both of them are probably concerned with important facts underlying the mode of action of hypnotic substances.

Moore and Eoaf have studied the solubilities of chloroform and certain other anaesthetics in solutions of blood serum and haemoglobin, and have found them to be considerably higher than in ordinary saline solution or in pure water. The curves connecting the vapour-pressure and concentration of chloroform in serum or haemoglobin solutions do not correspond to the normal behaviour of a simple solution, but show evidence of association between the solvent and the dissolved substance. Moore and Eoaf therefore consider that anaesthetics form unstable compounds or aggregates with the proteins of the tissue cells, which exist only so long as the partial pressure of the anaesthetic in the blood is maintained, and that anaesthesia is due to a paralysis of the chemical activities of the protoplasm as a result of the formation of such aggregations. The curves obtained do not point to the formation of definite chemical compounds, but are more of the nature of adsorption curves.

The action of certain substances, such as alcohol, cannot be explained by the foregoing theories. Alcohol is miscible with water in all proportions, and is only slightly soluble in fats, but this need occasion no surprise when it is borne in mind that alcohol does not really belong to the same class of substances as sulphonal and inert bodies of that type, as it exerts some action on proteins and is oxidized in the body. The action of alcohol is, therefore, probably specific, and of a different kind from that of sulphonal, etc.

A theory, according to which narcosis is due to deprivation of oxygen, has been suggested by Baglioni. Jt is based on the fact that in the case of various aromatic compounds, the paralysing action of the substance is inversely proportional to the amount of oxygen already in the side chain, and that deprivation of oxygen by breathing inert gases, such as carbon dioxide, produces symptoms not unlike those of chloroform narcosis. In support of this view, it might be pointed out that Herter has shown that chloroform, ether, and chloral-hydrate diminish the oxidizing capacity of the tissues. This hypothesis indicates a possible mode of action of the narcotics after they have once entered the cell, the preceding theories indicating the conditions which determine their entrance into substance.

Halogen Anaesthetics

The most widely used compound of this type is chloroform, which shares with ether the distinction of being the most widely used anaesthetic. The deaths that occasionally take place when chloroform is used may in some cases be due to decomposition products such as phosgene, but it was formerly thought that these were impurities present in the chloroform to start with, and hence methods were devised to purify the chloroform very carefully when it was manufactured. Pictet did this by freezing it and centrifuging away the liquid. Very pure chloroform was prepared by Anschtitz by taking advantage of the interesting fact that salicylide (obtained by treating salicylic acid with $POCl_3$ in a neutral solution) forms a crystal line compound with chloroform, giving it up on distillation

$$\left(C_6H_4\begin{matrix}CO\\ \\ O\end{matrix}\right)_4 + 2CHCl_3 \rightleftarrows \left(C_6H_4\begin{matrix}CO\\ \\ O\end{matrix}\right)_4 (CHCl_3)_2.$$

It has been found, however, that pure chloroform is rapidly decomposed by air and moisture, and this decomposition is best hindered by the addition of about 2 per cent, of ethyl alcohol. Oil of turpentine, thymol, and other substances may also be used for this purpose.

Its anaesthetic action appears to be rendered more certain by the addition of a small quantity of ethyl chloride. Ethyl chloride and ethyl bromide have sometimes been used as anaesthetics, and a mixture of these substances with chloroform has been used, under the name of Somnoform, but does not possess any special advantage.

The narcotic action of chloroform is closely connected with the amount of chlorine it contains, the entrance of chlorine into the molecule of many aliphatic compounds conferring narcotic properties upon the substances so formed. This fact is illustrated by the following series of compounds :

CH_4	Methane	Without narcotic effect.
CH_3Cl	Methyl-chloride	Weak narcotic effect
CH_2Cl_2	Methylene-dichloride	Stronger narcotic action.
$CHCl_3$	Chloroform	Strong narcotic action.
CCl_4	Carbon-tetra-chloride	Strong narcotic action.

In this series the intensity and the persistence of the action increase with the amount of chlorine. A similar connection can be traced in the derivatives of aldehyde, atetaldehyde, $CH_3.CHO$, having a slight narcotic action and trichloraldehyde (chloral), CCl_3. CHO, a very strong one.

Although carbon tetrachloride appears to have a stronger narcotic action than chloroform, it has no advantage over it, but on the contrary seems to be too toxic for safe use. Its use by barbers to cleanse the hair has led to accidents, at least one fatal case having been recorded. On the other hand, some of the other members of this series have been suggested as substitutes for chloroform; methylene dichloride, $CH_2\ Cl_2$, has been recommended, as it produces less vomiting, and methyl-chloroform, CH_3. CCl_3, is said to be less dangerous.

The bromine substitution products' of the lower hydrocarbons also have narcotic properties. Bromoform, ($CHBr_3$), is sedative, and has been used to suppress attacks of whoopingcough. Ethyl bromide, (C_2H_5Br), has also been used for producing short slight anaesthesia ; it is a better anaesthetic and is far less toxic than ethylene-dibromide, the halogen esters of monacid alcohols being generally better anaesthetics and safer than the corresponding esters of diacid alcohols.

Ethyl chloride is used as a general anaesthetic and also as a local one, the latter by spraying it on to the surface to be operated upon. Its use in this case has nothing to do with the presence of chlorine in the molecule, but is due simply to its low boiling-point, the rapid vaporization producing so great a cooling as to cause anaesthesia. Ether is often used in the same way, and methyl-ethyl ether, $CH_3—O—C_2H_5$, which has a very low boiling-point, would be very useful for this purpose, as also would be the lower paraffins.

The volatile general anaesthetics are too transient in their action to be used as hypnotics for continued action, but never theless, a compound closely related to chloroform has proved very useful for the latter purpose. It was known that chloral hydrate, (CCl_3 $CH(OH)_2$), is hydrolyzed by alkalies, giving chloroform and alkaline formate, and thinking that a similar action might take place in the body, Liebreich suggested its use as a hypnotic. It was found to possess strong hypnotic properties, but Mering showed that it does not form chloroform in the body, being reduced to trichlorethyl alcohol instead

$$CCl_3.CHO \rightarrow CCl_3.CH_2.\ OH.$$

It has been suggested that the action of the halogen narcotics is due to the liberation of free halogen in the body, but there are several facts not in harmony with this view. One of the chief of these is that after administering some of these substances, the amount of alkaline chlorides in the urine is not increased. For example, after taking chloroform, the amount of chloride is increased, but not after taking chloral-hydrate, carbon tetrachloride, and dichloracetic ester, $Cl_2CH—\ COOC_2H_5$, all of which have a hypnotic action. On the other hand, trichloracetic acid splits off chlorine in the body, and has no hypnotic action.

Although chloral-hydrate was the first artificial hypnotic to come into general use, it suffers from many drawbacks. Thus, it cannot be injected subcutaneously like morphine, and it has a harmful by-effect on the heart. These drawbacks cannot be got rid of, but attempts have been made to prepare derivatives in which the unpleasant taste and burning feeling in the stomach produced by chloral-hydrate should be absent. All these preparations depend for their use upon the regeneration of chloral itself, as it is found that those which are more stable have no hypnotic action, and therefore all of them must possess the injurious by-effects of chloral itself. For this reason these derivatives of chloral are at a disadvantage compared with the hypnotics of other classes, and they owe their origin rather to the cheapness of chloral and to the fact that it was the first synthetic hypnotic to be brought into use.

Most of these chloral derivatives depend for their produc tion on the reactivity of the aldehyde group in chloral. Conden sation products of choral with oximes have been prepared, but they have not come into practical use. Their formation is in accordance with the general equation—

$$CCl_3.CHO{+}X = NOH = CCl_2—\overset{\displaystyle H\ \diagup}{\underset{\displaystyle \diagdown\ O—N{=}X}{C}}—OH$$

Examples are

$$CCl_3-\underset{\displaystyle \backslash O-N=C\begin{matrix}CH_3\\ \\ CH_3\end{matrix}}{\overset{\displaystyle H}{C}}-OH$$

from acetoxime, and similar compounds from nitroso-naphthol, benzaldoxime, camphor-oxime, etc. Other derivatives include those in which the aldehyde group has been combined with basic radicals, e.g. chloral -ammonia, chloral imide, CCl_3—CHOH=NH_2, etc.

Chloral-formamide, known as Chloralamide, is used as a mild hypnotic and sedative.

Other compounds have been prepared by combining the aldehyde group in different condensations with sugars, but these are not now used in medicine.

Dormiol is a condensation product of chloral with the hyp notic substance, tertiary amyl alcohol:

$$CCl_3-CHO + HO-\underset{C_2H_5}{\overset{CH_3}{C}}-CH_3 = CCl_3-\underset{OH}{\overset{H}{C}}-O-\underset{C_2H_5}{\overset{CH_2}{C}}-CH_3$$

It is a liquid with a burning taste and is insoluble in water. It is as poisonous as chloral, and closely resembles it in its action.

A tasteless solid polymeride of chloral, which possesses strong narcotic properties, has been obtained by the action of aluminium chloride on chloral. Other tasteless compounds have been obtained by combining chloral with orthoform and new ortho form. Hypnal is formed by the condensa tion of chloral with antipyrine, but this compound, as well as many others of the same type, has no advantage over a mixture of their components, as they readily split into these, and hence act in the same way as the mixture. The only advantage which can sometimes be claimed for them is that the unpleasant taste of free chloral is masked.

Chloral-urethane, was first prepared to combine the hypnotic action of chloral with that of urethane. An ethylated derivative of this is soluble in water, and has been given the name of Somnal. Butyl-chloral,has a stronger hypnotic action than chloral, but its effect disappears more rapidly, and its irritant action on the stomach is stronger than that of chloral. Trigemin,is a compound of butyl-chloral hydrate and pyramidon.

Acetone-chloroform (tertiary-trichlor-butyl alcohol), dis covered by Willgerodt in 1886, has been introduced into medicine under the name of Chloretone. It is prepared by adding potash to a mixture of chloroform and acetone, and is a white crystalline solid, melting at 96-97°C—

$$(CH_3)_2\ CO + CHCl_3 = \begin{matrix} CH_3 & & OH \\ & C & \\ CH_3 & & CCl_3 \end{matrix}$$

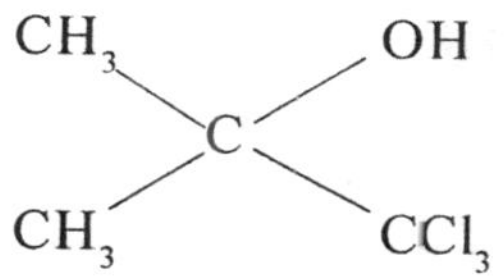

It has a great advantage over chloral in having no irritant action on the stomach. On the contrary, it has a sedative as well as an anaesthetic action, and very favourable results have attended its use in sea-sickness, vomiting, chorea, etc. It is the chief ingredient in the proprietary medicine known as Zotos. It also possesses antiseptic powers, for the sake of which it has been occasionally used, and it has sometimes found employment as a local anaesthetic for the mucous membrane of the larynx.

The corresponding bromine compound, $CBr_3.\ C(CH_3)_2.\ OH$, is known as Brometone, and is also used as a sedative. Bromal hydrate, $CBr_3.CH(OH)_2$, is of no use as a hypnotic, but in large doses it has an anaesthetic action. The corresponding iodine compound, $CI_3.\ CH(OH)_2$, is toxic, and affects the muscles and nerve-endings, but has only a slight action on the higher centres, and therefore has practically no hypnotic action.

In general, the aromatic halogen compounds have no hypnotic action, but tribromosalol (Cordal) appears to be an exception, as it is said to be a good hypnotic.

On the other hand, a large number of aliphatic bromine compounds have been brought into use as mild hypnotics and nervous sedatives. These are mostly derivatives of urea or of borneol or yvalerianic acid, and probably owe their sedative action partly to the presence of the bromine atom, and partly to the organic radicle present.

HYPNOTICS

In a large number of compounds the presence of an ethyl group seems to confer upon the substance the power of entering into a close connection with the nervous system. In the case of ethyl alcohol, an excessively large dose is required to produce sleep, but a number of compounds have been discovered which have a hypnotic action in much smaller doses, although their action appears to be due chiefly to the presence of ethyl groups. This difference in their behavfour is probably accounted for by the fact that alcohol is largely oxidized by the tissues, and so only a small fraction of it can produce a hypnotic effect, while these other compounds offer a certain amount of resistance to oxidation, and therefore can exert a more powerful hypnotic action.

Methyl alcohol has no narcotic action; in the other monohydric fatty alcohols, the narcotic action increases with increasing length of the unbranched side chain. In general, primary alcohols are less active than secondary, and these are less active than tertiary. In the case of tertiary alcohols, the strength of the action depends upon the nature of the radicles attached to the tertiary carbon atom. If the radicle is methyl, the action is relatively weak, but if it is ethyl it is stronger, and the strength increases with the number of ethyl groups.

Amylene hydrate, was introduced as a hypnotic in 1887, but the alcohols have not been used as general inhalation anaesthetics, owing to the slight volatility of the higher members, which are the only ones possessing marked narcotic properties. On the other hand, the ethers are far more volatile, and ordinary ethyl ether, $C_2H_5—O—C_2H_5$, is the most widely used general anaesthetic. If one of the hydrogen atoms in urea be replaced by a tertiary alkyl group, derivatives are obtained possessing a narcotic action. In accordance with the general rule, those containing an ethyl group united to the tertiary carbon atoms have a greater effect than those which contain only methyl groups in this position. For instance, substances containing tertiary amyl, or tertiary heptyl, are more active than those containing tertiary butyl.

Tertiary-butyl-urea produces sleep in doses of 4 grams; tertiary-amyl-urea, is an excellent hypnotic, which is both more active and more pleasant to take than amylene hydrate. It is, however, slower in its action on account of its smaller solubility. This substance is completely oxidized in the organism, but symmetrical diamyl urea is a very stable substance, which passes into the urine unchanged and has no physiological action. Tertiary-heptyl-urea is very slightly soluble. In doses ofgram it produces first intoxication and afterwards sleep.

Besides the urea derivatives already mentioned, many other amides possess hypnotic properties, but most of these are uncertain in their action. Of greater importance are the ureides of dibasic acids, many of which have strongly marked sedative properties. Thus, diethyl-malonyl-urea,

$$(C_2H_5)_2C\begin{matrix}\diagup CO—NH \diagdown \\ \diagdown CO—NH \diagup\end{matrix}CO$$

is a very stable sub and ethyl-propyl-malonyl-urea,

$$\begin{matrix}C_3H_7 \\ C_2H_5\end{matrix}\!\!>C\begin{matrix}\diagup CO—NH \diagdown \\ \diagdown CO—NH \diagup\end{matrix}CO$$

have a strong hypnotic action; whilst dipropyl-malonyl-urea has so intense a hypnotic action that it is too dangerous to be used.

Special mention must be made of Veronal, which has attained a very great clinical importance, and is now perhaps more widely used than any other synthetic hypnotic. It was formerly supposed to be practically free from toxic properties, but lately several cases of veronal poisoning have occurred. Nevertheless, in the usual doses, it is of very low toxicity and free from harmful by-effects, and has the advantage, compared with most synthetic hyp notics, in the matter of taste, solubility, and promptness of action. Its sodium salt

$$(C_2H_5)_2C\begin{matrix}\diagup CO—N—Na \diagdown \\ \diagdown CO—NH \diagup\end{matrix}CO$$

or more likely

$$\begin{array}{l} C_2H_5 \\ \quad\searrow \\ \quad\quad C \\ \quad\nearrow \\ C_2H_5 \end{array} \begin{array}{l} CO—N \\ \nearrow \quad\quad \searrow\!\!\!= \\ \quad\quad\quad\quad\quad C—O—Na \\ \searrow \quad\quad \nearrow \\ CO—NH \end{array}$$

known as VeronalSodium (Medinal), has the valuable property of being extremely soluble (1 in 5) in water. Veronal, which is a white crystalline solid, is prepared by a general method applicable to other dialkyl-barbituric acids. This consists in allowing dialkyl derivatives of malonic ester to react with urea or alkyl substituted ureas in presence of sodium ethoxide or other metallic ethoxides

$$R_2C\begin{array}{l} CO—OC_2H_5 \\ \\ CO—OC_2H_5 \end{array} + \begin{array}{l} H_2N \\ \quad\quad\searrow \\ \quad\quad\quad CO \\ \quad\quad\nearrow \\ H_2N \end{array} = R_2C\begin{array}{l} CO—NH \\ \quad\quad\quad\quad\searrow \\ \quad\quad\quad\quad\quad CO \\ \quad\quad\quad\quad\nearrow \\ CO—NH \end{array} + \begin{array}{l} HOC_2H_5 \\ \\ HOC_3H_5 \end{array}$$

Another method of preparing these compounds is to convert the dialkyl-malonic acids into chlorides, and then to heat these with urea

Instead of using alcoholic sodium ethoxide as a condensing agent in the first process, one can employ the metal sodium itself, or its amide or the dry powdered sodium ethoxide. Another method actually adopted for the manufacture of veronal is to heat the amide of diethyl-malonic acid with diphenyl carbonate to 200-220,

$$\begin{array}{l} C_2H_5 \\ \quad\quad C \\ C_2H_5 \end{array}\begin{array}{l} CO—NH_2 \\ \\ CO—NH_2 \end{array} + \begin{array}{l} C_6H_5—O \\ \quad\quad\quad\quad CO \\ C_6H_5—O \end{array} = \begin{array}{l} C_2H_5 \\ \quad\quad C \\ C_2H_5 \end{array}\begin{array}{l} CO—NH_2 \\ \quad\quad\quad\quad CO \\ CO—NH_2 \end{array} \begin{array}{l} +C_6H_5OH \\ \\ +C_6H_5OH \end{array}$$

Dial or diallyl-barbituric acid is said to have a stronger hypnotic cation than veronal. Luminal (phenylethyl-barbituric acid)

$$\begin{array}{l} C_2H_5 \\ \quad\quad\searrow \\ \quad\quad\quad C \\ \quad\quad\nearrow \\ C_6H_5 \end{array} \begin{array}{l} CO—NH \\ \nearrow \quad\quad\quad \searrow \\ \quad\quad\quad\quad\quad\quad CO \\ \searrow \quad\quad\quad \nearrow \\ CO—NH \end{array}$$

is also a hypnotic and has been recommended as a specific in epilepsy. A detailed description of its preparation has recently been published.

The presence of an alkyl group confers hypnotic properties on many of the urethanes. Ethyl urethane itself, is a mild hypnotic but methyl-propyl-carbinol-urethane, known as Hedonal, has a stronger hypnotic action. It is not much used, however. It can be prepared by the action of warm methyl-propyl-carbinol on urea nitrate

$$CO \begin{matrix} NH_2 \\ NH_2.HNO \end{matrix} = \quad + HO - CH \begin{matrix} CH_2 \\ C_2H_7 \end{matrix}$$

$$= CO \begin{matrix} NH_2 \\ O-CH \begin{matrix} CH_3 \\ C_3H_2 \end{matrix} \end{matrix} + NH_4NO_3$$

This method can be applied to the preparation of other similar compounds, using various secondary alcohols.The introduction of an acyl group (such as CO—CH_3) into the amino group of the urethanes lessens their toxicity, but does not otherwise alter their physiological action.

The pinacones show a narcotic action, which is least in the case of dimethyl-pinacone greater in the case of methylethyl-pinacone and greatest in the case of diethyl-pinacone

$$\begin{matrix} CH_3 \\ CH_3 \end{matrix} C(OH)—C(OH) \begin{matrix} CH_3 \\ CH_3 \end{matrix}$$

greater in the case of methyl-ethyl-pinacone—

$$\begin{matrix} CH_3 \\ C_2H_5 \end{matrix} C(OH)—C(OH) \begin{matrix} CH_3 \\ C_2H_5 \end{matrix}$$

and greatest in the case of diethyl-pinacone—

$$\begin{matrix} C_2H_5 \\ C_2H_5 \end{matrix} C(OH)—C(OH) \begin{matrix} C_2H_5 \\ C_2H_5 \end{matrix}$$

KETONES AND SULPHONES

Most of the ketones have hypnotic properties. Acetone is somewhat similar in its effect to ethyl alcohol; diethyl-ketone, $(C_2H_5)_2.CO$ (propion), is a stronger hypnotic-and anaesthetic, but its insolubility and unpleasant taste have prevented its extended use. Benzophenone, $(C_6H_5)_2CO$, has a slight hypnotic action, but much less than that of the aliphatic ketones. The mixed aromatic and aliphatic ketones have more marked hypnotic properties in virtue of the aliphatic portion of the molecule, acetophenone (hypnone), C_6H_5—CO—CH_3, being a fairly strong hypnotic, while phenyl-ethyl-ketone, C_6H_5—CO.C_2H_5, has a still more powerful action.

Of far greater practical importance are the sulphone deriva tives obtained from the ketones by the action of the mercaptans, and subsequent oxidation of the condensation product thus formed

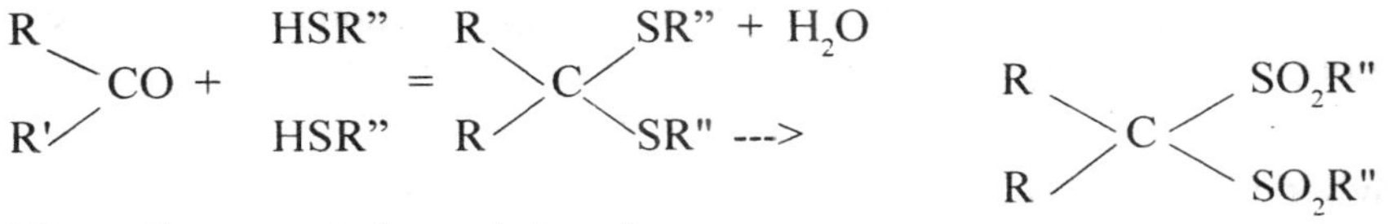

$$\begin{matrix} R \\ R' \end{matrix}\!\!>\!CO + \begin{matrix} HSR'' \\ HSR'' \end{matrix} = \begin{matrix} R \\ R \end{matrix}\!\!>\!C\!<\!\!\begin{matrix} SR'' \\ SR'' \end{matrix} + H_2O \dashrightarrow \begin{matrix} R \\ R \end{matrix}\!\!>\!C\!<\!\!\begin{matrix} SO_2R'' \\ SO_2R'' \end{matrix}$$

Baumann and East discovered that sulphonal

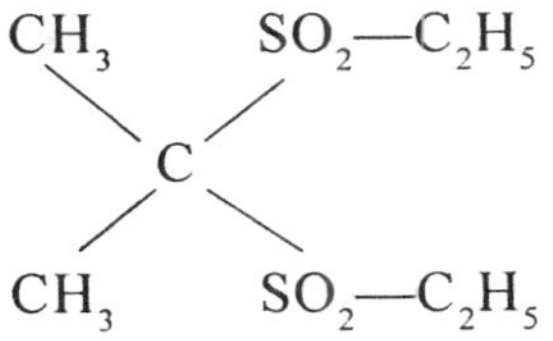

$$\begin{matrix} CH_3 \\ CH_3 \end{matrix}\!\!>\!C\!<\!\!\begin{matrix} SO_2—C_2H_5 \\ SO_2—C_2H_5 \end{matrix}$$

when administered to animals, produced a strong hypnotic effect. A large number of sulphones were then examined by these investigators, who foundthat di-sulphones in which the sulphone groups are united to two different carbon atoms, are inert; for example, ethylene-diethylsulphone.

$$\begin{matrix} CH_2—SO_2—C_2H_5 \\ | \\ CH_2—SO_2—C_2H_5 \end{matrix}$$

They also found that sulphones derived from methane are inert, as also are those containing methyl groups but no ethyl groups.

E.g. Methylene-dimethyl-sulphone, is inactive.

Methylene-diethyl-sulphone, is inactive. Ethylidene dimethyl sulphone, is inactive.

On the other hand, ethylidene-sulphone has a similar action to that of sulphonal, while the entrance of an ethyl group into the central methylene group also brings about the appearance of a narcotic action.

Trional, has a more powerful and prolonged action than sulphonal. Tetronal is very insoluble, and therefore is not so good a hypnotic, but for dogs it appears to have the most powerful action of all the sulphones.

The above examples indicate that the intensity of the hypnotic action appears to be conditioned by the number of ethyl groups in the molecule. A study of the fate of these compounds in the body has revealed the curious fact that those sulphones which are readily decomposed by ordinary chemical means are relatively stable in the organism, while those which are more resistant to ordinary chemical reagents are oxidized by the organism. For example, sulphonal, reversed sulphonal, trional and tetronal, though very stable to strong reagents, such as permanganate, are to a great extent decomposed in the organism, while $CH_2(SO_2.C_2H_5)_2$, which is easily decomposed by alcoholic potash, is found unaltered in the urine.

Sulphonal and trional are of great practical importance, as they are amongst the most widely used hypnotics. Technically, sulphonal is prepared by the condensation of ethyl mercaptan and acetone in the presence of zinc chloride, and oxidation of the mercaptol thus formed with excess of permanganate.

Trional is prepared in a similar manner, but the preparation is complicated by the necessity of introducing the extra ethyl group in place of methyl. This can be done in three different ways.

(1) Methyl-ethyl-ketone is condensed with ethyl mercaptan by means of dry HC1, and the resultant mercaptol oxidized

$$\begin{matrix} CH_3 \\ C_2H_5 \end{matrix} > CO \xrightarrow[HSC_2H_5]{HSC_2H_5} \begin{matrix} CH_2 \\ C_2H_5 \end{matrix} > C < \begin{matrix} S-C_2H_5 \\ S-C_2H_6 \end{matrix} \rightarrow \begin{matrix} CH_3 \\ C_2H_5 \end{matrix} > C < \begin{matrix} SO_2-C_2H_5 \\ SO_2-C_2H_5 \end{matrix}$$

This method is strictly analogous to the method of preparing sulphonal.

(2) Propionic aldehyde is condensed with ethyl mercaptan, and the product oxidized as before. The oxidation product is then methylated with methyl iodide and caustic soda to form trional

$$\begin{matrix} C_2H_5 \\ H \end{matrix} > C{=}O \xrightarrow[HSC_2H_5]{HSC_2H_5} \begin{matrix} C_2H_5 \\ H \end{matrix} > C < \begin{matrix} SC_2H_5 \\ SC_2H_5 \end{matrix}$$

$$\rightarrow \begin{matrix} C_2H_5 \\ H \end{matrix} > C < \begin{matrix} SO_2C_2H_5 \\ SO_2C_2H_5 \end{matrix} \rightarrow \begin{matrix} C_2H_5 \\ CH_3 \end{matrix} > C < \begin{matrix} SO_2-C_2H_5 \\ SO_2-C_2H_5 \end{matrix}$$

(3) By starting with ordinary aldehyde, instead of propionic aldehyde, and ethylating with ethyliodide and sodium ethylate in the last stage instead of methylating, trional can be obtained

$$\begin{matrix} CH_3 \\ H \end{matrix} > C{=}O \rightarrow \begin{matrix} CH_3 \\ H \end{matrix} > C < \begin{matrix} SC_2H_5 \\ SC_2H_5 \end{matrix} \rightarrow \begin{matrix} CH_3 \\ H \end{matrix} > C < \begin{matrix} SO_2-C_2H_5 \\ SO_2-C_2H_5 \end{matrix}$$

$$\rightarrow \begin{matrix} CH_3 \\ C_2H \end{matrix} > C < \begin{matrix} SO_2-C_2H_5 \\ SO_2-C_2H_5 \end{matrix}$$

Various alkaloids and many of their derivatives have a hypnotic action. Hyoscine, for example, is one of the most powerful narcotics and one of the most deadly poisons known, but in small doses and in conjunction with other alkaloids it has found useful application as a hypnotic and sedative. The hypnotic action of the opium alkaloids has already been referred to. A synthetic derivative of one of these which has recently been advocated as a good hypnotic is Eukodal, the hydrochloride of dihydrohydroxycodeinone.

Atropine And Tropeines Cocaine And The Local Anesthetics

Atropine was discovered in 1831 in the roots of the belladonna plant, and is a strongly poisonous alkaloid. Its chief use in medicine depends upon its action in dilating the pupil and paralysing the accommodation of the eye, and it is also used to check the inhibition of the heart arising from administration of chloroform and the depressant action of morphine. on the respiratory centre. Atropine is an ester, and on hydrolysis yields a basic sub stance, tropine, and optically inactive tropic acid. It has been shown that the alkaloid hyoscyamine, which is also obtained from belladonna and is laevo-rotatory, is the ester of tropine with laevo tropic acid, and therefore atropine appears to be racemic hyoscyamine. This view of the nature of atropine has been confirmed by Ladenburg, and dextro-hyoscyamine has also been prepared by the union of tropine with dextro tropic acid. The pharmacology of these three stereo-isomerides, *d*-hyos cyamine, *l*-hyoscyamine, and the racemic form, atropine, has been investigated by Cushny, using the frog as the subject of the experiments. It was found that all three were alike in certain respects, but that with regard to some aspects of their action dextro-hyoscyamine was the strongest and the laevo variety the weakest, while with other effects of the drug exactly the reverse was the case. In all cases the action of atropine was intermediate between that of the two optically active forms, and this fact is explained by Cushny by the assumption that atropine is probably decomposed in solution into its two active components.

As atropine is the ester of tropine with racemic tropic acid, it is obvious that a knowledge of the constitution of these two substances is necessary in order to know that of atropine. Tropic acid is a relatively simple substance, being indeed a homologue of mandelic acid, and having the constitution

$$\begin{array}{c} \quad\;\; H \\ \quad\;\; | \\ C_6H_5\text{—}C\text{—}COOH \\ \quad\;\; | \\ \quad\;\; CH_2 \\ \quad\;\; | \\ \quad\;\; OH \end{array}$$

This view of its structure has been confirmed by a synthesis of the acid. The question of the constitution of tropine is one which has presented far greater difficulties, but thanks to the researches of Ladenburg, Merling, Willstatter, and others, our knowledge of the constitution of this substance is as complete as that of any alkaloid, and there is no doubt that it is represented by the formula

$$\begin{array}{lll} CH_2\text{—}CH\text{—}CH_4 & & \\ |\qquad\quad | \qquad\; \diagdown & & \\ |\qquad\quad N{=}CH_2 & CH\text{—}OH & \\ |\qquad\quad | \qquad\; \diagup & & \\ CH_2\text{—}CH\text{—}CH_2 & & \end{array}$$

and this has bcen confirmed by Willstater's brilliant synthesis. The constitution of this substance is of great importance, as not only does it show that atropine and hyoscyamine are repre sented by the formula

```
CH2—CH—CH2                 C6H5
|    |                      |
|   N—CH3   CH—O—CO—CH
|    |     /                |
CH2——CH——CH2               CH2—OH
```

but it also explains the constitution of many other alkaloids which are derivatives of it. For example, when tropine is heated with sodium and amyl alcohol, it is converted into a substance which is stereo-isomeric with it, and which is identical with the substance pseudo-tropine, obtained by the hydrolysis of the coca alkaloid, tropa-cocaine (benzoyl-pseudo-tropine). Both tropine and pseudo tropine yield the same substance on oxidation, namely, tropi none, and this on reduction yields pseudo-tropine and not tropine itself

```
CH2——CH——CH2       CH2——CH——CH2
     |    |         |    |    |
     N—CH3 CO       N—CH3 CH . OH
     |    |         |    |    |
CH2——CH——CH2       CH2——CH——CH·COOH
```

Tropinone *Egconine*

It has been shown that cocaine is a derivative of ecgonine, which is in turn a carboxylic acid of tropine, and hence our knowledge of the constitution of this important alkaloid is also dependent on that of tropine. Having accomplished the synthesis of atropine by combining tropine with tropic acid, Ladenburg prepared other esters of tropine with various organic acids, to which he gave the name tropeines. Other tropeines have been prepared by Merck, Ladenburg, and others, and these substances are of great interest, as their mydriatic action can be easily compared, and hence they afford a convenient series of compounds for studying the relation between chemical constitution and physiological action. It was found that the tropeines derived from benzoic and cinna mic acids were without mydriatic action, as also was that from lactic acid, but] those derived from mandelic and atrolactinic acid possess mydriatic properties

```
               H                              CH3
               |                               |
C6H14N—O—CO—C—C6H5        C6H14N—O—CO—C—C6H5
               |                               |
               OH                              OH
```

Mandelic *Atrolactinic*

It will be noticed that the substances mentioned as having a mydriatic action, all contain both a benzene ring and an aliphatic hydroxyi group in the side chain containing the —O—CO group, and a statement has found its way into a good deal of the literature of the subject under the name of "Laden burg's Eule," that only those tropeines which possess these characteristics have a mydriatic action. The so-called rule was never enunciated by Ladenburg, and it has been shown by Jowett and Pyman to be incorrect. For example, they found that the following three tropeines all have a mydriatic action: α-hydroxy-β-2 pyridyl-propionyl tropeine

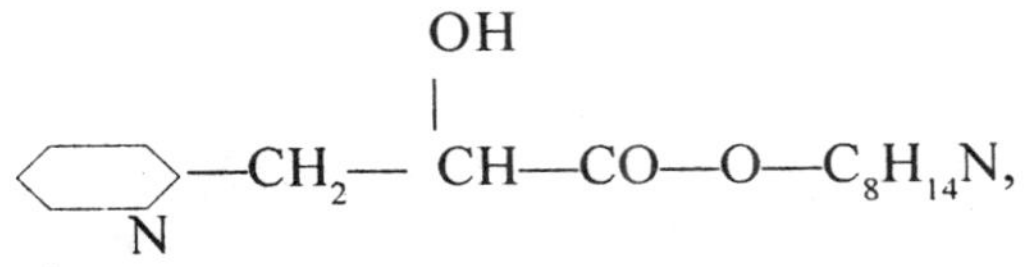

$$C_5H_4N—CH_2—CH(OH)—CO—O—C_8H_{14}N,$$

ortho-hydroxy-benzoyl tropeine (salicyl tropeine),

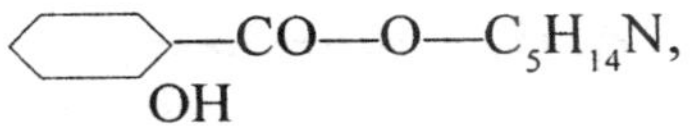

$$C_6H_4(OH)—CO—O—C_5H_{14}N,$$

meta-hydroxy-benzoyl tropeine,

$$HO—C_6H_4—CO—O—C_8H_{14}N,$$

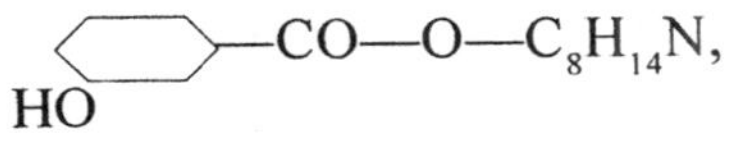

although the first contains no benzene nucleus, and the second and third no aliphatic hydroxyi group. The tropeine derived from mandelic acid is known as homa tropine, as it is the lower homologue of atropine

$$C_8H_{14}N—O—CO—CH(OH)—C_6H_5$$

Homatropine

$$C_8H_{14}N—O—CO—CH(CH_2\,.OH)—C_6H_5$$

Atropine

It is widely used in ophthalmic practice as a substitute for atropine, as its mydriatic action is nearly as great, and it has the advantages of being less toxic ; moreover, its mydriatic action develops and passes off more rapidly. Other esters have been prepared which resemble the tropeines to some extent in their chemical structure, and also have a mydriatic action, namely, the esters derived from triacetone methyl-alkamine, and its lower homologue vinyl diacetone alkamine

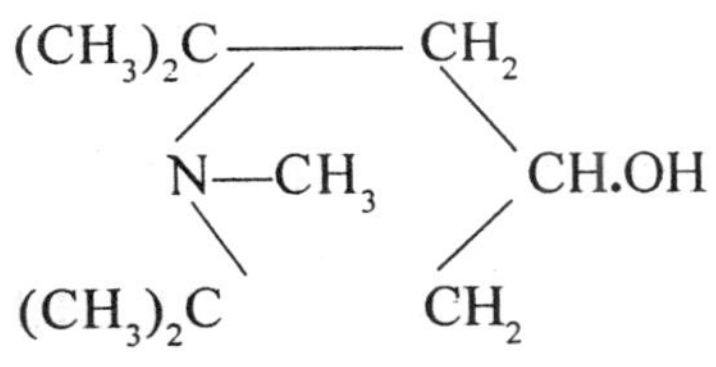

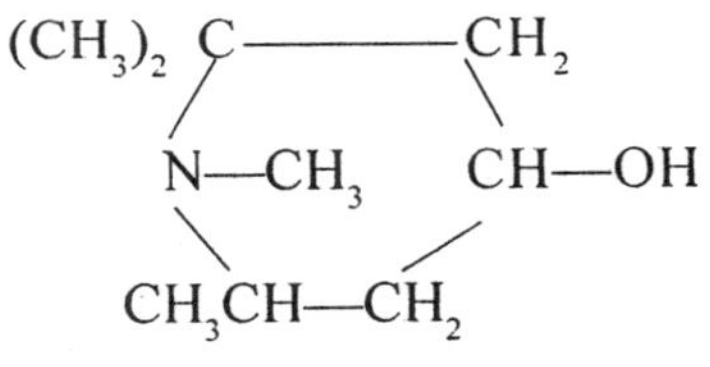

$(CH_3)_2C—CH_2—CH.OH—CH_2—(CH_3)_2C$, ring closed through $N—CH_3$

$(CH_3)_2C—CH_2—CH—OH—CH_2—CH_3CH$, ring closed through $N—CH_3$

The structure of these substances is closely related to that of tropine. The former is prepared by reduction and methylation of triacetonamine

$$
\begin{array}{ccc}
(CH_3)_2\,C\text{—}CH_2 & (CH_3)_2C\text{—}CH_2 & (CH_3)_2\,C\text{—}CH_2 \\
NH \quad CO \rightarrow & NH \quad CHOH \rightarrow & N\text{—}CH_3 \quad CHOH \\
(CH_3)_2C\text{—}CH_2 & (CH_3)_2\,C\text{—}CH_2 & (CH_3)_2\,C\text{—}CH_2
\end{array}
$$

a process carried out by Fischer, who also showed the relation between this substance and tropine. Its ester with mandelic acid resembles homatropine and atropine in having mydriatic action. The preparation of N-methyl-vinyl-diacetone-alkamine and some of its derivatives will be described later in connection with substitutes for cocaine, but for the present attention need only be drawn to the fact that various derivatives of the tropeine type have been prepared from it, some of which have a mydriatic action. This substance, methyl-vinyl-diacetonamine, exists in two forms, one, α, melting at 137-138 C., and the other, β, melting at 160-161°C. The existence of two asymmetric carbon atoms.

$$
\begin{array}{c}
(CH_3)_2\,C\text{——}CH_2 \\
CH_3\text{—}N \qquad CH\text{—}OH \\
C\text{——}CH_2 \\
CH_3 \quad H
\end{array}
$$

in the molecule explains the existence of these two forms as stereo-isomerides, and it is interesting to note that it is only the mandelic ester derived from the β-derivative which possesses mydriatic properties. This furnishes an interesting example of the difference in physiological action of stereo-isomerides.

Cocaine and the Local Anaesthetics

The alkaloid cocaine was discovered in coca leaves in 1860. It had long been known that the South American Indians were in the habit of chewing these leaves as a stimulant to enable them to stand great exertion without fatigue. The first use of cocaine in this country was for a similar purpose, but its great importance among alkaloids at the present time is due chiefly to Keller's important discovery that cocaine is a powerful and rapid local anaesthetic.

By hydrolysis with alkalies, cocaine yields ecgonine, benzoic acid, and methyl alcohol. Ecgonine was shown by Willstatter to be a carboxylic acid of tropine

$$
\begin{array}{ccc}
 & CH & \\
CH & | & CH\text{—}COOH \\
| & | & \\
CH_2 & N\text{—}CH_3 & CH\text{—}OH \\
 & | & CH_2 \\
 & CH &
\end{array}
$$

and by treatment with benzoyl chloride to yield benzoyl ecgonine, in which the hydroxyl group is converted into O—CO. C_6H_5; this on conversion into its methyl ester yields cocaine, which therefore has the formula

```
      / CH ─────────── CH—COOCH3
     /   |               |
  CH2    |               |
   |     N—CH3          CH—O—CO—C6H5
  CH2    |               |
     \   |               |
      \ CH ─────────── CH2
```

It is found that the free carboxylic acid, benzoyl-ecgonine itself, has no local anaesthetic action, but that any of its alkyl esters, such as ethyl, propyl, etc., resemble its methyl ester, cocaine, in having this action. This applies enly to the aliphatic esters, as the aromatic do not appear to have been prepared as yet. The effect of esterification is probably accounted for by an alteration of the anchoring group

```
      / CH ─────────── CH—COOH
     /   |               |
  CH2    |               |
   |     N—CH3          CH—O—CO—C6H5
  CH2    |               |
     \   |               |
      \ CH ─────────── CH2
```

Benzoyl-ecgonine (type of first series of esters).

```
      / CH ─────────── CH—COOH3
     /   |               |
  CH2    |               |
   |     N—CH3          CH—OH
  CH2    |               |
     \   |               |
      \ CH ─────────── CH2
```

Ecgonine methyl ester (type of second series of esters).

Ecgonine can be esterified in the usual way, leaving the hydroxyl group intact, and in this manner another series of esters can be obtained. Ecgonine methyl ester has no local anaesthetic action, but can be converted into cocaine by benzolyating the hydroxyl group. In this case, however, the nature of the group used to esterify the hydroxyl is important, for if the benzoyl group is replaced by others, the anaesthetic property is lost or greatly diminished.

Thus truxilline (isatropyl-cocaine) has no anaesthetic action, but is a strong cardiac poison, and Ehrlich found that, of several different cocaine derivatives, such as isatropyl-cocaine, valeryl-cocaine hydriodide, and phenylacetyl-cocaine hydriodide, the last named was the only one which had anaesthetic properties, but to a less degree than cocaine. All of these have a characteristic toxic effect on the liver, and differ from cocaine only in having the benzoyl group replaced by the one named.

Coca leaves, which are the only commercial source of cocaine, contain various other alkaloids, most of which are, however, de void of the useful physiological properties of cocaine. These other alkaloids are amorphous substances which yield ecgonine on hydrolysis, and therefore, owing to the high price of cocaine various methods have been devised to utilize them in improving the yield of cocaine obtainable from the leaves.

According to one method, the alcoholic solution of the amorphous bases is boiled with hydrochloric acid, filtered from the precipitated organic acids, and practically pure ecgonine hydrochloride obtained by evaporation of the filtrate. By means of benzoyl chloride or benzoic anhydride this is converted into benzoyl ecgonine, which is then esterfied with methyl alcohol giving cocaine. Various modifications of this method have also been devised.

Cocaine has several disadvantages when used for hypodermic injection, one of the most serious being that its solutions do not keep well, but become mouldy and decompose on boiling, so that they cannot be readily sterilized. For this reason, and also on account of the high price of cocaine, various attempts have been made to prepare analogous compounds which it was hoped would resemble cocaine in its useful physiological effects.

As cocaine is a derivative of ecgonine, which is closely related to tropine, and as atropine, one of the esters of tropine, has a slight anaesthetic action, various attempts have been made to prepare substances from tropine which should have an action resembling that of cocaine. Several synthetic tropeines have been prepared, and have already been discussed, but none of these are of value as substitutes for cocaine.

Strangely enough, however, a natural tropeine was discovered in Java coca leaves, which is a stronger lccal anaesthetic than cocaine. This substance, which is called tropacocaine, also has the advantage over cocaine in being less toxic and more resistant to microorganisms, and hence its solutions can be preserved for some length of time. It is the benzoyl ester of pseudo-tropine, which only differs from ordinary tropine in its space configuration. It differs from cocaine and atropine in having no mydriatic action, and in this respect it resembles the other pseudo-tropeines, such as those of mandelic acid and tropic acid.

Pseudo-tropine is obtained from tropine by heating it' with sodium amylate, and it is also obtained from tropinone

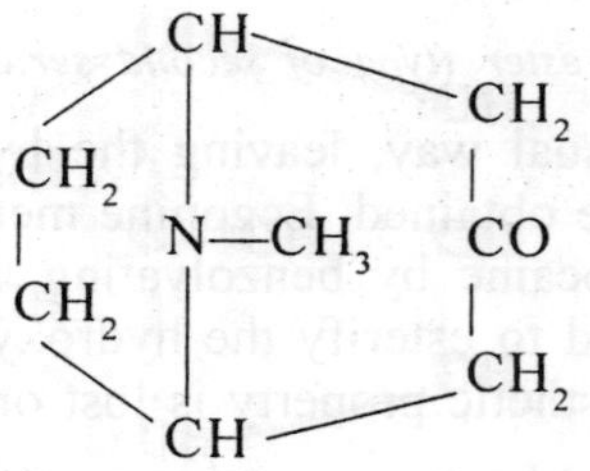

$$\begin{array}{ccc} & CH & \\ & & CH_2 \\ CH_2 & & \\ & N-CH_3 & CO \\ CH_2 & & \\ & & CH_2 \\ & CH & \end{array}$$

by electrolytic reduction in acid solution.

It should be pointed out that tropinone is obtained by oxidiz ing tropine with chromic acid, or with permanganate in cold strong acid solution, or with other oxidizing agents. Pseudo tropine is then easily converted into tropacocaine by means of benzoyl chloride

benzoyl chloride—

Tropinone

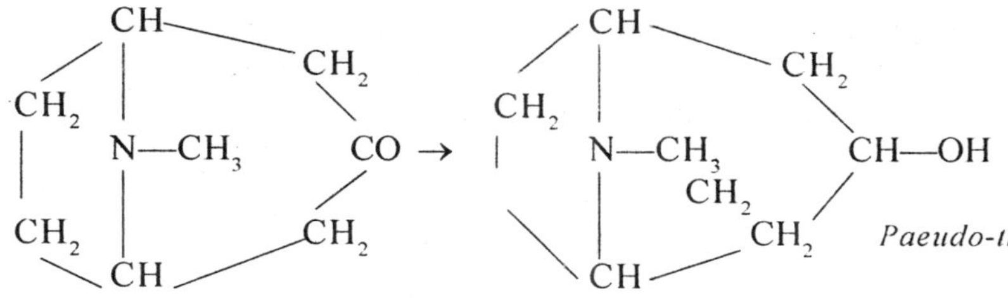

$$\begin{array}{l} \quad CH\text{——————}CH_2 \\ CH_2 \quad | \qquad\qquad \backslash \\ | \quad\;\; N\text{—}CH_3 \qquad CO \\ CH_2 \quad | \qquad\qquad / \\ \quad CH\text{——————}CH_2 \end{array} \rightarrow \begin{array}{l} \quad CH\text{——————}CH_2 \\ CH_2 \quad | \qquad\qquad \backslash \\ | \quad\;\; N\text{—}CH_3 \qquad CH\text{—}OH \\ \qquad\quad | \quad CH_2 \quad / \\ \quad CH\text{————}CH_2 \end{array}$$

Paeudo-tropine

Tropacocaine →

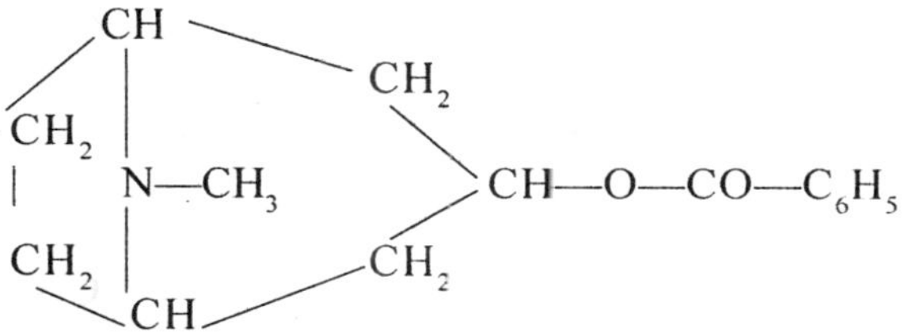

$$\begin{array}{l} \quad CH\text{——————}CH_2 \\ CH_2 \quad | \qquad\qquad \backslash \\ | \quad\;\; N\text{—}CH_3 \qquad CH\text{—}O\text{—}CO\text{—}C_6H_5 \\ CH_2 \quad | \qquad\qquad / \\ \quad CH\text{——————}CH_2 \end{array}$$

By the action of HCN on tropinone, Willstatter obtained the cyanhydrin

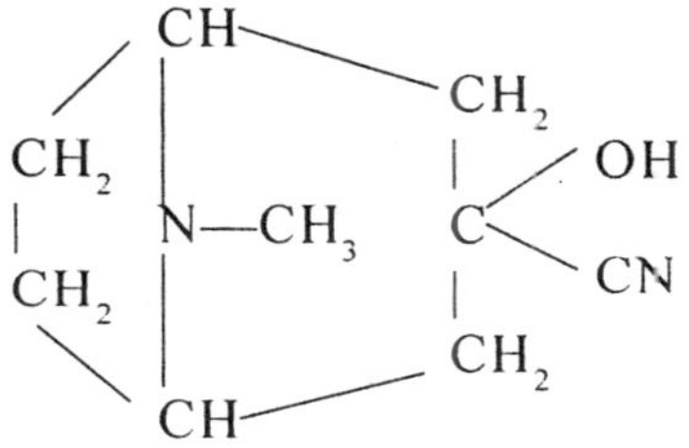

$$\begin{array}{l} \quad CH\text{——————}CH_2 \\ CH_2 \quad | \qquad\qquad | \quad / \; OH \\ | \quad\;\; N\text{—}CH_3 \qquad C \\ CH_2 \quad | \qquad\qquad | \quad \backslash \; CN \\ \quad CH\text{——————}CH_2 \end{array}$$

which on hydrolysis yielded

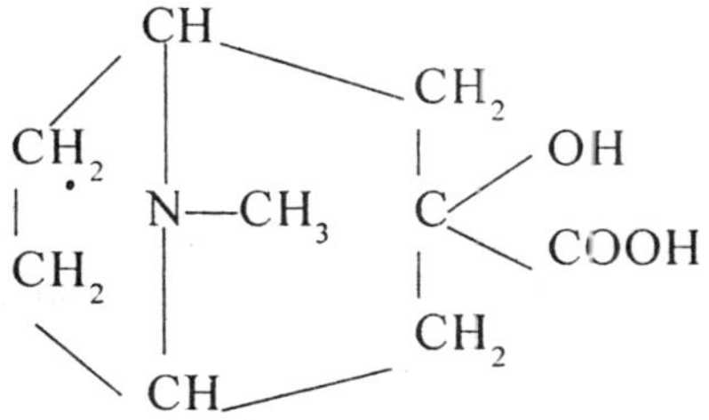

$$\begin{array}{l} \quad CH\text{——————}CH_2 \\ CH_2 \quad | \qquad\qquad | \quad / \; OH \\ | \quad\;\; N\text{—}CH_3 \qquad C \\ CH_2 \quad | \qquad\qquad | \quad \backslash \; COOH \\ \quad CH\text{——————}CH_2 \end{array}$$

which only differs from ecgonine in having the carboxyl group and the hydroxyl group united to the same carbon atom. He termed the substance α–ecgonine, and from it he prepared by benzoylation and methylation a substance, α-cocaine

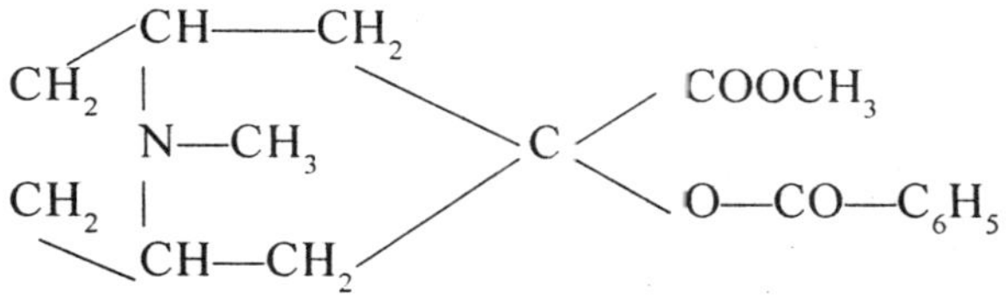

$$\begin{array}{l} \quad CH\text{———}CH_2 \\ CH_2 \quad | \qquad\qquad \backslash \quad / \; COOCH_3 \\ \quad\;\; N\text{—}CH_3 \qquad C \\ CH_2 \quad | \qquad\qquad / \quad \backslash \; O\text{—}CO\text{—}C_6H_5 \\ \quad CH\text{—}CH_2 \end{array}$$

which is, as would be expected from its close structural relation to it, very much like cocaine itself in its chemical and physical properties, such as the crystallizing power of its salts, but strangely enough, it is totally devoid of anaesthetic properties. Although α-cocaine does not possess anaesthetic properties, a similar a compound derived from N-methyl-triacetone-alkamine was found to be a valuable local anaesthetic

```
      CH ──────────── CH2            CH2 ─── CH ─── CH2
     /  |              |              |       |       |  / COOCH3
  CH2   |              |              |  CH3─N        C
   |    N─CH3        CH.OH            |       |       |  \ O─CO─C2H5
  CH2   |              |             CH2      |       |
     \  |              CH2              \    CH ─── CH2
      CH ────────/
```

Tropine — *α-cocaine*

```
         CH3                              CH3
          |                                |
  CH3 ─── C ──────── CH2          CH3 ──── C ──── CH3
          |           |                    |       |   / COOCH3
          N─CH3      CH . OH      CH3 ──── N       C
          |           |                    |       |   \ O─CO─C6H5
  CH3 ─── C ──────── CH2          CH3 ──── C ──── CH2
          |                                |
         CH3                              CH3
```

N-methyl-triacetone-alkamine — *α-eucaline*

The substance in question is known as a-eucaine, and was obtained by Merling in the following manner. Three molecules of acetone were allowed to react with one of ammonia, giving triacetonamine, which on treatment with HCN yields the cyanhydrin. This on hydrolysis yields triacetone-alkaminecarboxylic acid, which on benzoylation and methylation yields N-methyl-benzoyl-triacetone-alkamine-carboxylic acid methyl ester (α-eucaine)

```
3(CH3)2CO    (CH3)2C ── CH2          (CH3)2 C ── CH2
    →           |        |      →           |     |  / OH
+NH3           NH       CO  HCN            NH     C
                |        |                  |     |  \ CN
             (CH3)2C ── CH2          (CH3)2 C ── CH2

→  (CH3)2 C ── CH2                   (CH3)2 C ── CH2
          |     |  / OH     →               |     |  / O.CO.C6H5
         NH     C                     CH3 ─ N     C
          |     |  \ COOH                   |     |  \ COOCH3
   (CH3)2 C ── CH2                   (CH3)2 C ── CH2
```

This substance is a cheap substitute for cocaine, and it has the advantage of being less toxic and of being stable to boiling water, so that its solutions can be sterilized by boiling. It has the drawback of being somewhat painful and irritant when injected, and it has now been superseded by β-eucaine, the hydrochloride of benzoyl-vinyl-diacetone-alkamine.

```
            CH3
            |
     CH3—C——CH2
            |     |
     H—N    CH—O—CO—C6H5
            |
     CH3—CH—CH2
```

This substance is stable to boiling water and so can be readily sterilized. It is less toxic than α-eucaine, and is easily sol uble in water, the lactate, which is often used instead of the hydrochloride, being soluble up to 30 per cent. It is equal to cocaine in its anaesthetic properties, and is widely used in many branches of surgery. Vinyl-diacetonamine, the parent substance of β-eucaine, was obtained by Harries by the inter action of diacetonamine and acetaldehyde, but it is obtained in better yield by boiling the acid oxalate of diacetonamine with diethylacetal in alcoholic solution. The vinyl-diacetonamine, on reduction with sodium amalgam, yields a mixture of the two isomeric (cis and trans) vinyl-diacetone-alkamines.

This mixture is converted into the stable isomer of lower melting-point by boiling with sodium amylate, which is then converted into the base of β-eucaine by treatment with benzoyl chloride

```
       CO                  CO                          CO
      /  \                /  \                        /  \
  CH3| H |CH3          CH3   CH2                    CH2   CH2
         |      →             |       →              |     |
  +   | O | C(CH3)3           |      CH3.CHO         |     |
         |                  C(CH3)2        CH3—CH      C(CH3)2
                            /                     \   /
  +   | H |                NH2                      NH
        |
       NH3
```

2 (Acetone) + Ammonia *Diacetinamine* *Vinyl-diacetonamine*

```
        CH.O.CO.C6H5                 CH . OH
          /    \                      /  \
       CH2     CH2                 CH2    CH2
        |       |                   |      |
  CH3—CH       C(CH3)2       CH3—CH       C(CH3)2
         \     /                    \    /
           NH                         NH
```

Base of β-eucaine *Vinyl-diacetone-alkamine*

Recently, von Braun and Miiller 5 have prepared an interesting local anaesthetic from cocaine itself, namely N-λ-benzoxypropyl anhydronorecgonine ethyl ester.

Cocaine is converted into anhydronoregconine ethyl ester

$$
\begin{array}{l}
CH_2—CH—CH—COOC_2H_5 \\
\quad\quad\;\; | \quad\quad\; | \\
\quad\quad NH \quad CH \\
\quad\quad\;\; | \quad\quad\; \| \\
CH_2—CH—CH
\end{array}
$$

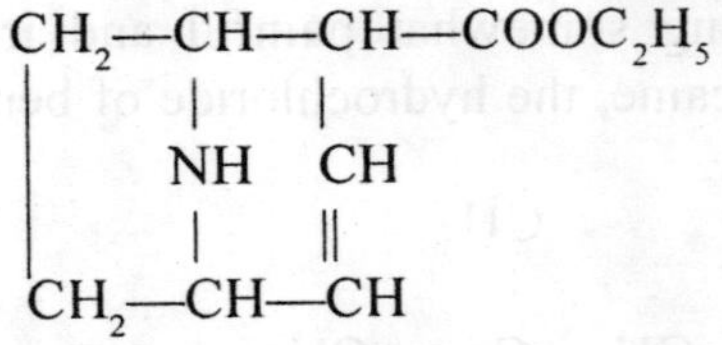

and this is then treated with λ-brompropyl benzoate,

$$Br.\ CH_2.\ CH_2.\ CH_2.\ O.\ CO.\ C_6\ H_5,$$

whereby the hydrogen of the imino group is replaced by the benzoxypropyl,

$$—CH_2\ CH_2—CH_2\ O.\ CO.\ C_6H_5\ ,$$

group. The compound so obtained is an oil, of which the hydrochloride is stable and soluble in water, and is advocated as a local anaesthetic under the name of Eccaine. It is said to be more active than cocaine and non-toxic, and is sufficiently stable for its solutions to be sterilized by boiling.

In the same paper, the above authors discuss the relation between chemical constitution and physiological action in the cocaine series, and they bring forward a good deal of evidence to show that the physiological activity of cocaine is in some way connected with the presence of an acylated hydroxyl group in the y-position to the nitrogen. This does not explain the lack of activity of a-cocaine which also contains this grouping.

Besides the members of the cocaine series and the acetone. alkamines, there are a large number of other substances pos sessing local anaesthetic properties. Many of the antipyretics of the aniline type have this property, and in the case of phenetidine derivatives, the anaesthetic character of the sub stance is greatly enhanced by combination with a second base. Holocaine, one of the best-known substances of this type, is the mono-hydrochloride of

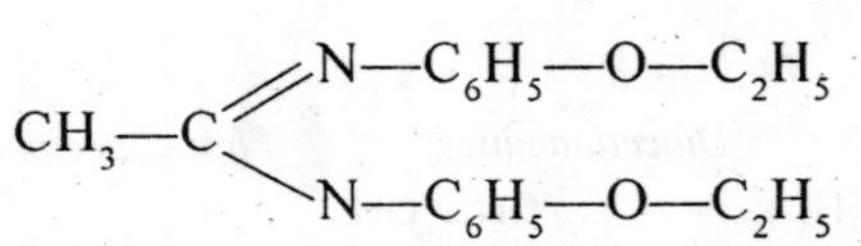

$$CH_3—C\begin{cases} =N—C_6H_5—O—C_2H_5 \\ \;\;N—C_6H_5—O—C_2H_5 \end{cases}$$

It is prepared by condensing phenetidine with phenacetin

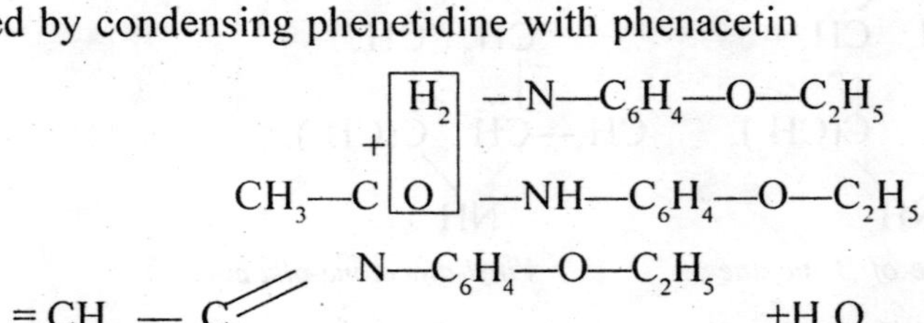

$$
\begin{array}{l}
\quad\quad\quad\quad \boxed{H_2} \; —N—C_6H_4—O—C_2H_5 \\
\quad\quad\quad + \\
CH_3—C\boxed{O} \; —NH—C_6H_4—O—C_2H_5
\end{array}
$$

$$= CH_3 — C\begin{cases} =N—C_6H_4—O—C_2H_5 \\ \;\;NH—C_6H_4—O—C_2H_5 \end{cases} + H_2O$$

Holocaine has the drawbacks of being more toxic than cocaine. 'and of being sparingly soluble in water; but, on the other hand, its aqueous solutions keep well, and have a rapid anaesthetic action. It is used in ophthalmic surgery.

In recent years attention has been directed chiefly to alkamine esters, which contain the grouping

```
    R
    |    |    |
  —N—C—C—O—CO—R'
    |    |    |
```

a grouping very similar to that indicated by asterisks (*) in the formula for cocaine

```
CH2—CH——  CH—COOCH3
|      |       |
|     *N—CH3CH—O—CO—C6H5
CH2—CH——CH2
      *      *
```

Stovaine, alypine, and novocaine are well-known members of the group. Stovaine

```
CH3          O—CO—C6H5
     \      /
        C
     /      \
C2H5         CH2—N(CH3)2, HCl
```

is a well-known synthetic anaesthetic, obtained by the action of magnesium-ethyl-bromide on diethyl-amino-acetone, and benzoylatibn of the product thus obtained

```
CH3            Br—Mg—C2H5     CH3
|                              |
CO                →      C2H5—C—OH
|                              |
CH2—N (CH3)2                   CH2—N(CH3)2
                             CH3
                              |
                    → C2H5—C—O—CO—C6H5
                              |
                             CH2—N(CH3)2
```

It is very widely used for producing spinal anaesthesia. Alypine is the hydrochloride of tetramethyl-diammo-dimethylethyl-carbinyl benzoate and is therefore similar to stovaine in its constitution, being the dimethyl-amino derivative of the latter. As a local anaesthetic, it is said to have the useful properties of cocaine without most of its drawbacks, producing rapid anaesthesia, and being free from injurious effects on the heart and respiration. A whole series of local anaesthetics which also possess anti septic properties has been discovered by Einhorn and Heintz. They found that the benzoyl derivative of amino-hydroxy-ben zoic ester possessed distinct anaesthetic properties, and contrary to the behaviour of cocaine, the removaj of the benzoyl group yielded a substance of which the anaesthetic' properties were greater than those of the benzoyl derivative. A large number of substances of this type were produced, of which *p*-amino-m-hydroxy-benzoic-methyl-ester, was introduced into practice under the name of Orthoform. It is a white powder, very slightly soluble in water, which is non toxic and has no action on the unbroken skin, but produces anaesthesia on coming into contact with the peripheral nerves themselves. It is used as a dusting powder for painful wounds, etc. The high price of orthoform led to the production of an isomeric substance, called New Orthoform, which has the same physiological action, but is cheaper. These compounds are not sufficiently soluble to be adminis tered hypodermically, and their more soluble hydrochlorides are too strongly acid for this purpose. To overcome this de fect, Einhorn has prepared various derivatives of glycocoll with different amino-hydroxybenzoic acids

$$\text{HO—Aryl group}\begin{cases}NH_2\\COOR\end{cases} + HO \cdot CO—CH_2—NR_2$$

Amino-bydroxybenzoic acid + Glycocoll derivative

$$= \text{HO—Aryl group}\begin{cases}NH \cdot CO \cdot CH_2 — NR'_2 + H_2O\\COOR\end{cases}$$

These compounds differ from the parent substances in being strongly basic, and hence they can form soluble salts (such as hydrochlorides) of neutral reaction. Their anaesthetic action is not closely related to that of the parent substance. A large number of these compounds have been prepared, and the diethyl-glycocoll derivative of,

NH_2

HO

$COOCH_3$

$NH.CO.CH_2—N(C_2H_5)_2$

HO

$COOCH_3$

having the formula has been introduced into therapeutics, in the form of its hydrochloride, under the name of Nirvanine. It is easily soluble, and is less toxic than orthoform, which it resembles in its general behaviour.

The simplest local anaesthetic of this type is ethyl paraamino-benzoate, known as Ancesthesine. It resembles orthoform in most of its properties, and is said to be free from any toxic action. It is insoluble in cold water, but its solution in olive-oil may be used for hypodermic in jection. Its salt w·'·h para-phenolsulphonic acid,

$$NH_2.C_6H_4.COOC_2H_5,\ HO.C_6H_4.SO_2H$$

is soluble in water, and has been used for hypodermic injection under the name of Subcutin.

The isobutyl ester of *p*-amino-benzoic acid $NH_2C_6H_4.\ COOC_4H_9$ has also been suggested as a local anaesthetic under the name of Cycloform.

Novocaine is the hydrochloride of the diethylamine derivative of anaesthesine, having the formula

$$NH_2\ \langle\bigcirc\rangle\ —CO—O—CH_2—CH_2N(C_2H_5)_2,\ HCl$$

which, it will be seen, contains the previously mentioned grouping

$$R'—CO—O—\overset{|}{\underset{|}{C}}—\overset{|}{\underset{|}{C}}—NR—$$

It is a non-irritant and powerful local anaesthetic, only oneseventh as toxic as cocaine, and of recent years has found very extended use, so that it is now the most valued of all local anaesthetics.

The preparation of novocaine can be carried out in various ways, but all the methods are somewhat difficult, involving the preparation of ethylene-chlorhydrin and of diethylamine, so that attention has recently been given to improvements in the production of the latter compound.

Ethylene chlorhydrin, for example, may be heated with para nitrobenzoyl chloride, and the resultant chlorethyl para-nitro benzoic ester, then heated with diethylamine for twenty-four hours in a closed vessel at 100-120°C. The diethylaminoethyl ester of para-nitrobenzoic acid so obtained is then reduced with tin and hydrochloric acid to form novocaine

$$\underset{NO_2}{\overset{COCl}{\bigcirc}} + HO—C_2H_4—Cl = \underset{NO_2}{\overset{CO\ .\ O—CH_2—CH_2—Cl}{\bigcirc}}$$

$$\downarrow HN(C_2H_5)_2$$

$$\underset{NH_2}{\overset{CO\ .\ O.\ CH_2\ .\ CH_2.\ N(C_2H_5)_2,}{\bigcirc}} \leftarrow \underset{NO_2}{\overset{CO.O—CH_2—CH_2—N(C_2H_5)_2}{\bigcirc}}$$

Another method is to condense ethylene chlorhydrin with diethylamine to form chlorethyl-diethylamine

$$Cl—CH_2—CH_2—OH + HN\ (C_2H_5)_2$$
$$= Cl—CH_2—CH_2—N(C_2H_5)_2 + H_2O.$$

This is then heated with sodium para-aminobenzoate to form the base of novocaine

$$H_2N\langle\ \ \rangle COONa + Cl —CH_2—CH_2—N(C_2H_5)_2$$
$$= NaCl + H_2N\langle\ \ \rangle COO—CH_2—CH_2—N(C_5H_5)_2$$

In a third variation, para-aminobenzoic acid is condensed with ethylene chlorhydrm by heating them to 100 in sulphuric acid solution. The compound so obtained is then heated in a sealed tube with diethylamine at 100-110 to form novocaine

$$\underset{NH_2}{\overset{COOH}{\bigcirc}} + HO—CH_2—CH_2—Cl = \underset{NH_2}{\overset{COO—CH_2—CH_2—Cl}{\bigcirc}} + H_2O$$

$$\downarrow HN(C_2H_6)_2$$

$$\underset{NH_2}{\overset{COO—CH_2—CH_2—N(C_2H_5)_2}{\bigcirc}}$$

Another process for preparing novocaine is by heating the ethyl ester of *p*-amino-benzoic acid with diethylamino-ethanol, $(C_2H_5)_2NCH_2.\ CH_2.OH$, and 2 per cent, of the aluminium compound of the latter.

Chapter 4

Organic Synthetic Drugs

Phenol Derivatives

Phenol, also known under an older name of carbolic acid, is a toxic, colourless crystalline solid with a sweet tarry odor. Its chemical formula is C_6H_5OH and its structure is that of a hydroxyl group (-OH) bonded to a phenyl ring; it is thus an aromatic compound.The word phenol is also used to refer to any compound which contains a six-membered aromatic ring, bonded directly to a hydroxyl group (-OH).

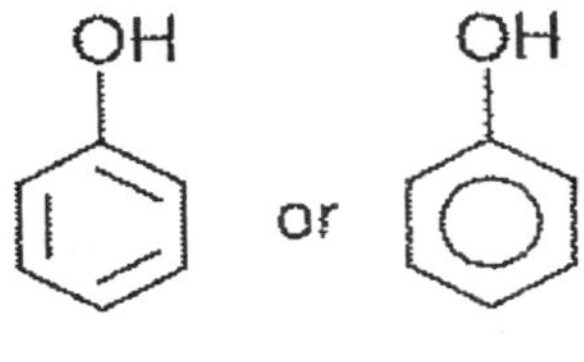

OH OH

or

Phenol has a limited solubility in water (8.3 g/100 ml). It is slightly acidic: the phenol molecule has weak tendencies to lose the H^+ ion from the hydroxyl group, resulting in the highly water-soluble phenoxide anion C_6H_5O-. Compared to aliphatic alcohols, phenol shows much higher acidity; it even reacts with aqueous NaOH to lose H^+, whereas aliphatic alcohols do not. One explanation for the increased acidity is resonance stabilization of the phenoxide anion by the aromatic ring. In this way, the negative charge on oxygen is shared by the ortho and para carbon atoms. In another explanation, increased acidity is the result of orbital overlap between the oxygen's lone pairs and the aromatic system. In a third, the dominant effect is the induction from the sp^2 hybridized carbons; the comparatively more powerful inductive withdrawal of electron density that is provided by the sp^2 system compared to an sp^3 system allows for great stabilization of the oxyanion. In making this conclusion, one can examine the pKa of the enol of acetone, which is 10.9 in comparison to phenol with a pKa of 10.0.

Phenol can be made from the partial oxidation of benzene or benzoic acid, by the cumene process, or by the Raschig process. It can also be found as a product of coal oxidation.Phenol has antiseptic properties, and was used by Sir Joseph Lister (1827-1912) in his pioneering technique of antiseptic surgery, though the skin irritation caused by continual exposure to phenol eventually led to the substitution of aseptic (germ-free) techniques in surgery. It is also the active ingredient in some oral anesthetics such as Chloraseptic spray. Phenol was also the main ingredient of the Carbolic Smoke Ball, a device sold in London designed to protect the user against influenza and other ailments. In the early part of the 20th century, it was used in the Battle Creek Sanitarium to discourage female masturbation.

It is also used in the production of drugs, weedkiller, and synthetic resins. Exposure of the skin to concentrated phenol solutions causes chemical burns which may be severe; in laboratories where it is used, it is usually recommended that polyethylene glycol solution is kept available for washing off splashes. Washing with large amounts of plain water (most labs have a safety shower or eye-wash) and removal of contaminated clothing are required, and immediate ER treatment for large splashes; particularly if the phenol is mixed with chloroform (a commonly used mixture in molecular biology for DNA purification). Notwithstanding the effects of concentrated solutions, it is also used in cosmetic surgery as an exfoliant, to remove layers of dead skin. It is also used in phenolization, a surgical procedure used to treat an ingrown nail, in which it is applied to the toe to prevent regrowth of nails.

Injections of phenol have occasionally been used as a means of rapid execution. In particular, phenol was used as a means of extermination by the Nazis during the Second World War. Phenol injections were given to thousands of people in concentration camps, especially at Auschwitz-Birkenau. Injections were administered either by medical doctors or by their assistants; such injections were originally given intravenously, more commonly in the arm, but injection directly into the heart, so as to induce nearly instant death, was later preferred . One of the most famous inmates at Auschwitz to be murdered by carbolic acid injection was St. Maximilian Kolbe, a Catholic priest who volunteered to undergo three weeks of starvation and dehydration in the place of another inmate and who was finally injected with carbolic acid so that the Nazis could make more room in their holding cells.

Phenol Derivatives

The entrance of a hydroxyl group into the benzene nucleus increases its solubility and its reactivity, and, as might be expected, these changes are accompanied by an increase in its physiological activity and its antiseptic powers. Phenol, C_6H_5OH, was the first antiseptic to be widely used, and its antiseptic powers are increased by the entrance of halogen or additional hydroxyl groups into the nucleus. The entrance of more hydroxyl groups adjacent to the first, also increases the toxicity of the substance, phenol, being less toxic than catechol, and this in its turn being less toxic than pyrogallol.

OH

OH

OH

Pyrogallol

On the other hand, the entrance of alkyl groups into the nucleus lowers the toxicity and increases the antiseptic properties, and for this reason the three isomeric cresols, $C_6H_4(OH)(CH_3)$, are better antiseptics than phenol. Unfortunately this advantage is marred by the fact that they are much less soluble in water than phenol, and hence many attempts have been made to obtain derivatives of cresol which should retain their antiseptic properties, and yet be more soluble in water.

The cresols form an emulsion with (hard) yellow soap, and an emulsion of this sort is known as Creolin, but it suffers from the drawback that it is demulsified by mineral acids, alkalies, or common salt. A solution of the cresols in soft soap is known as Lysol, and has attained wide use as an anti septic. It is prepared by mixing the crude coal-tar cresol (cresylic acid) with linseed oil in presence of alcohol until completely saponified, and the final product dissolves easily in water.

Lysol solutions suffer from the drawback that they vary in their antiseptic power according to the amount of cresol present, and hence they have to be tested bacteriologically, but they possess the advantage of being less poisonous than phenol or mercuric chloride.

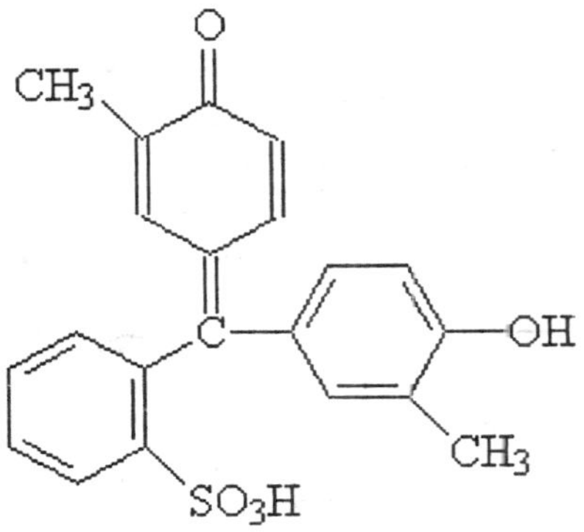

O
CH_3
C
OH
CH_3
SO_3H

Cresol

The cresols can also be rendered soluble in . water by mixing them with the sodium salts of organic sulphonic acids. Thus the cresols and other insoluble substances can be brought into solution by mixing them with the neutralized products obtained from the action of sulphuric acid on resinous oils, etc.

The sodium salts of cresotinic acid, salicylic acid, and of various fatty acids, can also be used to render the cresols soluble. The only substances of this type which are of practical importance are solutions of cresol in soap, but a solution of cresol in sodium cresotinate has been sometimes used internally under the name of Solveol, as a substitute for guaiacol and creosote.

Thymol, is used as an antiseptic and anthelminthic, but for the latter purpose thymol carbonate has been recommended instead. It is prepared from thymol by the action of carbonyl chloride,$COCL_2$ and is called Thymatol. The polyhydric phenols have not been much used in therapeutics. Resorcmol, has however, found considerable application in dermatology, and its acetyl derivative, $C_6H_4(OH)$ $(O.CO.CH_3)$ is used in the same way under the name of Euresol.Pyrogallol, is also used in some skin diseases on account of its reducing properties. The high price of phloroglucinol, has prevented it from coming into therapeutic use.

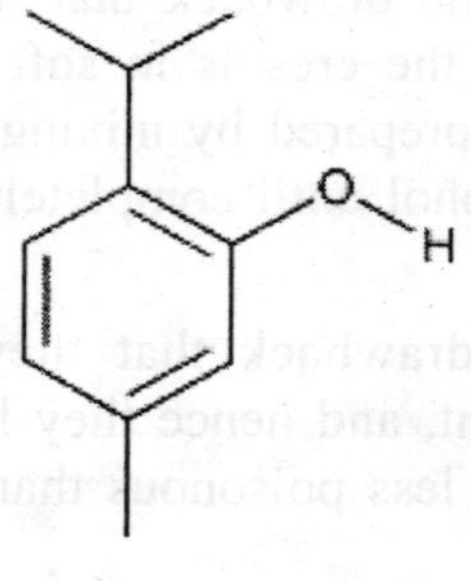

O
H

Thymol

A mixture of various phenyl-sulphuric acids, known as Aseptol, is obtained by allowing cold fuming sulphuric acid to act on phenol, and adding alcohol to the reaction product. The substance thus obtained is unstable and liberates phenol, but has no special value. The naphthols are not much used in therapeutics; a-naphthol is more poisonous than b-naphthol, and so only the latter finds any therapeutic application. Its sodium salt is soluble in water, and has received the name Microcidin.

An extended investigation on the effect of substituting halogen atoms or alkyl groups for hydrogen in the nucleus of phenol, and on the germicidal power of other phenolic derivatives, has been carried out by Bechhold and Ehrlich. The antiseptic power of these compounds was compared by finding the amount of the phenol required to prevent the growth of certain bacteria under standard conditions, the diphtheria bacillus being the one usually chosen. It was found that the entrance of chlorine or bromine into the nucleus of phenol is accompanied by an in crease in the antiseptic power. Trichlor-phenol was found to be twenty-five times and tri-brom-phenol forty-six times as active as phenol itself. Tetrachlor-, pentachlor-, and penta brom-phenols were increasingly active in the order given, the last-named being five-hundred times as powerful in its action as phenol itself.

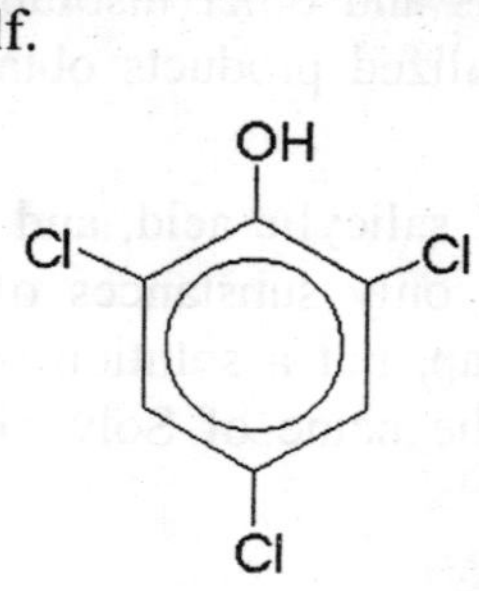

OH
Cl
Cl
Cl

Trichlor-phenol

The tetrabrom derivatives of all three cresols were found to be far more active in their germicidal properties than, tetrachloror tetrabrom-phenol, the derivative of ortho-cresol being slightly more powerful than the meta or para isomerides. A one-per-cent. solution of this substance takes less than two minutes to kill the diphtheria bacillus, whereas a corresponding solution of phenol requires more than ten. As the toxicity of this compound is stated to be comparatively slight, it might find useful application, but the toxicity is apparently still too great to permit of its being used internally.

In fact, although it was found that the entrance of a bromine atom reduces the toxicity and characteristic convulsive action of phenol itself, nevertheless the conclusion was arrived at that none of these compounds were suitable for use as internal disinfectants, as they were no more damaging to bacteria than to the animal body. The further introduction of halogen is accompanied by a rise of toxicity, that of the tribromand trichlorphenols being about equal to that of phenol itself, while the tetra-and penta-halogen derivatives are extremely toxic.

The simple phenolic compounds and their halogen derivatives are therefore not suitable for internal disinfection, but greater success has attended the use of phenolic derivatives containing a second group in the molecule, which lowers the toxicity of the compound.

The introduction of a carboxyl group, as is usually the case, lowers the toxicity. It is true that it greatly lowers the antiseptic power of phenol, but in spite of this, the ortho-carboxylic acid of phenol, (salicylic acid), has marked antiseptic properties, and it has proved to be of great value medicinally. Another type of phenolic derivatives which has proved of value as an internal antiseptic, is represented by guaiacol. Salicylic acid and its derivatives are used more on account of their value in lowering the temperature and diminishing the pain in rheumatism rather than for the sake of their antiseptic properties, but the acid and its ester with phenol (salol) are also used as antiseptics. The derivatives of salicylic acid will be discussed in the next section, and those of guaiacol in the one after that.

Salicylic Acid and Salols

The introduction of a carboxyl group into phenol lowers its physiological activity. Meta-hydroxyand para hydroxy-benzoic acids are practically inert physiologically, but salicylic acid, in addition to having a very slight toxicity, pos sesses special therapeutic properties which are of very great value.

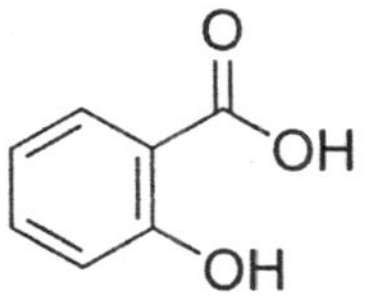

O
OH
OH

Salicylic acid

The most important of these is a powerful action against most of the symptoms of acute rheumatism, which is marvellous in its intensity. Salicylic acid also possesses marked antiseptic properties, for which it is often used to check gastric fermentation, as its irritant action on the stomach is much less than that of phenol, and for the same reason it has been used to prevent putrefaction in food, beer, etc. In the body it is rapidly absorbed, and circulates as the sodium salt, which is often used therapeutically instead of the free acid, as of course its action is the same, and it has the advantage of being far more soluble in water.

$COOCH_3$

OH

Salol

Salicylic acid was first synthesized by Kolbe l by the action of carbon dioxide on phenol. Originally this was carried out by passing carbon dioxide into hot phenol in the presence of sodium, but it was found better for technical purposes to prepare dry sodium phenate, C 6 H 5 ONa, and to pass carbon dioxide into this. Even by this method, however, only a 50 per-cent. yield of salicylic acid was obtained, and if potash were used instead of soda, para-hydroxybenzoic acid was the chief product. This process was greatly improved by Schmitt, who heated sodium phenyl carbonate under pressure at 140°C., by which means a quantitative yield of sodium salicylate was obtained. This process is also applic able to the preparation of naphthol-carboxylic acid and oxyquinoline-carboxylic acid. Salicylic acid and its sodium salt frequently produce unpleasant gastric symptoms, and to overcome this defect various derivatives have been prepared, one of which, acetyl-salicylic acid is of very great importance. This substance is known under various trade names, such as Aspirin, etc.

It has the characteristic action of salicylic acid, being hydrolyzed in the intestine with liberation of sodium salicylate, and it also has a slight sedative action of its own. It is a favourite remedy for feverish colds, headaches, etc. It was first obtained by heating salicylic acid with excess of acetic anhydride or acetyl chloride, but a better yield is obtained by carrying out the acetylation in the presence of a condensing agent, such as sulphuric acid, zinc chloride, or sodium acetate. Propionyl, butyryl, and other acyl derivatives of salicylic acid have been obtained in the same way. Of these, methylenecitrylsalicylic acid is known as Novaspirin, salicylosalicylic acid as Diplosal, and succinyl salicylic acid as Diaspirin. The calcium salt of acetylsalicylic acid is known as Soluble Aspirin or Kalmopyrin, and the sodium salt as Tylnatrin. These, and also the lithium salt, Hydropyrin, are more soluble in water than is Aspirin itself, which they resemble in their therapeutic effect, whilst the last named has also the characteristic action of lithium salts.

A substance which is isomeric with aspirin has been obtained by the action of acetyl chloride on salicylic acid in presence of ferric chloride. It is an aceto-salicylic acid of the formula $(CH_3.CO).C_6H_3.(OH)(COOH)$, and is not toxic, but it has far less antiseptic power than salicylic acid itself. Salicylacetic acid in which the hydroxylic hydrogen is replaced by an acetic acid residue, (CH_2COOH), instead of by the acetyl group, $(CO.CH_3)$, was first obtained by the oxidation of the ortho-aldehyde of hydroxyphenyl-acetic acid. Subsequently, improved methods of preparing it were devised, but this substance does not appear to have come into use as a drug.

Acetyl-salicylic acid, is the only one of real practical value, but other substances have been used instead of salicylic acid itself, chiefly on account of the distrust with which synthetic salicylic acid was at one time viewed. This distrust arose from the fact that the synthetic acid used often to contain the therapeutically useless para-hydroxybenzoie acid, as well as sometimes being contaminated with the posi tively harmful cresols. To make sure of obtaining a natural product, some physicians preferred to prescribe the glucoside salicin instead of salicylic acid itself. This glucoside is hydro lyzed by the organism, with liberation of saligenin (salicyl alcohol), which then fornıs salicylic acid by slow oxidation. In this way, the use of saligenin or salicin ensures a gradual action of the salicylic acid. Saligenin may be synthesized by the action of formaldehyde on phenol.

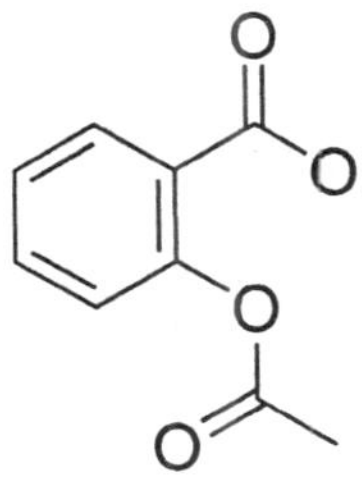

Acetyl-salicylic acid

Ortho-coumaric acid, bears the game relationship to cinnamic acid, as salicylic acid does to benzoic acid, and as cinnamic acid is more physiologically active than benzoic acid, it was to be expected that o-coumaric acid would have even more powerful antiseptic properties than salicylic acid. This was found to be the case, all three coumaric acids having a marked germicidal action, which is strongest in the case of the ortho acid.

Incidentally, it should be noted that sodium cinnamate has been found to be active in promoting leucocytosis, and has been recommended in cases of tuberculosis. Its esters with guaiacol, phenol, ortho and para-cresol are too irritant to be of use, but its meta-cresol ester is free from these drawbacks, and is used under the name of Hetocresol as a dusting powder for tubercu lous wounds. A similar compound of cinnamic acid with thymol has also been prepared.

A dilute aqueous solution of sodium cinnamate has been used in Germany under the name Hetol, and a glycerol solution of this substance which has certain advantages over the aqueous one was advocated by Morgan in 1902. The use of drugs, such as sodium cinnamate and sodium ortho-coumarate, which produce leucocytosis, appears to have given promising results in the treatment of inoperable cancer when combined with local treatment with substances such as copper oleate and antimony oxide.

The acetyl derivative of o-coumaric acid which is suitable for being taken by the mouth has been introduced by Martindale under the name of Tylmarin.

Salol and Esters of a Similar Type

The first ester, both the components of which are physiologically active, to be used in medicine was salol. The introduction of this substance by Nencki marked an important development in pharmacology, and many other attempts have been made to convert substances which are too toxic for ordinary use into esters from which the active component is liberated so slowly that it produces no injurious by-effects. In the case of salol itself the ester on hydrolysis in the intestine liberates phenol and salicylic acid, the hydrolysis taking place so gradually that the former component can exert its antiseptic effect without, under ordinary conditions, giving rise to its characteristic toxic symptoms. In this case both components of the ester are active, but this " salol principle," as it is called, can be extended to esters in which only the acid or the alcohol is active, but the use of which in the free state is hindered owing to the possession of toxic or corrosive properties.

Derivatives of this kind may be classed as partial salols, and comprise two types:

(1) Esters in which an active (aromatic) acid is esterified with an inert hydroxylic substance (alcohol), and which therefore bear a general resemblance in their physiological action to the acid from which they are derived, but which may differ from it in being free from harmful by-effects.

(2) Esters in which an active hydroxyl compound (alcohol or phenol) is esterified by an inactive acid. In this case the action of the substance resembles that of the alcohol or phenol, the sodium salt of the acid being inert.

It must not be lost sight of, however, that the ester itself may exert a specific action in the unhydrolyzed state, and therefore in every case the physiological action of the ester must be experimentally demonstrated before it can be used in therapeutics, as a priori reasoning based on the behaviour of the component acid and alcohol might be upset by the specific action of the ester.

Dealing first with the true salols themselves, it may be said with confidence that no other ester of this class is nearly as important as salol itself. It was found by Nencki that fatty or aromatic acids when allowed to react with phenols in presence of zinc chloride, aluminium chloride, etc., yielded ketones, but if $POCl_3$ were employed as the condensing agent, esters were formed.

Thus salol itself, is produced by heating two molecules of phenol, two molecules of salicylic acid and one of $POC1_3$ at 120°C. This method has also been applied to the preparation of the esters of salicylic acid with many other phenols, naphthols, resorcin, etc., and also to the preparation of the esters of other acids, such as nitro-salicylic and oxynaphthoic. A cheaper and simpler modification of this method of preparation is to allow carbonyl chloride to react with an equi-molecular mixture of the sodium salts of the phenol and of the acid. The ester thus formed can usually be separated from the reaction mixture by distilling it off with steam. In this way an enormous number of esters of a similar type to salol have been prepared.

Salol is also obtained by heating salicylic acid alone at 160°-240°C., provided that the water split off by the reaction be removed by distillation, and that the access of air be prevented.

Another method for the preparation of salol is by heating polysalicylide $(C_7H_4O_2)x$ (obtained by heating salicylic acid with $POG1_3$ alone with phenol. The higher members of the salol series may be prepared by heating salol itself with the higher phenol, whereby the lower phenol is replaced by the higher. This method is especially suitable for use with phenols which are too reactive to be treated with the vigorous condensing agents used in the first methods ; e.g. hydroquinone, eugenol, carvacrol, etc.

Partial Salols of First Type

Methyl salicylate (oil of wintergreen) is the best known of these. The synthetic pro duct is superior to the natural in being free from the irritant effects of the latter. It is slower in its action than salicylic acid itself. Ethyl salicylate appears to have a specific action of its own, and produces undesirable effects, and hence is not used in therapeutics. Mesotan or Ericin is the methoxy-methyl ester of salicylic acid.

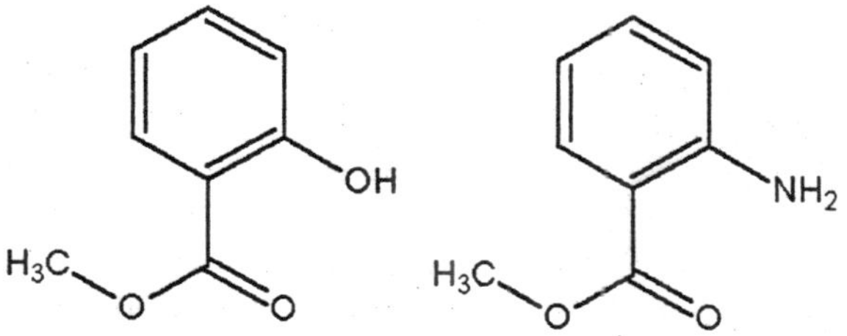

OH

H_3C O O

NH_2

H_3C O O

Methyl salicylate

Salacetol is obtained by the action of monochloracetone on sodium salicylate and is therefore the salicylic ester of acetol.It is rapidly saponified in the intestine, but has no particular advantage over salicylic acid. The methyl esters of the acyl salicylic acids have also been used medicinally. Methyl acetylsalicylate is known as Methylrhodin, and methyl benzoylsalicylate as Benzosalin.

Partial Salols of Second Type

Partial Salols of the Second Type, in which only the phenol is the active part, comprise carbonates and esters of fatty acids.

True Salols

Of esters of the type of salol in which both components are active, salol itself is the most important, and next to it comes β-naphthol salicylate, which is known as Betol, and as Naphtholsalol. The benzoate of ftnaphthol is also used to a considerable extent under the name of Benzonaphthol. Menthyl salicylate is used medicinally under the names of Salimenthol and Salit. In addition to the esters of salicylic acid itself, the phenyl esters of acetylsalicylic and of salicylosalicylic acid are used, and are known respectively as Vesipyrin and Disalol.

In general it is only the monobasic acids which lend them selves in an economical manner for use as salols, as the esters of polybasic acids usually only split off one molecule of the phenol, the rest being excreted still in combination with the acid. Thus, for example, triphenyl phosphate $(C_6H_5O)_3$ PO, is hydrolyzed in the intestine into diphenyl-phosphoric acid, and phenol, the former not being further decomposed in the organism, and thus only a third of the phenol is utilized.

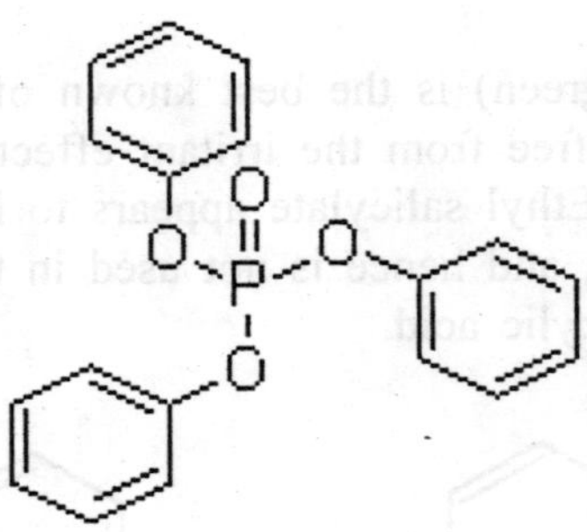

Triphenyl phosphate

Another type of salicylic acid derivative comprises those in which it is combined with basic radicles. Salicyl-amide, is more soluble than salicylic acid, and has a more marked analgesic action. Many compounds have been prepared in which salicylic acid is combined with substances having antipyretic and analgesic properties, especially the derivatives of acetyl-para-amino-phenol. The salicylate of acetyl-para-amino-phenol, is known as Salophen. In the same manner lactylamino-phenol can be condensed with salicylic acid.

Guaiacol, Creosote, and their Derivatives

Beechwood tar has long been used in the treatment of phthisis, and the chief active ingredients appear to be guaiacol and creosol.

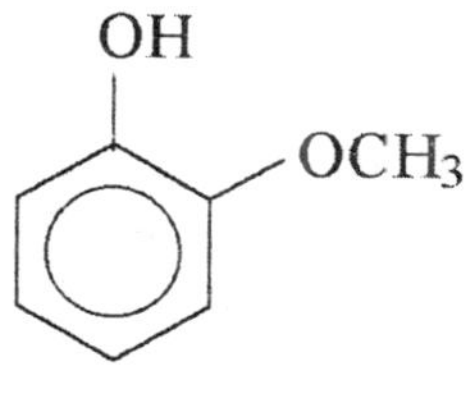

OH

OCH_3

guaiacol

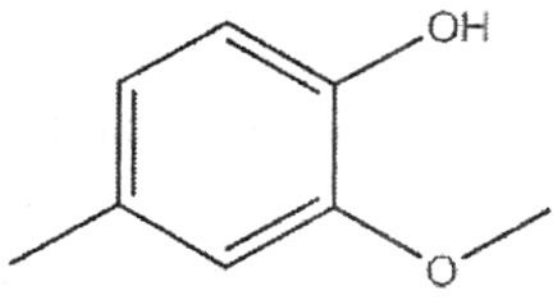

OH

O

creosol

These can be obtained from the beechwood tar, but guaiacol is also largely produced by synthetic methods. It can be obtained by the partial methylation of catechol, , but this method is too expensive for technical use, and a cheaper synthesis is that which starts from ortho-anisidine.

This substance is pre pared from ortho-nitro-phenol by methylation and subsequent reduction. It is then diazotized, and the solution poured into sulphuric acid containing a large quantity of sodium sulphate, heated to 135°-160°C., whereupon the guaiacol distils over with the steam as it is formed. In this way the production of by-products is diminished. Guaiacol, like phenol, is corrosive and poisonous, and it is probable that creosol would be less toxic, though more strongly antiseptic, in the same way as cresol and phenol.

Guaiathol, and other higher homologues have been prepared by heating catechol and the required alcohol under pressure with zinc chloride at 160°-220°C., but they are more expensive than guaiacol and have no advantage over it. With the object of diminishing the unpleasant and toxic properties of guaiacol and creosote, various derivatives have been prepared in which the hydroxyl group is esterified with an acid according to Nencki's salol principle.

Carbonic acid was the first used, and various other organic and inorganic acids have also been tried with varying degrees of success. In every case, the action of the compound is due to the guaiacol split off in the intestine. Guaiacol carbonate, known as Duotal, is prepared by the action of phosgene on an alkaline solution of guaiacol Duotal is thus obtained as an insoluble preparation, which has less toxicity and unpleasant taste than guaiacol itself, but which is not quite tasteless.

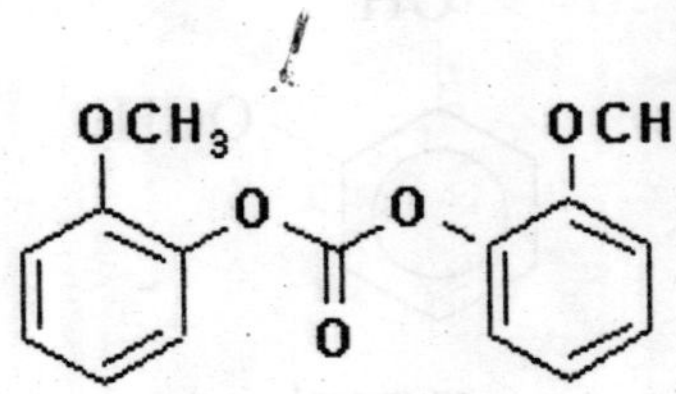

Guaiacol carbonate

In the same way, carbonates of menthol, borneol, carvacrol, thymol, creosol, eugenol, etc., have been prepared, and also the corresponding esters of carbamic acid using chlorocarbonic amide instead of $COCl_2$.

$$XONa + Cl.CO.NH_2 = NaCl + XO.CO.NH_2$$

This process, besides being used for the pure substances, may also be applied to the mixture of phenols comprising creosote, by which means a mixture of neutral carbonic esters free from corrosive properties is obtained, which is called creosote carbonate. Phenolic substances, such as guaiacol, can also be converted into alkyl carbonates instead of carbonates by the action of chloroformic ester on the hydroxyl compound or its sodium salt.

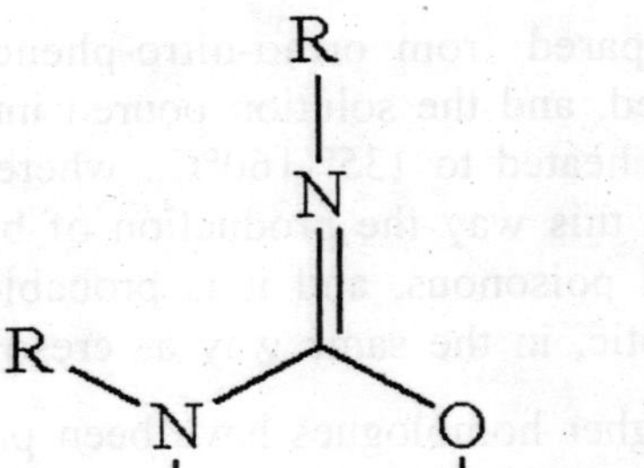

Hydroxyl compound

The carbonate of catechol, forms addition products with compounds containing an alcoholic hydroxyl group or a primary or secondary amino group, whereby one of the hydroxyl groups of the catechol is regenerated with the formation of a mixed ester of carbonic acid.

In the preparation of carbonates of phenols by the phosgene method, it may happen that those which are easily decomposed, such as the derivatives of iso-eugenol or menthol, are broken up, and therefore the method requires modification. In these cases, diphenyl or diethyl carbonate is first prepared, and this then made to react with the desired phenol. For example, with iso-eugenol

As the phosphites are said to be useful in the treatment of tuberculosis, it is not surprising that guaiacol phosphite, has been prepared. It is a crystalline powder, which has the advantage over the carbonate and phosphate of guaiacol in being soluble in fatty oils, and it is prepared by suspending guaiacol and an equivalent of soda in alcohol and adding one molecular proportion of PCl_3 to the cooled mixture. It is then heated to boiling and the alcohol distilled off. Substances such as this are known under the names of Phosphatol, Creosote-phosphite, etc.

Guaiacol benzoate is known as " Benzosol," and the acetate as " Eucol" ; the cinnamate, which is known under the name of " Styracol," is said to be a very good antiseptic and antitubercular, free from harmful effects. Cacodyliagol is , $(CH_3)_2As.O.O.\ C_6H_4(OCH_3),H_2O$.

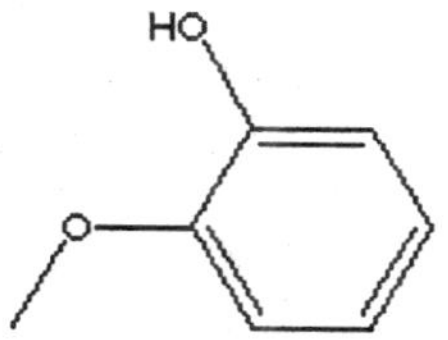

HO

O

Guaiacol benzoate

Another type of guaiacol derivatives is that in which the hydroxylic hydrogen is replaced by alkyl groups or other groups in which the hydroxylic oxygen is directly united to carbon. Veratrole may be included in this class, but it does not show the physiological action of guaiacol in any great degree, and probably does not split off any guaiacol in the organism. If, however, instead of replacing the hydrogen by a methyl group as in veratrol, one replaces it by a larger group, then the resulting compound iserol ether of guaiacol known as " Guaiamar," is one of the most important of this type. It is soluble in water and is prepared by the action of mono-chlorhydrin, $CH_2(OH)CH(OH)CH_2Cl$, on an alka line salt of guaiacol, or by the interaction of guaiacol and glycerol in presence of condensing agents. Glycerol ethers of other phenols have been prepared in the same manner. Taste less soluble derivatives of guaiacol and other phenolic substances are obtained by the action of alloxan on phenols in presence of certain condensing agents, such as H_2SO_4, HCl, $ZnCl_2$, etc.

An insoluble guaiacol preparation, known as " Cetiacol" or "Palmaicol," is obtained by the action of guaiacol in sodiun ethylate on spermaceti oil at 80°C. It is said to be without irritant action on the alimentary canal.

Guaiaperol is an addition product of guaiacol and piperidire, having the formula $C_5H_{11}N,(C_7H_8O_2)_2$. The object aimed at is to combine the action of guaiacol with the tonic action of piperidine on the heart and circulation. It is said to be without irritant action.

Creosoform is a product obtained by the condensation of formaldehyde and creosote, and has been recommended as an internal antiseptic.

N
H–C–H
H–C–H H–C–H
N
H–C–H H–C–H
H
N C N
H

Hexamethylene-tetramine

A substance of this type recently introduced is " Hexamecoll," a preparation of guaiacol and hexamethylene-tetramine. It is a powder which readily splits up into its components, and is recommended as an external disinfectant.

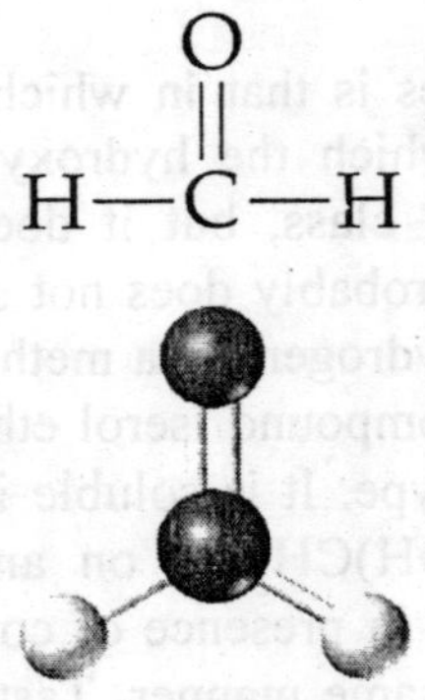

Formaldehyde

Another class of guaiacol derivatives comprises substances which do not regenerate guaiacol in the organism. It was to be expected that the entrance of a sulphonic acid group into guaiacol would lower its activity, and such is found to be the case. By sulphonating guaiacol at 70°-80°C. Guaiacol is obtained, which is therapeutically useful, its potassium salt known as " Thiocoll" being soluble and non-irritant. If the sulphonation be carried out at 140-150°C., The sodium salt of (OH, $—O—CH_2—COOH$) known as " Guaiacetin," is soluble in water and nearly tasteless, and can be obtained by the action of chloracetic acid on catechol in the presence of alkali.

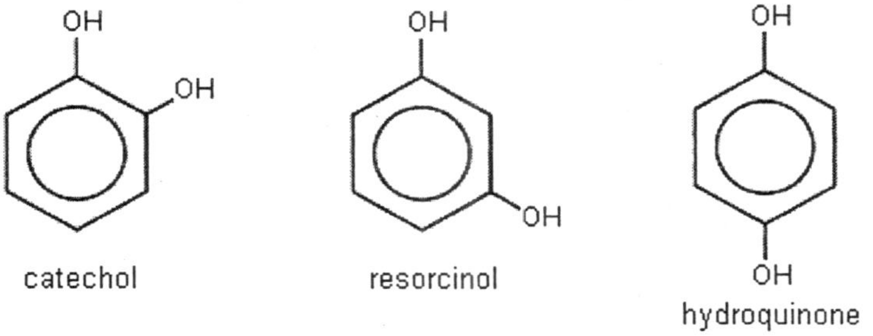

catechol resorcinol hydroquinone

ORGANIC DYESTUFFS

The organic dyestuffs are of interest both from the practical point of view on account of their use in therapeutics, and also on account of their importance in the development of the theories concerning the mode of action of drugs. Many of the diseases caused by microbial or parasitic infection can be successfully treated by a suitable serum, but there are a large number, especially those of protozoal origin, which do not lend themselves to serum-therapy.

In these cases a cure has been sought by chemical means, and at one time great hopes were entertained of curing many diseases of microbial origin by the use of general antiseptics, such as those which have proved so valuable in surgery. Generally speaking, however, the results obtained by the introduction of antiseptic methods into medicine have been disappointing when compared with the success which has attended their use in surgery.

Various processes such as the intravenous injection of form aldehyde solutions, the inhalation of antiseptic vapours such as creosote for consumption, and many others, have not fulfilled the hope that was placed in them. This is not surprising when it is borne in mind that most antiseptics are quite as injurious to the tissues of the higher organism as they are to the bacteria.

It has been pointed out that a greater measure of success may be hoped for in the case of those diseases which are caused by protozoa rather than by bacteria. The latter are probably amongst the oldest living organisms and have probably acquired a high degree of resist ance, but the former do not appear to have been able to adapt themselves so successfully to their environment, and the diseases they give rise to are very likely of more recent date than those caused by bacteria.

In some of these diseases favourable results have been obtained by the use of a chemical specific, which is not equally fatal to all forms of life, but which reacts specifically with certain micro-organisms in a manner somewhat analogous to the action of the aforementioned curative sera. This is expressed by saying that a substance of this type should have a strong affinity for the parasites (parasitotropic), but should be only slightly active to the organism (organotropic). Instances of this are afforded by syphilis and malaria, which yield far more readily to specific antisepsis, the former to mercury and arsenic and the latter to quinine, than do diseases of bacterial origin.

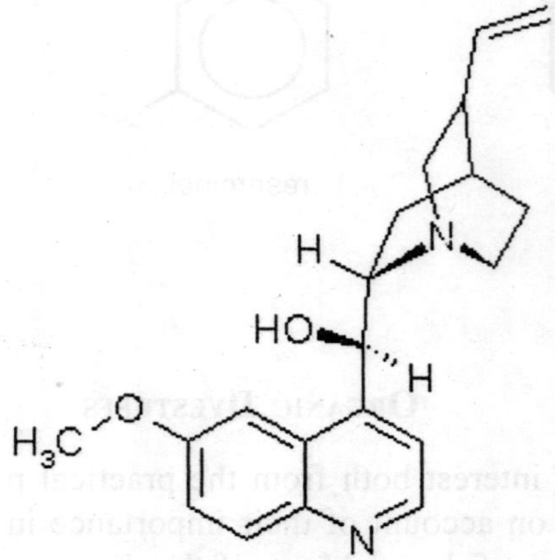

Quinine

Our knowledge of the chemical specifics and their probable mode of action, we mainly owe to Ehrlich, and his success in this field was largely due to a careful study of the reactions between the parasites and the dyes which reveal them in their surroundings. He realized that the relation between a dye and a particular type of cell is a chemical fact of great importance, and that such a dye must contain an anchoring group that can enable it to attach itself to the particular type of cell in question.

To vary the simile, the chemical specific may be likened to a poisoned arrow, the point being the particular dye which has a selective affinity for the parasite, and therefore fixes it. If a poison can be attached to such a dye or to the anchoring group of such a dye, the arrow will be not merely a dye, but also the desired chemical specific. This procedure is not so simple as might be gathered from the above, as the attachment of a poisonous element or radicle may destroy the anchoring effect of the other groups.

On the other hand, many dyes are themselves poisonous to certain parasites, and act as chemical specifics in these particular infections. For example, some of the triphenylmethane dyes, many of the azo dyes derived from benzidine, and various thiazine dyes, such as methylene-blue, have a strong parasiticidal action. Thus, amongst the dyes derived from benzidine, Trypan Bed and Trypan Blue have a strong trypanocidal action.

The former is obtained by tetrazotizing benzidine-orthosulphonic acid, and coupling with b-naphthylamine 3-6-disulphonic acid, and therefore has the formula the latter is obtained by tetrazotizing o-tolidine, and coupling with 8-amino-naphthol (1), 3-6 disulphonic acid.

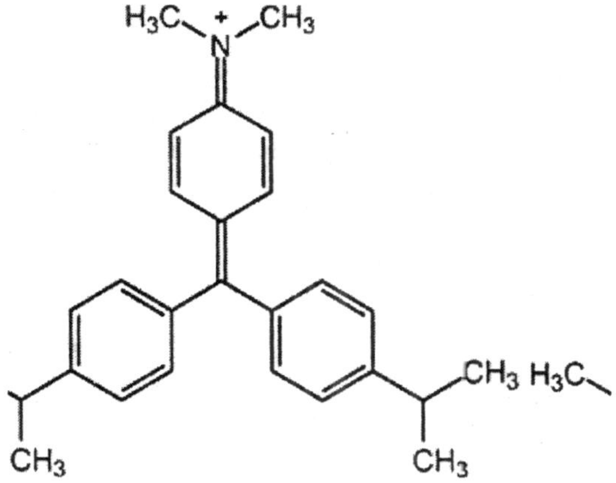

Triphenylmethane dyes

It is noteworthy that all the azo dyes of this series which are effective trypanocides have the sulphonic acid groups in the 3-6 position. Naga Eed which differs from Trypan Eed only in the absence of the sulphonic acid group in the benzidine portion of the molecule was found to have a strongly poisonous action on the spirilla of relapsing fever, but it also acts strongly on the red blood-corpuscles. Trypan Eed and Trypan Blue have shown good results in the treatment of nagana, especially when used in conjunction with certain triphenylmethane dyes such as diand tri-hydroxy malachite greens, and para-rosaniline.

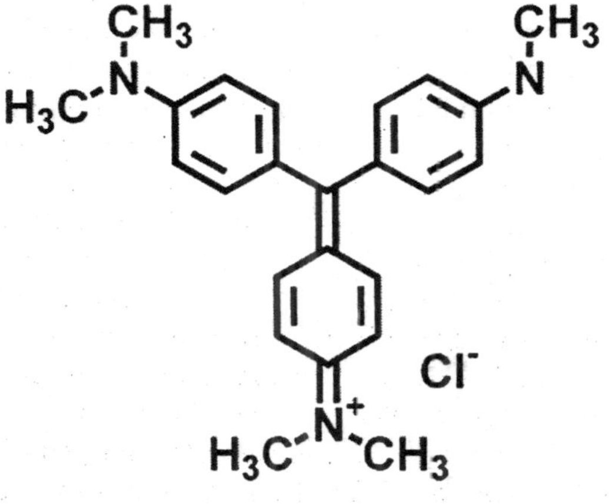

Pararosaniline

Malachite Green has the formula and Brilliant Green, which is the sulphate of the corresponding ethyl compound, is used fairly extensively as a general antiseptic. Methylene blue is used internally in a variety of conditions such as rheumatism, cystitis, nephritis, etc. This dye was found to have strong parasiticidal action on the spirilla of relapsing fever in testtube experiments, but failed when used on infected animals.

A dye which is largely used as a paste for external application for stimulating the growth of epithelium over granulating wounds, is Scarlet Bed, an azo dye derived from amidoazo toluene and β-naphthol, and therefore having the formula, $CH_3.C_6H_4.N:N.C_6H_3.N:N.C_{10}H_6O$. Some of the dyes derived from acridine have recently attracted considerable attention as antiseptics both for external and intravenous use, namely, Proflavine, diaminoacridine sulphate and Trypaflavine or Acriflavine, 3-6 diaminomethylacridinium chloride.

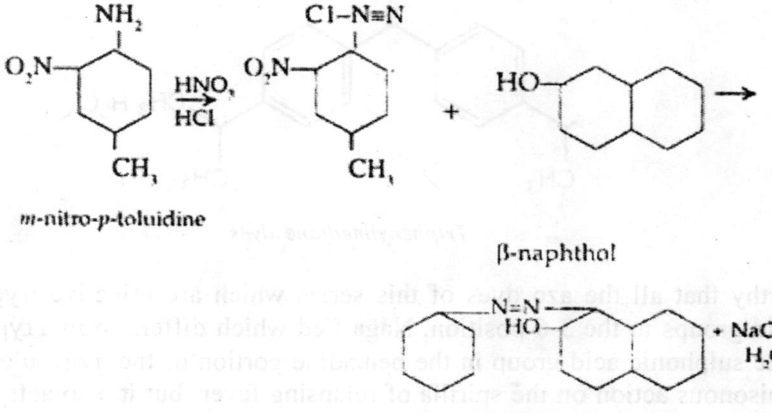

NH₂ / O₂N / CH₃ —HNO₂, HCl→ Cl–N≡N / O₂N / CH₃ + HO (β-naphthol) →

m-nitro-*p*-toluidine

β-naphthol

N=N / HO + NaCl / H₂O

Toluidine red

The latter was originally introduced by Bhrlich as a trypanocide. Proflavine is prepared by the interaction of aniline, form aldehyde and caustic potash, and heating the resulting product with aniline hydrochloride, whereby diaminodiphenylmethane is formed. This is then nitrated, and the product reduced with tin and hydrochloric acid, and the reduction product which contains the tin double salt of tetra-aminodiphenylmethane is heated in an autoclave at 140 to form 3-6 diaminoacridine.

To prepare Acriflavine from Profiavine, the amino groups in the latter are. protected by acetylation, and the diacetyl compound is then methylated by methyl sulphate or methyl toluenesulphonate in nitrobenzene solution. The acetyl groups are then hydrolyzed from the resulting compound by heating with hydrochloric acid, and on cooling the desired hydrochloride crystallizes out in red needles. Browning and his colleagues have recommended Proflavine and Acriflavine in the treatment of wounds, as they have high antiseptic power, together with freedom from irritant or toxic action, and no inhibiting effect upon the phagocytic action of the leucocytes or upon the process of healing. Up to the present, the clinical evidence appears to be very conflicting, but it is probable that good practical results will be obtained with these compounds.

Naphthalene, Pyridine, and Quinoline Derivatives

The antiseptics derived from other cyclic systems, such as naphthalene, pyridine and quinoline, are none of them so important as certain benzene derivatives, such as phenol. The naphthols and their derivatives have already been mentioned in connection with the phenols, but naphthalene itself has antiseptic properties, and on account of its volatility is a useful insecticide, and has largely replaced camphor as a means of protecting clothes and other fabrics from moths. It is often sold for this purpose under the name of carbon. All pyridine carboxylic acids have strong antiseptic properties. Uvitonicacid, is said to be a stronger antiseptic than salicylic acid, but the expense of its preparation has prevented it from being used for this purpose. It is produced by the action of alcoholic ammonia on pyruvic acid. Quinoline has an antiseptic action, which is increased by the entrance of methyl groups. A quinoline derivative has been placed on the market under the name of Chinosol, and is prepared by boiling an alcoholic solution of ortho-hydroxyquinoline, III with acid potassium sulphate. The product appears to consist of ortho-hydroxyquinoline sulphate and potassium sulphate. It has been used as an antiseptic both internal and external.

Formaldehyde

The strong antiseptic properties of formaldehyde have only recently been made use of in medicine, as the application of this substance has been hindered by its corrosive and toxic action. It readily polymerizes, forming tri-oxy-methylene which is also strongly antiseptic, but gives rise to bad effects when taken internally. Formaldehyde has long been used for the disinfection of rooms, for which purpose its volatility renders it eminently suitable, but its use as a drug depends on the manufacture of compounds which slowly split off formaldehyde under the influence of the secretions of the organism.

One of the earliest compounds of this class was Glutol, obtained by the action of formaldehyde on gelatine; other compounds of this type have been prepared with formaldehyde and casein, and formaldehyde and nucleic acid, the latter compound yielding soluble alkali salts. The compounds of formaldehyde and carbohydrates are of greater importance.

Classen found that formaldehyde reacts with starch, dextrin, etc., yielding insoluble, odourless, non-irritant substances, which slowly split off formaldehyde, and are antiseptic without being poisonous. Soluble condensation products of dextrin and formaldehyde (Dextroform) can be obtained by a modification of this process. A condensation product of formaldehyde with lactose has, under the name of Formamint, attained great popularity. This substance is used in the form of tablets, which, when allowed to dissolve in the mouth, liberate formaldehyde, and so check and prevent septic conditions in the throat and mouth. Formaldehyde reacts readily with ammonia, forming hexamethylenetetramine, $(CH_2)_6N_4$, discovered by Butlerow in 1860

$$6CH_2 + 4NH_3 = (CH_2)_6\ N_4 + 6H_2O.$$

It is a basic substance, very soluble in water, and its constitution is probably represented by the formula. This substance has strong antiseptic properties, but its aqueous solutions can be taken internally without causing poisonous or irritant symptoms. It is largely used in medicine, under the names of Hexamine, Urotropin, Cystamine and Cystogen, as a urinary antiseptic, for which purpose it is stated to be better than salicylic acid or any of the older preparations. It has some solvent action on uric acid, but it is improbable that it is capable of dissolving uric acid to any extent in the concentra tions in which it can be present in the organism. It is also said to be of value in laryngitis, pharyngitis, etc. The anhydromethylene-citric-acid derivative of urotropine, has been introduced under the names of New Urotropin and Helmitol. A large number of other derivatives and additive products of hexamine have been introduced into medicine as urinary antiseptics.

Tannic and Gallic Acids

Tannic acid is distinguished by its astringent and styptic action, and its freedom from toxic effects ; gallic acid is more irritant and five times as antiseptic as tannic acid. Gallic acid has the structure, but that of tannic acid is not known with certainty. Its empirical formula is $C_{14}H_{10}O_9$, from which it appears to be a digallic acid. In order to avoid the unpleasant taste of tannic acid itself, various insoluble derivatives have been prepared, which pass unchanged through the stomach, and are decomposed in the intestine, where they liberate free tannic acid. Tannigen is an insoluble acetyl derivative of tannic acid, soluble in alkalies. Tannoform, one of the best examples of this class, is a con densation product of tannic acid and formaldehyde

$$2C_{14}H_{10}O_9 + HCHO = CH_2\ (C_{14}H_{10}O_9)_2 + H_2O.$$

It combines the antiseptic properties of formaldehyde with the astringent action of tannic acid, and is used internally as an astringent and antidiarrhoeic ; externally it is used as an antiseptic for wounds, etc. Similar preparations of gallic acid have been devised, but the compounds of this acid containing bismuth are of more importance. Bismuth has a beneficent action on wounds, causing the surface to dry without impairing the healing power, and therefore its insoluble compounds with antiseptic and astringent substances readily lend themselves for use as iodoform substitutes.

Purine Derivatives

Caffine, the best-known drug of the purine group, is used as a cardiac tonic and cerebral excitant, but in addition it has a diuretic action, as also have other members of the purine group. Caffeine sodium cinnamate is known as Hetol. Although the action of theobromine on the nervous system is far weaker than that of caffeine, its diuretic action is as strong. Both substances suffer from the disadvantage of sparing solu bility and small power of resorption. To overcome the draw back of slight solubility, double salts of the sodium derivatives of caffeine and theobromine, with sodium salicylate, benzoate, and acetate, have been prepared. Compounds of sodium theobromine with sodium salicylate and acetate are known as Diuretin and Agurin respectively.

The acyl-amino derivatives of caffeine are said to have a strong diuretic action without the by-effects of caffeine. Monoacetyl-amino-caffeine, diacetyl amino-caffeine, etc., have been prepared. According to the investigations of Ach, the dimethylxanthines have a stronger diuretic action than trimethylxanthine (caffeine). Of these, theobromine (3-7 dimethyl) has the weakest, and theophylline (1-3 dimethyl) the strongest action, but that of paraxanthine (1-7 dimethyl) is the most persistent.

Theophylline has been synthesized by Fischer and by Traube, the technical preparation being based on the latter method. The synthesis of theophylline was carried out by Traube according to the following method.

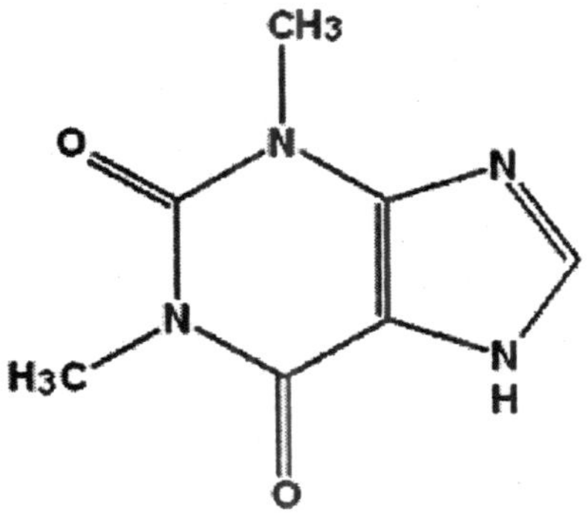

Theophylline

Dimethyl urea was first condensed with cyanacetic acid by means of $POC1_3$, and the resulting compound (II.) converted into the cyclic base (III.) by the action of alkali. This base, when treated with sodium nitrite and acetic acid, yields the isonitroso compound (IV.) which is reduced to the corresponding amine (V.) by means of ammonium sulphide. This amine is then converted, by formic acid, into its formyl derivative (VI.), which loses H_2O and yields theophylline (VII.) when heated with alkali. 8-Amino-theophylline is obtained by the action of ammonia on 8-chloro-theophylline, and has strong diuretic action. 8-Aminoparaxanthine and its alkyl derivatives have been obtained in a similar manner.

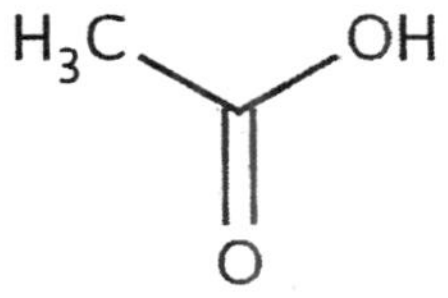

Acetic acid

3-Monomethyl-xanthine has an appreciable diuretic action, but in 7-monomethyl-xanthine it is vanishingly small. The diuretic action of xanthine itself is extremely small, and that of iso-caffeine (1-7-9 trimethyl-xanthine) is also very weak.

Uric Acid Solvents

In addition to the diuretics, the various compounds that have been advocated as remedies for gout may be divided into two classes those which are designed to diminish the formation of uric acid in the body, and those which are intended to act as solvents for the uric acid after it has been formed. Of the latter class it may be said that many substances have been obtained, such as piperazine and urotropine, which are capable of dissolving uric acid in vitro, but are probably quite incapable of dissolving it in the highest possible concentrations that can be present in the body. Naturally many compounds have been prepared which are intended to combine both functions.

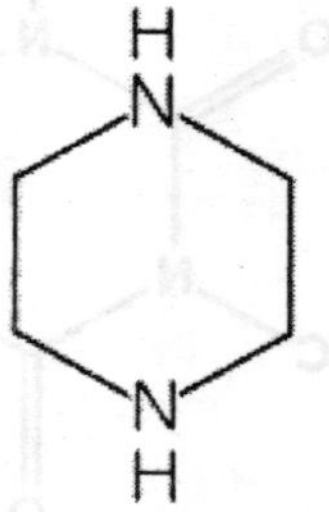

Piperazine

For example, quinic acid, $C_6H_7(OH)_4$. COOH, which exists in cinchona bark and coffee beans, is said to diminish the formation of uric acid, and its lithium and piperazine salts have been introduced under the names of Urosin and Sidonal respectively. In these cases the lithium or piperazine is intended to act as a uric acid solvent.

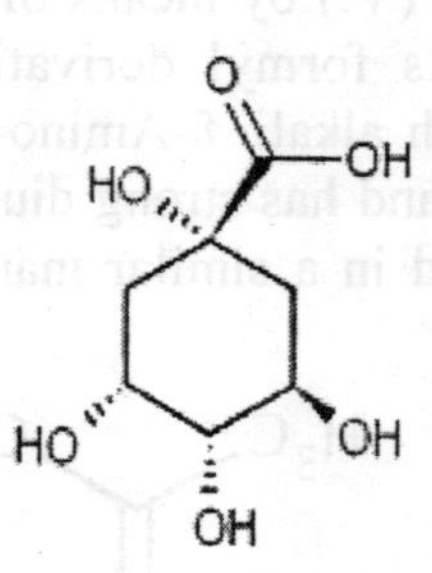

Quinic acid

It has also been noticed that the presence of organic acids in the organism generally decreases the amount of uric acid formed, and that this effect is greater in proportion to the number of carbon atoms present in the acid. For this reason the use of diphenyl tartrate has been suggested and salicylic acid has also been used, especially in the form of a condensation product of saligenin and tannic acid. A salicylate of urea has been recommended under the name of Ursal. Other compounds used for this purpose are hippuric acid, methylene-hippuric acid and its meta-nitro compound.

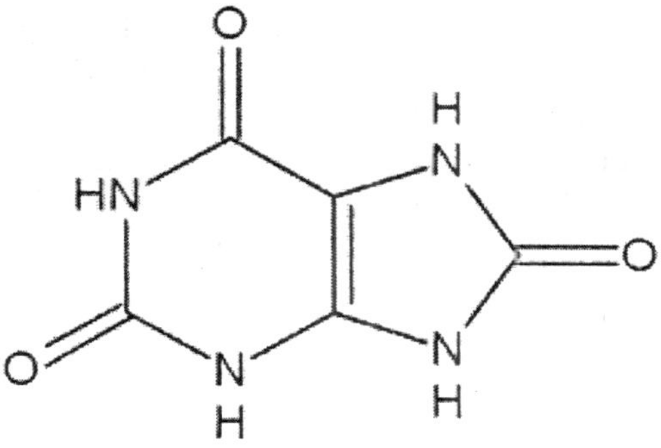

Uric acid

Drugs of the second class, namely, those the function of which is to prevent the deposition of uric acid rather than to prevent its formation, are more numerous. The alkaline carbonates have been widely used for this purpose, and especially lithium carbonate, as it has been found that the lithium salt is by far the most soluble of all the inorganic salts of uric acid. The alkaline tartrates, citrates, etc., are also used as well as the carbonates. The bad effects of lithium on the nervous system have led to the introduction of various organic bases which form even more soluble salts with uric acid.

The employment of lithium and these bases as uric acid solvents is fallacious, because there is always sufficient sodium present in the body to form the sparingly soluble sodium urate, and a double decomposition between salts takes place with the formation of the least soluble. If we designate sodium urate for the sake of simplicity as NaU, then an equation such as LiU + NaCl ⇔ NaU + LiCl, will run from left to right under the conditions present in the organism. Nevertheless, some considerable benefit is often derived from these remedies, although it may not be due to their solvent action on the uric acid.

Of the various organic bases that have been used as substitutes for lithium, piperazine is the most important. It was obtained by Hofmann by the action of ammonia on ethylene dichloride or dibromide. It is most readily purified by treating the reaction mixture with nitrous acid, whereby nitroso-piperazine is obtained, from which piperazine can be regenerated by the action of hydrochloric acid or reducing agents.

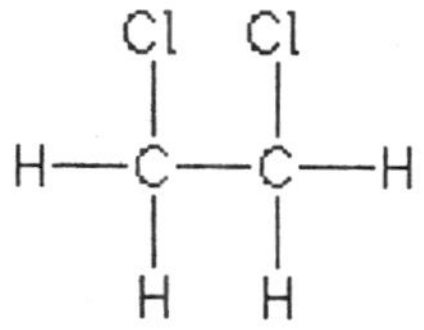

Ethylene dichloride

Many different modifications of this synthesis have been devised, but only one of them calls for mention. By the action of aniline on ethylene dibromide, diethylenediphenyldiamine is obtained, which yields a nitroso compound on treatment with nitrous acid. It has been shown that this, on treatment with alkalies, yields piperazine and nitroso-phenol. Aniline with ethylene dibromide gives just as ammonia gives piperazine. This with nitrous acid yields Piperazine is also known under the name of Dispermin, and its salt with quinic acid is used under the names of Urol and Sidonal.

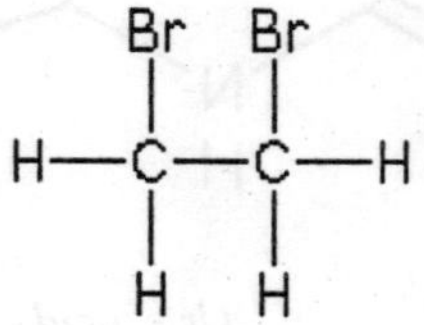

Br Br
H—C—C—H
H H

Ethylene dibromide

The tartrate of dimethyl-piperazine, is known as Lysetol, and has the advantage of being non-poison ous and non-hygroscopic. Dihydroxy-piperazine has also been investigated, and it resembles piperazine in its properties of a uric acid solvent. It is obtained by polymerizing aminoacetal dehyde with cold hydrobromic acid. An ethylene-ethenyl-diamine, has been prepared by heating ethylene diamine hydrochloride with sodium acetate and introduced under the name of Lysidin. It is said to be eight times as strong as piperazine in its solvent action on uric acid in vitro. Similar propenyl and butenyl deriva tives have been prepared.

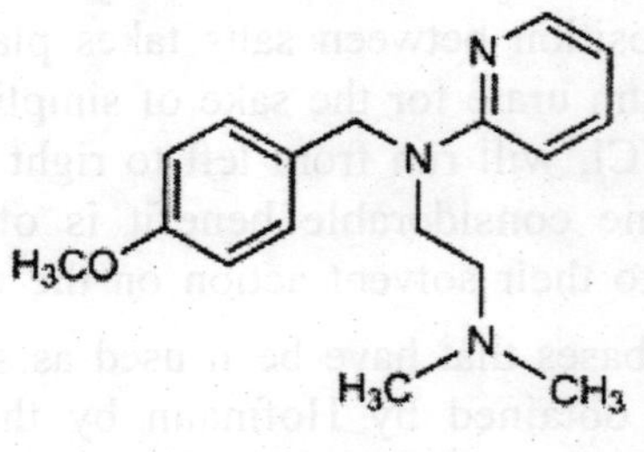

N
N
H_3CO
N
H_3C CH_3

Ethylene diamine hydrochloride

Urotropm (hexamethylene-tetramine) has been recommended as a uric acid solvent, its salts with quinic acid being known as Quinotropin, and its salicyl derivatives as Saliformin. Helmitol or New-urotropin, is the anhydro methylene-citrate of hexamethylene-tetramine. The acid, is obtained by heating citric acid with paraformaldehyde, or in better yield by the action of chlor-methyl alcohol, Cl. CH_2—OH, on citric acid at 130°-140°C.

The sodium salt of the acid has also been used by itself as a uric acid solvent, under the name of Citarin. According to Tunnicliffe and Rosenheim, the solubility of uric acid in blood serum is raised by the presence of piperidine, and they suggested the use of piperidine tartrate. A substance of unknown composition which is used a good deal as a uric acid solvent is thyminic acid, a complex substance prepared by a lengthy process from the thymus gland. It is also known by the trade name of Solurol. On boiling with aqueous sulphuric acid, it breaks down into Thymin which shows that it is chemically related to piperazine, lysidine. etc. Various other cyclic bases of this type, such as also have a strong solvent action on uric acid. Certain quinoline derivatives have been found useful in the treatment of gout, sciatica, etc., as they have analgesic pro perties in addition to being uric acid eliminants and urinary antiseptics. The simplest of these is Atophan, or 2-phenyl quinoline-4 carboxylic acid Its ethyl ester is known as Acitrin, and the methyl derivative of this is known as Novatophan.

ANTHRAQUINONE DERIVATIVES

Many aperient drugs, such as cascara, rheum (rhubarb), senna, and aloe, contain hydroxyl derivatives of methyl-anthraquinone, but unfortunately our knowledge of their constituents and of their physiological action is by no means in proportion to their great clinical importance. A large number of definite chemical compounds have been described as having been isolated from these drugs, but a good deal of the work published stands in need of confirmation, and there is still room for a great deal more research in this field. The position of the methyl group in these substances is not quite certain, but it is probable that most of them are derivatives of α-methyl-anthraquinone. These substances are very valuable as purgatives, owing to the fact that they have but little effect on the stomach and do not cause inflammation of the intestine.

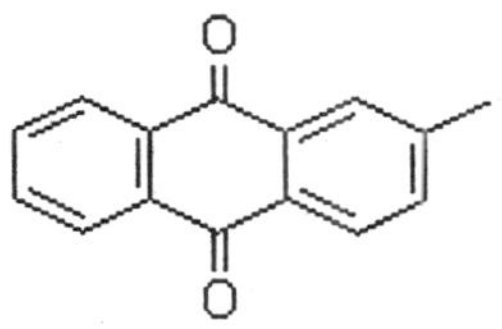

O

O

2-Methyl-anthraquinone

On the other hand, some of the drug is absorbed from the intestine into the system, and is liable to have an undesirable action on the kidneys. This effect is usually very slight, however. Chrysophanic acid, dihydroxy-methyl-anthraquinone is one of the milder of the natural purgatives of this group, and is present in rhubarb and many other purgative drugs, usually as a glucoside, chrysophan. The number and position of the hydroxyl groups has a powerful influence on the physiological action of the substance. For example, emodin, trihydroxymethyl-anthraquinone, $C_{15}H_{10}O_5$, probably having the constitution has a considerably more powerful action than chrysophanic acid. This substance is obtained together with rhamnose by the decomposition of the pentoside frangulin, obtained from the bark of Bhamnus frangula, and from other similar sources.

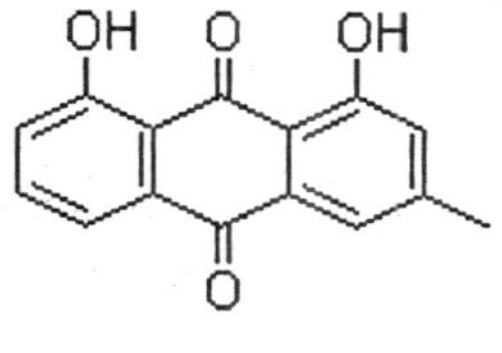

OH O OH

O

Chrysophanic acid

The active principles present in different varieties of aloes are known as aloin, $C_{17}H_{18}O_7$- + 1/ $2H_2O$, and barbaloin, $C_{16}H_{18}O_7$. The structure of these compounds is not known with certainty, but they are undoubtedly derivatives of polyhydroxy-anthraquinones.

The purgative action of the synthetic hydroxy-anthraquinones has been investigated by Vieth, and his results indicate that the number and position of the hydroxyl groups is of importance, but that the presence of the methyl group seems to have little influence on the physiological action. He found that the most active was anthrapurpurin, 1-2-7 trihydroxy-anthraquinone. A number of compounds, such as alizarin (1-2 dihydroxyanthraquinone), rufigallic acid (hexa-hydroxyanthraquinone), etc., are inactive. Some of. the active compounds contain a methyl group, and others do not.

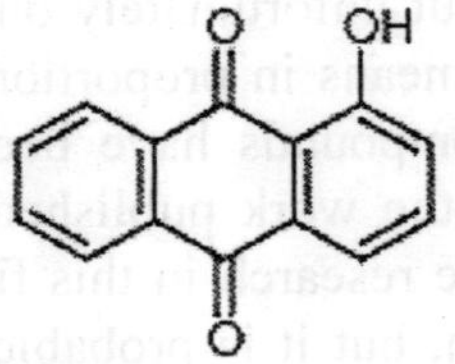

1-hydroxy-anthraquinone

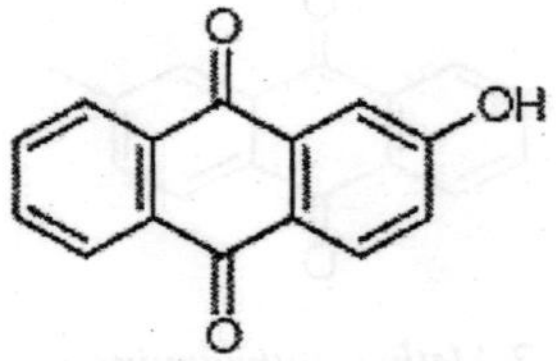

2-hydroxy-anthraquinone

An important factor in determining the purgative action is the length of time that the substance remains in the intestine, for when it is absorbed it can no longer exert a purgative action. The superiority of the glucosides and acetyl derivatives over the parent substances, and of the natural drugs over the synthetic, is due to the slower absorption of the former. Chrysophanic acid when quite pure is no longer a purgative, owing to its rapid absorption. Rapid absorption also increases the risk of undesirable effects on the kidneys, and therefore it is to be avoided on this account also. For these reasons, synthetic substances such as anthrapurpurin are inferior to the natural compounds, but synthetic substances of a more complex nature which are slowly absorbed have also been obtained. Anthrapurpurin diacetate has been tried as a mild laxative under the names Purgatin and Purgatol, but it is a kidney irritant. The acetyl derivatives of the tetra-alkyl ethers of rufigallic acid have been recommended as purgatives. Exodin is said to be the tetramethyl ether of the diacetyl compound, but according to Zernik it is a mixture of several substances, of which the hexamethyl ether is the one to which the purgative action is due, the diacetyl tetramethyl ether being inert. Exodin is a mild purgative.

Many derivatives of aloin have been prepared which are in tended to be free from the bitter taste of the parent substance whilst retaining its purgative action ; these being only slowly decomposed in the intestine, should be more active. A deriva tive of aloin and formaldehyde has been prepared in which the methylene group enters into two hydroxyl groups. Its action resembles that of aloin itself. H. Meyer has prepared tribromaloxin, $C_{17}H_{15}O_{7}Br_{3}$, which has a milder action than aloin, and triacetyl aloxin, $C_{1}7H_{15}O_{7}(CO.CH_{3})_{3}$, which is as powerful as the parent substance in its action, and has the advantage of being tasteless and keeping well.

An oxidation product obtained by the action of persulphate on aloin has a weak purgative action, and is free from harmful by-effects. A glucoside having a purgative action has been obtained from cascara sagrada, and named Peristaltin. Phenolphthalein, although not an anthraquinone derivative, has a purgative action without harmful effects on the kidneys, and has attained wide use as a laxative under the names Purgen and Laxin. It sometimes causes griping, and to overcome this drawback its acetyl-valeryl derivative has been introduced, under the name of Aperitol.

DRASTIC PURGATIVES

Very little is known of the chemical nature of these substances, which act by reason of their generally irritant properties. In large doses they give rise to bad effects, but in their specific doses they act more promptly than the anthraquinone derivatives, though they are apt to cause nausea and vomiting. Croton oil contains a large number of substances, but its intense purgative action is probably due to a resin, $C_{13}H_{18}O_{4}$. The action of jalap has also been attributed to a resin, which is, however, a mixture of several substances. The active principle of podophyllin is podophylltoxin, $C_{15}H_{14}O_{6}$, a neutral substance, which when heated with alkali forms podophyllinic acid, $C_{15}H_{16}O_{7}$.

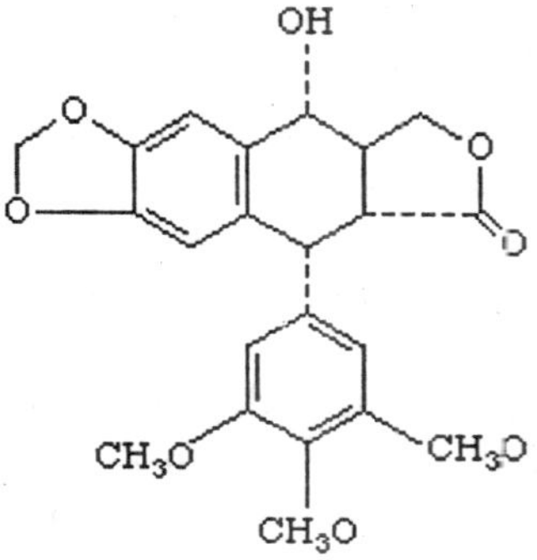

Podophylltoxin

By loss of water this acid forms picropodophyllin, $C_{15}H_{14}O_{6}$, an isomeride of podophyllin, and on further treatment with alkali yields acetic acid and orcine, Picropodophyllin is probably the lactone of podophyllinic acid, these substances being regarded as derivatives of g-pyrone. These two substances are without definite purgative action.

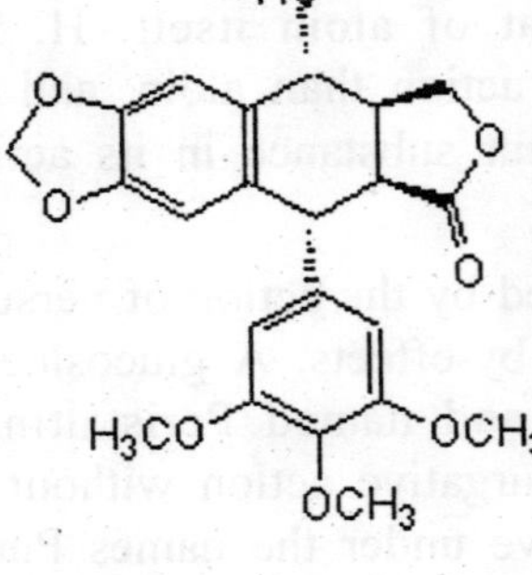

Picropodophyllin

Substances acting on Qastro-Intestinal Tract

Phenyl-dihydroquinazoline was accidentally found to possess a bitter taste, and to produce an early feeling of hunger, and hence it has been introduced as an aperitive, under the name of Orexin. Other substances of this type have also been investigated, but none of them are so satisfactory as this one. Dihydroqumazolines of the type are formed by the reduction of or thonitrobenzylform-anilides, -toluides, etc.

Orexin is also prepared by the action of formanilide on orthoaminobenzyl alcohol. Orexin is used in the form of the hydrochloride, owing to the bad taste of the free base. The tannate is insoluble in water, and therefore is quite tasteless.

It is soluble in hydrochloric acid, and therefore dissolves in the gastric contents, and can exert the required action; It is now largely used instead of the hydrochloride.Coto-bark has a specific effect on the walls of the intestine, dilating them and helping resorption, which renders it of use in diarrhoea. Its active constituent is coto'in, $C_{14}H_{12}O_4$, a derivative of benzophenone and phloroglucinol, having the formula

$$CH_3\ O.\ C_6H_2(OH)_2—CO—C_6H_5$$

This substance has a sharp taste, and in order to overcome this disadvantage, various derivatives have been prepared one of which, known as Fortoin, is a methylene-dicotom, $CH_2(C_{14}H_{11}O_4)_2$, formed by the action of formaldehyde on cotoin. It is free from the sharp taste of cotoin, and is said to have a stronger action, having especially an enhanced bactericidal effect.

Glucosides

Several glucosides which owe their physiological activity to their hydrolytic cleavage products have been mentioned in other parts of this volume. It will suffice to recall such compounds as salicin and chrysophan. Besides these, there are numerous glucosides which have a powerful and characteristic physiological action of their own. The most important of these are the active principles of digitalis and strophanthus, drugs having a powerful action in increasing the strength and diminishing the frequency of the heart-beat.

Digitalis contains several active glucosides, but unfortunately very little is known of their chemical constitution beyond the fact that most of them are derivatives of substances allied to cholesterin. The most important are digitalin and digitoxin; the latter, on warming with caustic soda, yields the physiologically inert digitoxic acid.

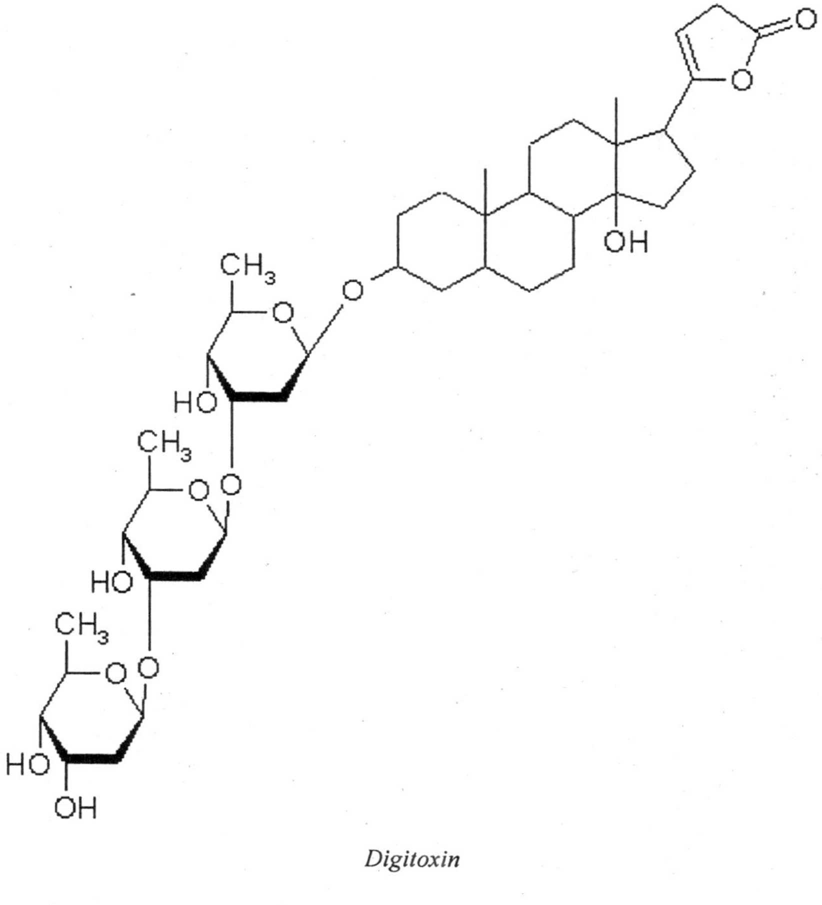

Digitoxin

The active constituent of strophanthus is a glucoside, strophanthin, $C_{40}H_{66}O_{19}$, which on hydrolysis yields strophanthidin, $C_{27}H_{38}O_7$, and a carbohydrate, $C_{13}H_{24}O_{10}$. Strophanthidin contains two lactone groupings, a benzene nucleus, an unsaturated linkage, CH=CH—, and probably three hydroxyl groups. The carbohydrate, $C_{13}H_{24}O_{10}$, is the methyl ether of a biose, $C_{12}H_{22}O_{10}$.

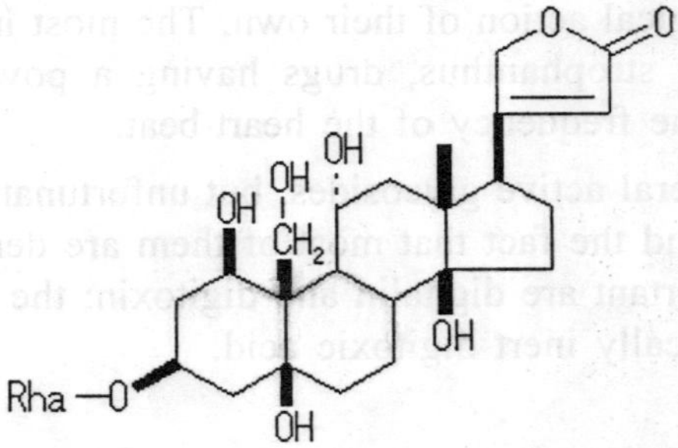

Strophanthin

Many physiologically active compounds are "glucosides" derived from pentoses instead of hexoses. Frangulin is an example of a pentose derivative, and more recently it has been shown that the poisonous haemolytic substance present in the fungus Amanita phalloides is a pentoside. Within recent years a considerable number of glucosides have been obtained synthetically, some of these being identical with natural products. For example, E. Fischer has obtained dl mandelonitrile glucoside, identical with prulaurasin ; this has been resolved into its dextro and laevo components, and the former found to be identical with sambunigrin, a glucoside occurring in the leaves of the elder. Unfortunately, however, the glucosides of greatest importance medicinally have not yet been synthesized, and are mostly of unknown structure.

Camphor and the Terpenes

The substances of the camphor group show a general resemblance to one another in their physiological action . All three have an antiseptic and slight local anaesthetic action. Borneol, known also as Borneo camphor, has less local irritant action, and can be tolerated in larger doses than camphor. Menthol is not much used internally, but it finds employment as a mild local anaesthetic. Camphor is the most important of these substances, and, as is well known, is obtained from the camphor-laurel, growing in China, Japan, and the East Indies, but within recent years it has also been produced artificially from turpentine.

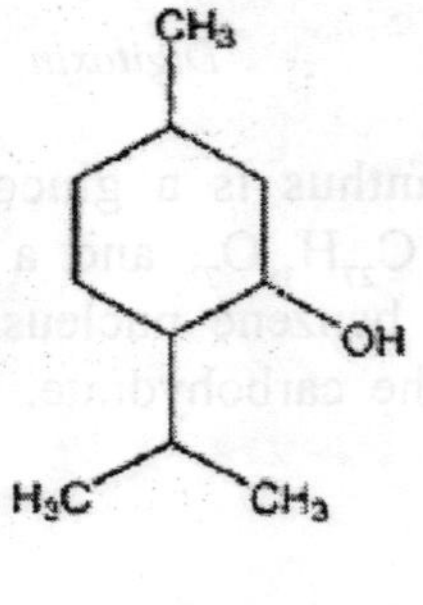

Menthol

It is a carminative and is used to check colds in the head, and for a very large variety of other medicinal purposes. It is injected as a stimulant in cases of collapse. It is necessary to distinguish between the so-called "artificial" or "synthetic camphor," and true camphor produced artificially. The former is i- reality not camphor at all, but is "pinene hydrochloride," obtained by the action of dry hydrogen chloride on turpentine. Turpentine consists chiefly of pinene, and on treatment with dry hydrochloric acid, the crystalline " pinene hydrochloride" is formed. This substance, which resembles camphor in many of its properties, is bornyl chloride, an intramolecular change taking place when hydrogen chloride acts on pinene.

True camphor has been completely synthesized, but the synthesis is an extremely lengthy and laborious one, and is impracticable as a commercial process. Camphor can, however, be economically prepared from turpentine (pinene) by a simple series of reactions. Turpentine on heating with anhydrous oxalic acid at 120°-130°C. yields camphor, borneol, and the oxalic and formic esters of borneol The borneol esters are hydrolyzed, and the resultant borneol oxidized to camphor by means of dichromate and sulphuric acid. Various other methods of oxidizing borneol to camphor have been devised, using permanganate, nitric acid, chromic acid, air, or ozone as the oxidizing agent.

Valerianic Acid Derivatives

Valerianic acid derivatives are used as nervous sedatives, and some of them, containing bromine and iodine. The esters of borneol and menthol with isovalerianic acid are known as Bornyval and Valid-ol respectively. Gynoval is the iso-borneol ester of the same acid, and Valyl is the diethyl-amide. It will be seen that this also contains the grouping—CO. N(C.,H r))., which is present in many other hypnotic and sedative compounds. New Bornyval is bornyl isovalerylglycollate.

Qlycerophosphates

One of the chief constituents of nervous tissue is lecith n, a complex ester of choline and distearylglycerophosphoric acid.m In reality it is somewhat more complex, for it contains also the corresponding derivatives of various other fatty acids such as palmitic, $C_{16}H_{32}O_2$, and of unsaturated acids such as oleic, $C_{18}H_{34}O_2$, and linoleic, $C_{18}H_{32}O_2$ as well as that from stearic acid, $C_{18}H_{36}O_2$. On hydrolysis it yields a glycerophosphoric acid which is probably a mixture of the α and β acids.

```
       O   H   H   H   H   H   H   H   H   H   H   H   H   H   H   H
       ||  |   |   |   |   |   |   |   |   |   |   |   |   |   |   |
H—O—C—C—C—C—C—C—C—C—C—C—C—C—C—C—C—C—H
           |   |   |   |   |   |   |   |   |   |   |   |   |   |   |
           H   H   H   H   H   H   H   H   H   H   H   H   H   H   H
```

Ppalmitic acid

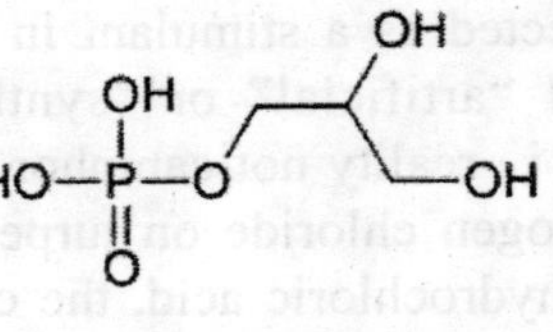

α-glycerophosphoric acid

As lecithin is said to exert a favourble influence on growth and metabolism, glycerophosphoric acid and its salts have been introduced into therapeutics. It is, however, improbable that lecithin is built up in the body from glycerophosphoric acid, and it is doubtful whether glycerophosphates have any advantage over the inorganic hypophosphites.

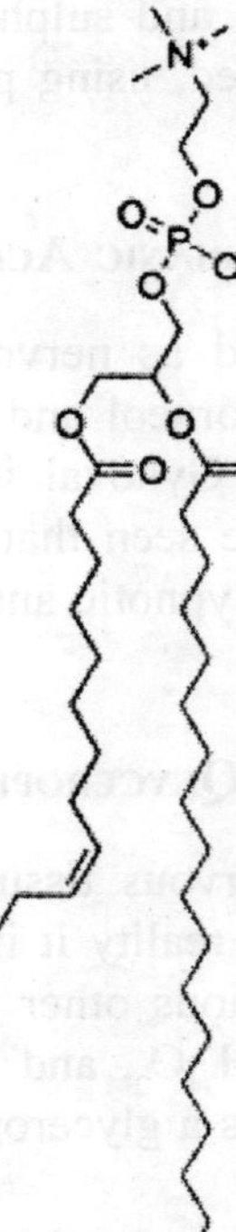

Lecithin

Glycerophosphoric acid is prepared by the action of phosphorus pentoxide on glycerol, or by heating phosphoric acid with glycerol for six days. A better result is obtained if the glycerol and syrupy phosphoric acid are heated for twenty-four to twenty-eight hours at 100°-105° in a vacuum.

Sodium glycerophosphate may be obtained by heating two molecules of glycerol with one molecule of monosodium or monoammonium phosphate in a vacuum, whereby a diglycerylmonometallic phosphate is formed. This is then saponified by caustic soda solution, and disodium glycerophosphate separates out on concentration. The crystalline sodium salt thus obtained has been shown to be identical with that of the β-glycerophosphoric acid obtained synthetically, and this conclusion as to its constitution has been confirmed by an indirect method. The sodium salt is very soluble in water, the calcium salt moderately soluble in cold water, but almost insoluble in hot. The ferric salt is soluble, and therefore can be used to combine the tonic properties of iron and phosphorus.

In the preparation of the crystalline sodium salt, there is always some uncrystallizable mother liquor, rich in glycerophosphate. This probably contains some of the sodium salt of the α-acid. The salts of this acid have been obtained synthetically by the action of trisodium phosphate on α-monochlorhydrin, $CH_2C1.CH(OH).CH_2.OH$, and by the oxidation of sodium monoallylphosphate. Bromo-lecithin is said to differ from lecithin in being absorbed in the intestine as such, without being split up into glycerophosphoric acid.

Sulphur Compounds

Sulphur is an abundant, tasteless, multivalent non-metal. Sulphur, in its native form, is a yellow crystalline solid. In nature, it can be found as the pure element or as sulphide and sulphate minerals. It is an essential element for life and is found in two amino acids, cysteine and methionine. Its commercial uses are primarily in fertilizers, but it is also widely used in gunpowder, matches, insecticides and fungicides. Elemental sulphur crystals are commonly sought after by mineral collectors for their brightly colored polyhedron shapes.

Sulphur is an essential component of all living cells. Sulphur may also serve as chemical food source for some primitive organisms: some forms of bacteria use hydrogen sulphide (H2S) in the place of water as the electron donor in a primitive photosynthesis-like process. Inorganic sulphur forms a part of iron-sulphur clusters, and sulphur is the bridging ligand in the CuA site of cytochrome c oxidase, a basic substance involved in utilization of oxygen by all aerobic life.

In plants and animals the amino acids cysteine and methionine contain sulphur, as do all polypeptides, proteins, and enzymes which contain these amino acids. Homocysteine and taurine are other sulphur-containing acids which are similar in structure, but which are not coded for by DNA, and are not part of the primary structure of proteins. Glutathione is an important sulphur-containing tripeptide which plays a role in cells as a source of chemical reduction potential in the cell, through its sulphhydryl (-SH) moiety. Many important cellular enzymes use prosthetic groups ending with -SH moieties to handle reactions involving acyl-containing biochemicals: two common examples from basic metabolism are coenzyme A and alpha-lipoic acid.

Disulphide bonds (S-S bonds) formed between cysteine residues in peptide chains are very important in protein assembly and structure. These strong covalent bonds between peptide chains give proteins a great deal of extra toughness and resiliency. For example, the high strength of feathers and hair is in part due to their high content of S-S bonds and their high content of cysteine and sulphur (eggs are high in sulphur because large amounts of the element are necessary for feather formation). The high disulphide content of hair and feathers contributes to their indigestibility, and also their odor when burned.

In traditional medical skin treatment which predates modern era of scientific medicine, elmental sulphur has been used mainly as part of cremes to alleviate various conditions such as psoriasis, eczema & acne. The mechanism of action is not known, although elemental sulphur does oxidize slowly to sulphurous acid, which in turn (though the action of sulphite) acts as a mild reducing and antibacterial agent.

Carbon disulphide, carbon oxysulphide, hydrogen sulphide, and sulphur dioxide should all be handled with care. Although sulphur dioxide is sufficiently safe to be used as a food additive in small amounts, at high concentrations it reacts with moisture to form sulphurous acid which in sufficient quantities may harm the lungs, eyes or other tissues. In organisms without lungs such as insects or plants, it otherwise prevents respiration. Hydrogen sulphide is toxic. Although very pungent at first, it quickly deadens the sense of smell, so potential victims may be unaware of its presence until death or other symptoms occur.

There are a number of drugs which are having sulphur as major component. For the most part, they are not related to one another, but are brought together here for convenience. Ichthyol is one of the best known of these. It is a substance of unknown composition, containing sulphur, obtained by the destructive distillation of "stink-stone" or "oil-stone," a deposit of fossil fish found in the Tyrol and southern Bavaria. By the action of sulphuric acid on the volatile basic oil thus produced, a sulphonic acid is formed, the salts of which are used medicinally. The ammonium salt of this ichthyol-sulphonic acid is a reddish-brown syrup, soluble in water, and is often known simply as " Ichthyol ". It is used both externally and internally for skin diseases, and is also used internally in tuberculosis and many other complaints, and as an intestinal antiseptic. The lithium and sodium salts are also used, and the zinc salt is used externally.

The disagreeable smell and taste of ichthyol are said to be largely due to impurities, and it can be obtained in a purer form by steam distillation in a vacuum.

Ichthalbin is an odourless and tasteless powder, obtained by precipitating an albumen solution with a solution of ichthyolsulphonic acid. It is used in the treatment of eczema and of intestinal catarrh. A compound of ichthyol-sulphonic acid and formaldehyde, is known as Ichthoform. It is a dark coloured powder insoluble in water, and with very little taste or odour. It is used internally as an intestinal antiseptic, and externally as a dressing for wounds (iodoform substitute).

Triphenylstibine sulphide, $(C_6HO_5)_3SbS$, is used under the name of Sulphoform in skin diseases as it liberates nascent sulphur on the skin.

Allyl-thiourea, also known as Thiosinamine, is a substance of some importance as a drug. It is formed by the action of ammonia on allyl -mustard oil and is a white crystalline substance moderately soluble in water. It is administered by hypodermic injection for the treatment of lupus, and for softening scar-tissues. Its sparing solubility in water is a drawback, and more concentrated solutions can be obtained by dissolving it with half a molecular proportion of sodium salicylate. This solution is known as Fibrolysin, and it is claimed that the presence of the sodium salicylate not only increases the solubility, but also renders the injectiqn less painful. It has been extensively used in relaxing scar-tissue. The additive compound of thiosinamine with ethyl iodide, $C_3H_5.NH.CS.NH_2,C_2H_5I$, is known as Tiodine, and is readily soluble in water giving solutions said to be painless on injection. lodolysin is a similar soluble, non-irritant preparation of thiosinamine. It is said to contain 43 per cent, of thiosinamine, and 47 per cent, of iodine.

Recently successful results have been claimed for a sulphur compound in the treatment of syphilis. This is di-o-aminophenyl disulphide known as Intramine. It is obtained by the oxidation of orthoaminophenyl mercaptan with ferric chloride, and is a pale yellow crystalline compound insoluble in water.

Chapter 5

Inorganic Synthetic Drugs

Inorganic chemicals are substances of mineral origin that do not contain carbon in their molecular structure. There are several basic types of products. Examples include naturally-occurring minerals, metals and their alloys, compounds that contain non-metallic elements, and metal complexes. Inorganic chemistry, the scientific discipline devoted to the study of inorganic chemicals, has developed inorganic chemical compounds sucii as inorganic fertilizer compounds, inorganic phosphates, and inorganic pigments. An inorganic fertilizer compound is designed for agricultural applications and typically contains a mixture of nitrogen, phosphorous, and potassium. In a biochemical setting, an inorganic phosphate is a free-phosphate ion in solution. An inorganic pigment is used to color plastics. Inorganic synthesis, the process of synthesizing inorganic chemical compounds, is used to produce many inorganic chemicals.

Most inorganic chemical reactions fall into four broad categories: combination reactions, decomposition reactions, single displacement reactions, and double displacement reactions. Specifications for inorganic chemicals include synonyms, molecular weight, formula/structure, functional group, Inorganic chemicals are widely used in many applications, including electroplating, dye and precision casting, alcohol distillation, and papermaking. Inorganic compounds are also used as feed additives, insecticides, wood preservatives, and antiseptics. Inorganic pigments have outstanding heat-stability, light-stability, weather resistance, and migration resistance. Inorganic chemical contaminants may be harmful if inhaled or ingested.

Most of the non-metallic inorganic antiseptics owe their antiseptic properties to their oxidizing power. Practically the only exception is boric acid, H_3BO_3, which is a very weak antiseptic. Hydrogen peroxide, H_2O_2, is a very useful antiseptic and disinfectant. In the presence of wounds and septic conditions it is decomposed into water and nascent oxygen, and has the great advantage of being odourless, non-poisonous, and almost free from irritant action. Other advantages of hydrogen peroxide are that it does not precipitate protein, and leaves only water when it has exerted all its antiseptic action. A 30 per cent, neutral solution of hydrogen peroxide is known as Perhydrol. The peroxides of the alkaline-earth metals may be regarded as derivatives of hydrogen peroxide, and MagnesiumPerhydrol or Biogen (a mixture of MgO and MgCy and ZincPerhydrol or Ektogen (a mixture of ZnO and ZnO_2) are used as antiseptics ; the former is for internal and the latter for external use.

Another compound of this type which may be regarded as a derivative of hydrogen peroxide is sodium perborate, $NaBO_3,4H_2O$, known as Perborax. Various organic derivatives of hydrogen peroxide have been prepared, but as they owe their antiseptic properties solely to the liberated oxygen, they have not been considered in the chapters devoted to the true organic antiseptics. An example of this type is benzoyl-acetyl hydrogen peroxide

$$C_6H_5 \cdot CO—O—O—CO \cdot CH_3 \quad \text{or} \quad \begin{matrix} C_6H_5 \cdot CO \searrow \\ \qquad\qquad O = O \\ CH_3 \cdot CO \nearrow \end{matrix}$$

known as Acetozone.For a similar reason, various metallic salts of organic acids are also included in this chapter, rather than in that on the organic antiseptics, as their action depends on the metal and not on the organic acid radicle.

Mercury Compounds

Mercury compounds are used very considerably as antiseptics, and also in the treatment of syphilis. As antiseptics, mercury salts suffer from the drawback that metallic instruments cannot be immersed in them, and for the treatment of syphilis, their corrosive, irritant and toxic nature is a serious disadvantage. For this reason, many efforts have been made to prepare soluble mercury salts which can be injected without giving rise to the bad effects of the simple soluble salts, such as the chloride. In another direction, attempts have been made to obtain soluble mercuric salts in which the electronegative character of the mercury is masked, so that solutions may be used for the sterilization of metallic instruments without the former being reduced with liberation of metallic mercury. Another drawback of mercuric chloride is the fact that it is not very soluble, and only dissolves slowly.

A method of administering mercury which has found a good deal of favour is by intramuscular injections of calomel (mercurous chloride, Hg_2Cl_2), and of metallic mercury. For example, Lambkin's Cream consists of an emulsion of 5 grams calomel and 20 grams creocamph made up to 100 c.c. with palmitine or carbolized liquid paraffin. The creocamph is a mixture of equal parts creosote and camphoric acid, and is used to minimize pain at the site of the injection.

Sometimes the older method of using mercurial ointment (the oleate or the free metal) is preferred, in spite of the indefinite amount of mercury that is absorbed through the skin, as com pared with the exact amount administered by intra-muscular injection. Colloidal metallic mercury has been tried as an injection medium, and various phenolic derivatives of mercury have been prepared, such as the mercury salts of phenol, naphthol, resorcinol, and tribromophenol, but these latter do not appear to be suitable for injection. The dimethyl, diethyl, and diphenyl derivatives of mercury are too dangerously toxic to be of any therapeutic value. Most satisfactory results seem to have been obtained with mercury derivatives of acid amides, and ammo-acids. Mercury derivatives of formamide, $(H. CO. NH—)_2$ Hg, and of succinimide as well as those

$$\left(C_2H_4 \begin{matrix} CO \\ \\ CO \end{matrix} N \right)_2 Hg,$$

of asparagine, $[NH_2. CO\ CH(NH_2). COO]_2$ Hg, and alanine, $[CH_3\ CH(NH_2)\ COO]_2$ Hg, have been prepared. Of these, mercury succinimide has been used in medicine under the name of Hydrargol.

Good results are said to have been obtained with secondary mercury salicylate,

$$C_6H_4 \begin{matrix} CO\text{—}O \\ \\ O \end{matrix} Hg,$$

a substance from which the mercury cannot be precipitated with H_2S. Although this compound is itself insoluble in water, it yields soluble double salts with the alkali chlorides. Mercurous tartrate, prepared by Lustgarten, is decomposed by the intestinal alkali, with liberation of finely divided mercury.

Of the various injection media that have been tried, the best results, so far, appear to have been obtained with various protein preparations. Some of these, obtained from glue, are soluble in water in any quantity, and are not precipitated by protein, nor acted on by alkalies with liberation of mercury.

The mercury salt of para-phenol-sulphonic acid, known as Hydrargyrol, $(HO\langle\ \ \rangle SO_2)_2Hg$, is said not to precipi tate proteins, and not to attack metallic instruments. A double salt of this substance and ammonium tartrate is more stable, $C_{12}H_{10}O_8S_2Hg_4[C_4H_4O_6(NH_4)_2] + 8H_2O$ and has been introduced into practice under the name of Asterol. It is soluble in water, giving a clear neutral solution, from which H_2S does not precipitate HgS. It is stated that aqueous solutions of asterol do not precipitate albumen, and though the bactericidal action is only about half that of mercuric chloride, the solutions have the advantage, of not losing any of their power in protein fluids, and of penetrating further into the tissues than solutions of sublimate. Asterol does not cauterize wound surfaces, and does not injure metallic instruments, and on account of these numerous advantages it is recommended for general surgical use.

Thymegol is a substance of a similar type to Hydrargyrol, and is the mercury potassium salt of thymol *p*-sulphonic acid. Similar preparations have been obtained by the action of mercuric oxide on alkaline solutions of phenol-disulphonic acid in molecular proportions. They are termed Hermophenyl, and are soluble in five parts water. Potassium mercury, hyposul phite is also said to be non-irritant and easily soluble. Hydrargotin is mercury tannate, and Toxynone is sodium m -acetylaminomercuribenzoate

$$HO\text{—}Hg\text{—}C_6H_3(NH.\ CO.\ CH3)\ (COONa).$$

Soluble mercury preparations which do not attack metallic objects have long been used in this country in the form of potassium mercury iodide, to which alkaline carbonate has been added. Similar compounds have been introduced in Germany by adding mercuric cyanide, cyanate, or p-phenol-sulphonate to alkaline carbonates.

Silver Compounds

Silver nitrate, besides its well-known caustic action, possesses strong bactericidal properties, but as it is precipitated both by proteins and by chlorides, its use is confined to the surface of the body. It would be desirable to obtain a com pound of silver which, while retaining its bactericidal proper ties, should be free from the caustic action of silver nitrate, and not be precipitated by chlorides or proteins. Brief mention need only be made of colloidal silver chloride and colloidal silver solutions, which were at one time in great vogue for disinfection of the tissues by intravenous injection. Such methods are useless, as the damage done to the tissues is greater than to the bacteria, and treatment of this kind is only justifiable for local infections.

These colloidal solutions, as also solutions of silver phosphate in ethylene-diamine, fulfil most of the chemical requirements outlined above, but they suffer from various drawbacks which have prevented their general application. The compound of silver nitrate and ethylene-diamine, however, is used as a substitute for silver nitrate under the name Argentamin.

As in the case of mercury, better results have been obtained with protein compounds. Argonin is a compound of silver and casein, but is difficultly soluble in water and sensitive to light. Protargol is a compound of this type which has a higher percentage of silver, is free from caustic effects, does not precipitate chlorides, and has the bactericidal effects of silver. To prepare it, a peptone solution is precipitated with silver nitrate solution or shaken with moist silver oxide, and the insoluble compound thus obtained, digested with protalbumose. Soluble protargol is formed in this way, and is separated from the solution by evaporation in vacuo. Similar compounds can be obtained from plant globulins, and the corresponding compounds have also been obtained from mercury, iron, copper, lead, zinc, and bismuth. Other silver albumin compounds have been obtained from gelatoses (hydrolytic products from glue).' Albargin, a powder readily soluble in water and of neutral reaction, is a substance of this type. It is recommended in various septic conditions as a non-irritant antiseptic.

Protosil is another silver protein compound, very soluble in water, and is not precipitated by chlorides or by protein. It contains about 20 per cent of silver.

Nargol is a silver nucleinate, soluble in warm water, and contains about 10 per cent, of silver. A silver compound of Methylene Blue has been recommended under the name of Argochrome, for intravenous injection in puerperal fever and general infections.

Antiseptic Properties of Zinc

Many salts of zinc possess astringent and antiseptic properties. A soluble zinc salt, recommended for various different complaints as a non-irritant antiseptic, is Nizin, a zinc salt of sulphanilic acid. Zinc perhydrol combines the antiseptic properties of hydrogen peroxide with the astringent properties of zinc.

Aluminium

The astringent properties of ordinary alum are well known. The aluminium salt of β-naphthol-disulphonic acid, is an example of the various organic aluminium salts that have been prepared. It is known as Alumnol and is recommended as a mild antiseptic, etc.

Iron

Ferric salts resemble those of aluminium in many of their properties, and they are used as astringents and styptics. For use as a styptic, a double compound of ferric chloride and antipyrine has been introduced under the name of Ferripyrin, but it has no advantages compared with ferric chloride itself. The chief therapeutic use of iron is, however, in anaemia and chlorosis. For this purpose, the ferrous salts are usually preferred, as they do not have so great a caustic action as the ferric, and hence do not disturb the stomach so much. The unpleasant by-effects of iron on the teeth and stomach have led to many attempts at obtaining compounds which should be free from these disadvantages.

By the reduction of haemoglobin, Robert obtained a substance, hamol, which contains the iron in the same form as haemoglobin, and Bunge obtained from egg-yolk a substance from which the iron could not be precipitated by ammonium sulphide. This substance was termed h&matogen by Bunge, and a similar sub stance has been isolated from the liver. A synthetic albuminate of iron was obtained by Schmiedeberg, but this differs from Bunge's haematogen in certain respects. Various other albumin ates of iron have been prepared, and many of these, as well as some of the foregoing substances, have been placed upon the market. None of these organic compounds are any more powerful agents in the treatment of anaemia than the simple inorganic salts, their advantage being simply that they are less irritant to the gastro-intestinal tract. Recently, iron has been used with success in the treatment of syphilis. The substance which has been found most useful in this respect is ferric sulphanilate, $Fe(SO_3C_6H_4NH_2)_3$, known as Ferrivine. It is a stable substance readily soluble in water. Compounds of arsenic and antimony, as well as certain bismuth preparations, are dealt with elsewhere.

Arsenic and Antimony Compounds

Arsenious oxide and the salts derived from it have for many years been used as tonics and in the treatment of anaemia, while the irritant properties of antimony have long been made use of in potassium antimonyl-tartrate. For the treatment of many diseases, the organic derivatives of arsenic, especially those containing an aromatic nucleus, have proved the most useful. In the case of antimony, its close resemblance to arsenic indicates the probability of its useful application in the same way, but up to the present, antimony compounds analogous to the most successful arsenic compounds have not been used. Within recent years it has been shown that the " sleeping sickness " of tropical Africa is caused by a protozoal parasite, the Trypanosoma gambiense, and this disease was found to react to various specifics, such as arsenic, although unfortunately their clinical use has not been completely successful. It was found that certain organic preparations, of which atoxyl was one of the first to be extensively tried, appeared to be more satisfactory than the older inorganic preparations, such as Fowler's solution (potassium arsenite). Nevertheless atoxyl is devoid of action on the trypanosomes in vitro, and it can only become specific for them when it undergoes some change in the body. Most of the atoxyl passes through the body unchanged, but the small portion which becomes changed is apparently capable of destroying a large proportion of the parasites. Probably the arsenic in this portion is changed into some compound of an unknown nature, the same or a similar compound being also formed from part of the arsenic in inorganic preparations. Although the active therapeutic agent is therefore probably the arsenic content, it is no doubt true that atoxyl possesses advantages over inorganic preparations by virtue of its physical properties, such as solubility. It is possible that it is able to penetrate into tissues which cannot be readily reached by inorganic arsenic.

Organic arsenic preparations of this type have also been used for the treatment of syphilis, but atoxyl does not appear to be so useful as mercury in this respect, although excellent results have been obtained with some of the newer preparations such as Salvarsan and its derivatives, especially when used in conjunction with mercury.

Antimony possesses somewhat similar pharmacological properties to arsenic, and on investigation antimony was found to have an even stronger trypanocidal action than arsenic, but its inorganic preparations labour under the disadvantage of being strongly irritant, while organic preparations of antimony analogous to atoxyl are difficult to prepare, and are not effective therapeutically.

Bismuth also has a powerful trypanocidal action, but it appears to be too toxic 'to the host to be of clinical value. All three metals, arsenic, antimony, and bismuth, appear to be fatal to trypanosomes in concentrations of one in two hundred thousand or even less, but Cushny has shown that a concentration of arsenic of one in three thousand four hundred is necessary to kill the harmless non-parasitic protozoa, such as Paramecium and Colipidium, while still stronger solutions of antimony and bismuth, namely 1/970 and 1/98 respectively, were necessary to effect the same result. These substances therefore appear to be true specifics against the trypanosomes, in the same way as quinine is a specific against the malaria organism.

Some varieties of trypanosomes seem to be more sensitive to a given drug than others, and it therefore is desirable to use as many different specifics as possible in order that the parasites which resist one drug may be destroyed by another. It has also been established that the trypanosomes rapidly acquire a tolerance for these drugs, and therefore it is necessary to use large doses at the commencement of the treatment.

Arsenic Compounds

It seems highly probable that atoxyl does not of itself exert a trypanocidal action, but that it gives rise to other substances which have this specific property. This problem has been the subject of extended investigations by Ehrlich and others, but the question still remains to be definitely settled. Atoxyl was first thought to be an anilide of arsenic acid, C_6H_5—NH—$AsO(OH)_2$, but it was shown by Ehrlich and Bertheim to be a sodium salt of para-aminophenylarsinic acid, $NH_2C_6H_4AsO(OH)_2$, an observation of very great importance, as it opened up the way for the preparation of a series of different compounds, by which means it was hoped to obtain compounds of a lower toxicity to the host (slightly organotropic) and a higher toxicity to the parasites (highly parasitotropic). For example, the amino-group may be acetylated, benzoylated, etc., or replaced by halogen, hydroxyl, or other groups by means of the diazo reaction, whilst by starting with substituted amines instead of aniline, derivatives can be prepared containing groups of the types —NHR and —NR_2, instead of NH_2. Of the derivatives thus obtained, some are $ and some 60 or 70 times as toxic as atoxyl. The entrance of a sulphonic acid group in the molecule yielded a substance the toxicity of which is of the same order as sodium chloride, but it is also quite inert towards the parasites. This is in accordance with the general physiological inertia of the sulphonic acids.

The acetyl derivative

$$CH_3 \cdot CO \cdot NH - C_6H_4 - AsO\begin{matrix} OH \\ ONa \end{matrix}$$

which is known as arsacetin, has certain marked advantages compared with atoxyl, being more stable and less toxic to some animals, while equally toxic to the parasites. Other acid groups have no advantage over acetyl, and with increasing length of the side chain the toxicity becomes far greater.

In order to make a decided advance, it is necessary to gain some idea of how atoxyl and its derivatives act in the animal body, as it would obviously be an advantage to use a substance as nearly as possible identical with the product of the meta bolism of the body. It is found that arsenious oxide, certain triphenylmethane dyes, etc., can kill parasites in blood serum in vitro, but that atoxyl and its derivatives do not, though in the body they exert a trypanocidal effect in very high dilutions, such as 1/120000. This difference between action in vitro and in vivo may be explained by various different hypotheses, such as

1. Atoxyl may be decomposed in the body into aniline and arsenic acid, the effect on the parasites being due to the inorganic arsenic.
2. According to Uhlenhuth and Woithe, atoxyl and compounds of a similar type may stimulate the cells to the production of derivatives (amboceptors) which kill the parasites.
3. The body may produce new products of a synthetic type leading to the production of more active compounds.

The first of these hypotheses is the simplest, and is in accordance with many of the observed facts, but its probability is weakened by the fact that no arsenic acid is excreted, and preparations have been obtained which are from ten to twenty times as active as inorganic arsenic. We have no knowledge of any actual facts in support of the second theory, and with regard to the third, it seems probable that some new highly active compounds are produced, but

Ehrlich considers that such products of metabolism are simpler than the original compounds, and are not very complex synthetic bodies.

Some significant facts have been brought to light which seem to indicate something of the nature of these metabolic changes. For example, there seems to be a close connection between the therapeutic efficiency of atoxyl and the resistance offered to it by the organism. Thus, a mouse which can readily tolerate 1/150 is no better influenced therapeutically by this dose than is an average mouse which can only stand 1/300 by that dose. One which was very sensitive and was poisoned by 1/400, showed a very marked trypanocidal effect. This seems to indicate that the organism changes atoxyl into a more toxic substance, which also acts very strongly on the parasites. A change of this kind seems probably to be connected with the reduction of the arsenic from the pentavalent to the trivalent state.

Atoxyl, on reduction, gives *p*-aminophenyl-arsenious oxide, NH_2⟨ ⟩AsO, and with stronger reducing agents *p*-diaminoarsenobenzene, NH_2⟨ ⟩As = O As⟨ ⟩NH_2. The reduction products of this type are generally more toxic and more active against trypanosomes than the corresponding derivatives containing pentavalent arsenic.

p-Hydroxy-phenyl-arsenious oxide, HO⟨ ⟩As = O, is the strongest, and in a dilution of one in ten million kills trypanosomes in one hour. When it is realized that atoxyl does not kill trypanosomes even in five per cent, solution, and that the hydroxy compound, HO⟨ ⟩AsO(OH)(ONa), does not do so in one or two per cent, solutions, it will be seen what an enormous change is produced by reduction. This is an example of the increased toxicity of unsaturated compounds The increased trypanocidal action in vitro is accompanied by an increased action in vivo. One cubic centimetre of a one in forty thousand solution of *p*-hydroxyphenyl-arsenious oxide caused the parasites to vanish from the blood of a mouse and for it to remain free for seven days. In the other members of the series, action in vivo is parallel to action in vitro, and therefore it indicates that the action of the derivatives of phenyl-arsinic acid is due to a reduction process, and indeed most of the evidence from all sides points to the probability of the trypanocidal action of arsenic and antimony being dependent on the presence of these elements in the trivalent state.

Some of these trivalent aromatic arsenic compounds are now extensively used in therapeutics. These are of the type R—As = As—R, rather than of the type R—As = O, and are distinguished by their comparatively low toxicity. The reduction product of atoxyl itself, NH_2⟨ ⟩As = As⟨ ⟩NH_2, has very high trypanocidal powers, but the compounds that have had the greatest success are dihydroxy-diarnino-arsenobenzene, HO⟨ ⟩As = As⟨ ⟩OH, and its derivatives, and the reduction product from phenylglycine arsinic acid
NH_2

$$(HO)_2AsO \cdot C_6H_4 —NH—CH_2 \cdot COOH,$$

namely

HOOC—CH_2NH— ⟨ ⟩As = As ⟨ ⟩ NH . CH_2—COOH.

The aliphatic arsenic compounds are, at the present day, of no very great therapeutic importance. Cacodyl

$$\begin{matrix} CH_3 \\ CH_3 \end{matrix} \!\!>\! As—As \!<\!\! \begin{matrix} CH_3 \\ CH_3 \end{matrix}$$

is poisonous, but the very soluble cacodylic acid, is comparatively harmless and inert. By reason of their excessive stability, its salts do not show a strong enough arsenical effect, and therefore do not find any therapeutic application. The same applies to "new" -cacodyl (Arrhenal), which is monomethyl-arsinic acid, $CH_3—AsO(OH)_2$, all these aliphatic compounds having been practically superseded by the aromatic.

Aromatic Arsenic Compounds

Atoxyl was the first aromatic arsenic compound to attain an extended use in therapeutics, and it is still one of the best known. By heating aniline arsenate, Bechamp in 1863 obtained a substance which he took to be the anilide of arsenic acid, $C_6H_5—NH—AsO(OH)_2$, and described the sodium, potassium, barium, and silver salts.

A sodium derivative of this substance was introduced into therapeutics under the name of atoxyl, and its correct empirical formula was established by Fourneau, who attributed to it the structure.

$$C_6H_5—NH—As \begin{matrix} \diagup OH \\ = O. \\ \diagdown ONa \end{matrix}$$

It is, however, a neutral substance, and the sodium-free product obtained from it was found to be acid, and to be identical with the product obtained by Bechamp. It seemed unlikely that this compound, $C_6H_8O_3NAs$, was really an anilide, and Ehrlich and Bertheim were able to show conclusively that it was para-amino-phenyl-arsinic acid, having the structure

$$NH_2 \langle\;\rangle As \begin{matrix} \diagup OH \\ = O. \\ \diagdown OH \end{matrix}$$

The following were the chief reasons in support of this view:

1. Atoxyl cannot be hydrolyzed into arsenic acid and aniline by any of the ordinary agents used to hydrolyze anilides.
2. It contains a primary amino group, which is easily diazotized and coupled with phenols and amines. It is readily acetylated, yielding a stable acetyl derivative.
3. Atoxyl corresponds in its properties with the arylarsinic acids of the type previously described by Michaelis and Reese.

4. Hydriodic acid reacts with atoxyl, replacing the arsenic group by iodine with formation of para-iodo-aniline, NH⟨ ⟩, thus proving that the arsenic and the amino group are in the para position to each other.

Atoxyl is prepared by heating arsenic acid with an excess of aniline at 180°-190°, or by heating aniline with aniline arsenate. The sodium salt is then obtained by extracting with sodium carbonate and recrystallizing. The acid is often known as arsanilic acid, and its sodium salt as sodium arsanilate, by analogy to sulphanilic acid, NH_2⟨ ⟩SO_2—OH.

These terms, on account of their convenience, will hereafter be used to denote these compounds.

The sodium salt of this acid contains water of crystallization, the amount varying somewhat in the different commercial products. " Soamin" is the name given to a pure product of sodium arsanilate crystallized with five molecules of water. Other commercial products of this salt are known as Arsamin, Atoxyl, etc. These are used for the treatment of sleeping sickness, syphilis, and other diseases of protozoal origin, but unless they are used with caution, unpleasant and even dangerous by-effects, of which blindness is one of the worst, may arise.

The acetyl derivative,

$$CH_3 \cdot CO — NH\ As(OH)(ONa) = O,\ 5H_2O,\ \text{is}$$

said to be less toxic to many animals than atoxyl itself, and has the advantage of being more, stable, so that its solutions can be sterilized by boiling. It is known as Arsacetin or Acetylatoxyl, and the acid is readily prepared by the action of acetic anhydride on arsanilic acid. Arsacetin is then obtained by neutralizing with soda.

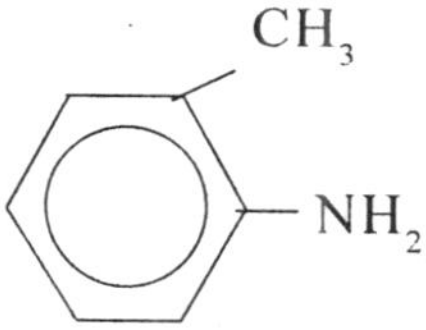

[Structure: benzene ring with CH_3 and NH_2 on adjacent positions]

Ortho-toluidine yields derivatives exactly analogous to those obtained from aniline. For example, 2-amino-tolyl-5 arsinic acid

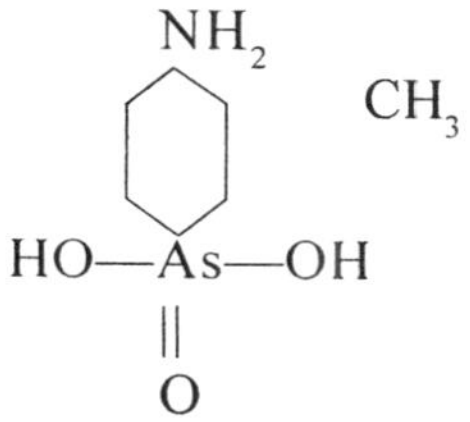

[Structure: benzene ring with NH_2 at top, CH_3 beside it, and HO—As(=O)—OH at bottom]

is obtained by heating o-toluidine arsenate with twice its weight of o-toluidine at 180°-185°. The sodium salt is then obtained by treatment with sodium carbonate.

The free acid can be acetylated, and the acetyl-compound yields a sodium salt,

$$CH_3—CO—NH—C_6H_4—AsO\begin{matrix}OH\\ONa\end{matrix}$$

analogous to arsacetin, and known by the trade name of " Orsudan". It is soluble in two and a half parts of water at body temperature, and resembles arsacetin in its action, being used in the same way for protozoal diseases.

Bis-2-amino-tolyl-5 arsinic acid

$$NH_2—C_6H_3(CH_3)—As(OH)(=O)—C_6H_3(CH_3)—NH_2$$

is obtained as a by-product of the action of o-toluidine on o-toluidine arsenate, and it also yields the corresponding symmetrical diacetyl derivative by treatment with acetic anhydride. In the same way, *bis-p*-aminophenyl-arsinie acid is obtained as a by-product in the preparation of arsanilic acid, and this compound also yields a diacetyl derivative. By the preparation of the following compounds

NH$_2$, CH$_3$, AsO(OH)$_2$ — *2-amino-tolyl-5 arsinic acid*

NH$_2$, AsO(OH)$_2$ — *1 amino naplithalene-4 arsinic acid*

NH$_2$, CH$_3$, AsO(OH)$_2$ — *3-amino-tolyl-6 arsinic acid*

NH$_2$, CH$_3$, CH$_3$, AsO(OH)$_2$ — *1-4-dimethyl-2-amino-phenyl-5 arsinic acid*

NH$_2$, Cl, AsO(OH)$_2$ — *1-chloro-2-amino-phenyl-5 arsinic acid.*

Benda and Kahn showed that the reaction of arsenic acid with aromatic amines is a general one, if the para position to the amino group is free. They also noticed the formation of arsinic acids of the type

$$(NH_2—R)_2As(=O)OH$$

as well as those of the type NH_2—R—$AsO(OH)_2$, but this had not been noticed by Bechamp or by A. and E. Adler, who had just previously prepared 2-amino-tolyl-5 arsinic acid and 1-amino-naphthalene-4 arsinic acid. By acetylation and subsequent oxidation of the former they obtained acetyl-anthranil-arsinic acid, which by hydrolysis yields anthranil-arsinic acid, which was then converted into salicyl arsinic acid by the diazo reaction

$$C_6H_3(NH_2)(CH_3)AsO(OH)_2 \rightarrow C_6H_3(NH.CO.CH_3)(CH_3)AsO(OH)_2 \rightarrow C_6H_3(NH.CO.CH_3)(COOH)AsO(OH)_2$$

$$\rightarrow C_6H_3(NH_2)(COOH)AsO(OH)_2 \rightarrow C_6H_3(OH)(COOH)AsO(OH)_2$$

Anthranil arsinic acid *Salicyl-arsinic acid*

Para-hydroxy-phenyl arsinic acid, $HO\langle\quad\rangle AsO(OH)_2$, is obtained from arsanilic acid by the diazo reaction, and it has also been obtained by the action of phenol on arsenic acid at 150, and the homologues of this acid may be obtained in a similar manner from ortho or meta cresol, or by the diazotization of the corresponding amino acids. As obtained from phenol it is a syrup, and it can be purified by recrystallization of its sodium salt. Para-hydroxy-phenyl arsinic acid is not used therapeutically, but is the starting-point for the production of some very important derivatives. The arsinic acids of the type $R_2N\langle\quad\rangle AsO(OH)_2$ and $RHN\langle\quad\rangle$—$AsO(OH)_2$ do not appear to have found any therapeutic application. Para-dimethyl-amino-phenyl arsinic acid, $(CH_3)_2N.C_6H_4.\ AsO(OH)_2$, is obtained by the action of arsenic trichloride on dimethylaniline, whereby $(CH_3)_2N.C_6H_4.\ AsCl_2$ is formed, which is hydrolyzed, yielding the acid $(CH_3)_2N.C_6H_4.$ —$As(OH)_2$, which gives the arsinic acid by oxidation with hydrogen peroxide.

An important derivative of arsanilic acid is phenylglycine para arsinic acid, which is obtained by mixing solutions of sodium arsanilate with chloroacetic acid in hot aqueous solution

$$(HO)(NaO)As(=O)—\langle\quad\rangle—N(H)—[H + Cl] — CH_2 — COONa$$

$$= (HO)(NaO)AsO—\langle\quad\rangle —NH—CH_2—COONa + HCl.$$

This substance is of importance, owing to the fact that its reduction product containing the arsenic in the trivalent condition is of great therapeutic value. This and other derivatives containing trivalent arsenic will be considered in the following section.

The benzene-sulphonyl derivative of atoxyl

$$C_6H_5 \cdot SO_2 \cdot NH \langle \bigcirc \rangle As \begin{matrix} \diagup ONa \\ = O, \\ \diagdown OH \end{matrix}$$

is known as " Hectine," and has been successfully used in the local treatment of syphilis. A compound of arsanilic acid with allyl-thiourea has been prepared, and is said to have valuable therapeutic properties and low toxicity.

The arsinic acids and their derivatives described in the pre ceding pages are mostly prepared by the action of arsenic acid on a primary amine or a phenol, and this is the simplest method in those cases where it is applicable. A more general method of preparing arsinic acids has, however, been devised by Bart. It consists in diazotizing a primary amine, and treating the diazo solution with a metallic arsenite, and warming the product whereby nitrogen is evolved and the salt of the arsinic acid formed

$$R \cdot NH_2 \rightarrow R, N_2X$$

$$R \cdot N_2X + As\ (OM)_3 = R.\ N_2AsO\ (OM)_2 + MX$$

$$R \cdot N_2AsO(OM)_2 \rightarrow R \cdot AsO\ (OM)_2 + N_2.$$

By a suitable choice of the radicle represented by E, practically any aromatic arsinic acid can be prepared.

Another widely applicable method of attaching arsenic to the aromatic nucleus is through the mercury compounds. Mercuric chloride readily reacts with many aromatic com pounds, giving para-substituted mercury compounds

$$R \langle \bigcirc \rangle + HgCl_2 = R \langle \bigcirc \rangle HgCl + HCl$$

On treatment with arsenic trichloride, the mercury is replaced by arsenic

$$R \langle \bigcirc \rangle HgCl_2 + AsCl_3 = HgCl_2 + R \langle \bigcirc \rangle AsCl_2$$

The dichlorarsines so obtained are easily converted into arsinic acids by hydrolysis and oxidation.

Trivalent Arsenic Compounds

The therapeutically valuable trivalent aromatic arsenic compounds are all of them obtained by reduction of the corresponding compounds containing pentavalent arsenic.

Reducing agents convert arsanilic acid into para-aminophenyl arsenious oxide, $NH_2—\langle \bigcirc \rangle AsO$, or into diaminodihydroxy-arsenobenzene, $NH_2—\langle \bigcirc \rangle As(OH) \langle \bigcirc \rangle NH_2$ or into para-diamino-arsenobenzene $NH_2 \langle \bigcirc \rangle As = As \langle \bigcirc \rangle NH_2$.

The first is produced by weak reducing agents such as hydriodic acid, or sulphurous acid, the second by sodium amalgam and methyl alcohol, and the third by stronger reducing agents, such as sodium hydrosulphite, etc.

Still stronger reducing agents, such as zinc and hydrochloric acid, reduce it to para-aminophenyl arsine, NH_2⟨⟩AsH_2.

Arsenophenol

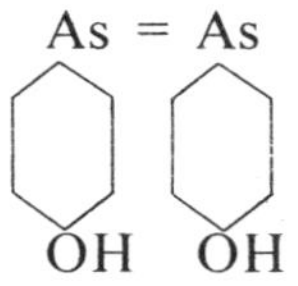

$$\begin{array}{cc} As & = & As \\ \bigcirc & & \bigcirc \\ OH & & OH \end{array}$$

is obtained by the reduction of para-hydroxyphenyl arsinic acid with a solution of sodium hydrosulphite, caustic soda, and magnesium chloride. The sodium derivative of this substance is soluble in water, from solutions in which it is precipitated by alcohol. Phenylglycine arsinic acid when reduced in this way gives arseno-phenylglycine

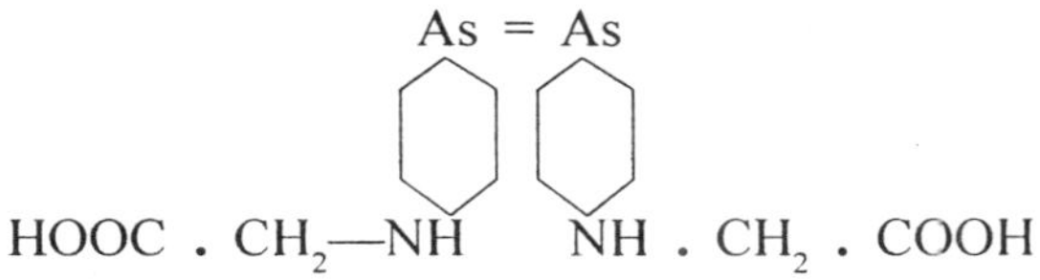

$$\begin{array}{ccc} & As = As & \\ & \bigcirc \quad \bigcirc & \\ HOOC \cdot CH_2—NH & & NH \cdot CH_2 \cdot COOH \end{array}$$

a reddish-brown powder, soluble in aqueous sodium carbonate. Very favourable results are said to have been obtained with this substance in the treatment of trypanosomiasis in rats, but it is not so well tolerated by larger animals, such as the horse or the donkey. 8 Favourable results are also said to have attended its use in syphilis, and it is claimed that in all cases it is free from the danger of harmful effects on the eyes. In recent years another derivative of arsenobenzene introduced by Ehrlich has been widely and successfully used in the treatment of syphilis.

This preparation is dihydroxy-diamino-arsenobenzene

$$\begin{array}{ccc} & As = As & \\ H_2N & \bigcirc \quad \bigcirc & NH_2 \\ & OH \quad OH & \end{array}$$

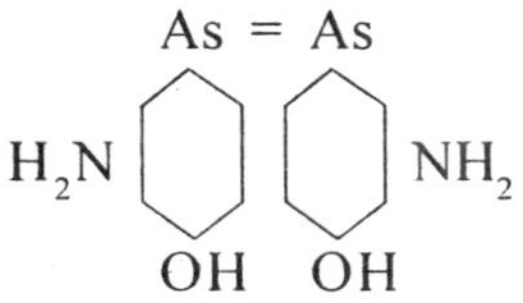

better known as "606" or Salvarsan. It is prepared from *p*-hydroxy-phenyl arsinic acid, HO⟨⟩$AsO(OH)_2$, which on treatment with nitric and sulphuric acids yields a mono-nitro compound, the nitro group entering into the ortho position to the hydroxyl and meta to the arsinic group. This compound, on reduction with caustic soda, sodium hydrosulphite, and magnesium chloride, gives dihydroxy-diamino-arsenobenzene

$$HO\langle\rangle AsO(OH)_2 \rightarrow HO\langle\overset{NO_2}{}\rangle AsO(OH)_2 \xrightarrow[\text{molecules of nitro compound}]{\text{From two}} \begin{array}{ccc} & As = As & \\ H_2N & \bigcirc \quad \bigcirc & NH_2 \\ & HO \quad HO & \end{array}$$

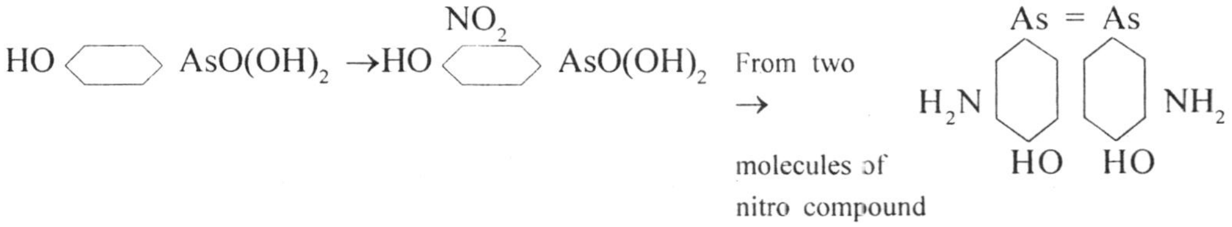

An alternative method of preparing 3-nitro-4-hydroxyphenyl arsinic acid starts from p-dimethylaminophenyl arsinic acid, prepared as described on p. 206. This on nitration gives 3-nitro4-dimethylaminophenyl arsinic acid, which on treatment with alkali gives 3-nitro-4-hydroxyphenyl arsinic acid

$AsO(OH)_2$ / $N(CH_3)_2$ → $AsO(OH)_2$ / NO_2 / $N(CH_3)_2$ → $AsO(OH)_2$ / NO_2 / OH.

In addition to the method already given, this can be reduced to dihydroxy-diamino-arsenobenzene by treating it with zinc and acetic acid at 25°-30°, and then with hydrochloric acid and sulphurous acid at 50°-60°. The addition of the sulphurous acid appears to prevent the reduction from going beyond the " arseno " stage. Although the composition of Salvarsan is usually represented by the above formula, the technical product always contains sulphur, and probably it contains a certain amount of the sul phamic acid in which one or both the amino groups contain the sulphonic acid group in the form

$$—NH—SO_3H_4.$$

Dihydroxy-diamino-arsenobenzene is obtainable as the hydro chloride under the trade names of Salvarsan, Kharsivan, and Arsenobenzol, but before use it is generally transformed into the sodium salt by adding the correct amount of caustic soda solution to its aqueous solution. Great care has to be taken in making up these solutions, which do not keep well, so that they have to be prepared immediately before use. They are administered either by intravenous or intramuscular injection. Splendid results have been obtained with this compound, especially when used in conjunction with mercury, though some of the more extravagant claims, as for example, that syphilis could be cured by a single injection, have not been substantiated.

Although its toxicity is low, it is by no means negligible, and fatal results have sometimes attended its use, but in the majority of cases these can be attributed to faulty technique or to its use in cases where the condition of the patient was already very bad. Its action on spirochaetes in vitro is very weak, and it was therefore presumed that it underwent some change in the organism with the formation of more active products. According to a recent investigation, these changes are very complex, a large number of different compounds being formed.

To overcome the drawbacks due to the laborious technique of administering Salvarsan, a derivative soluble in water was soon afterwards introduced. This substance is known as Neo salvarsan, Neokharsivan, Novarsenobenzol, etc., and has the structure represented by the formula

As = As

H_2N (ring) (ring) $NH—CH_2—SO_2Na$.

OH OH

It is prepared by adding an aqueous solution of formaldehyde sulphoxylate to an aqueous solution of Salvarsan. A precipitate is formed which redissolves in sodium carbonate solution to form a clear yellow solution of Neosalvarsan.

Not only does Neosalvarsan possess the advantage of ready solubility in water and in saline solutions, giving neutral solutions, but it is also said to be better tolerated by patients than is Salvarsan itself. It is administered by intravenous or intramuscular injection, and its therapeutic properties are similar to those of Salvarsan.

Another derivative of Salvarsan which has recently attracted a good deal of favourable attention is Galyl, a substance having the composition shown by the formula

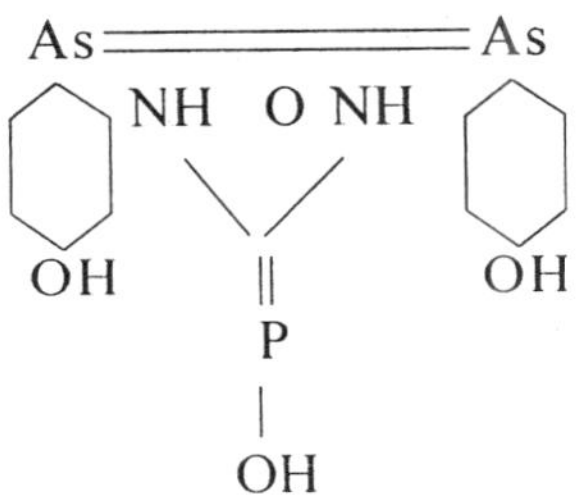

As=As
NH O NH
OH OH
P
OH

It is prepared as follows : 3-nitro-4-hydroxyphenyl arsinic acid is electrolytically reduced to 3-amino-4-hydroxyphenyl arsinic acid. This is then treated with phosphorus oxychloride in presence of caustic soda solution, and the product reduced with sodium hydrosulphite

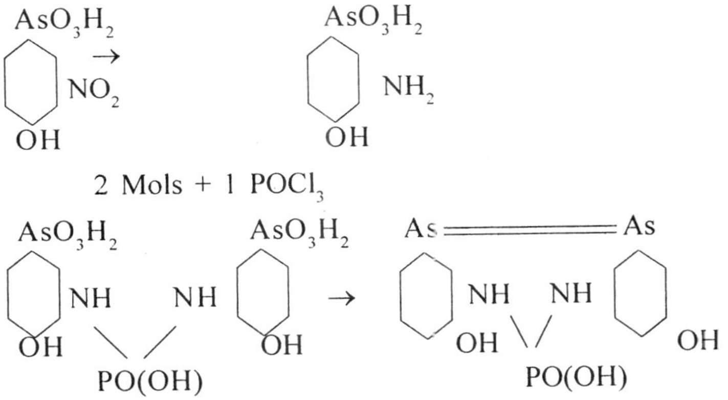

AsO_3H_2 → AsO_3H_2
NO_2 NH_2
OH OH

2 Mols + 1 $POCl_3$

AsO_3H_2 AsO_3H_2 As=As
NH NH → NH NH
OH OH OH OH
PO(OH) PO(OH)

It is used in the form of its sodium salt which is readily soluble in water, yielding solutions suitable for intravenous injection. Its solution in a dilute solution of glucose is also used for intramuscular injection.

Considerable attention has been paid recently to the metallic compounds of Salvarsan. These are of the additive type, and are formed by virtue of the residual affinity of the arseno As = As group. A compound of one molecule of Salvarsan with one molecule of cupric chloride is said to be remarkably effective against sleeping sickness. The preparation is described of copper and of silver compounds, by mixing a solution of Salvarsan with a copper or a silver salt, and precipitating the mixture with a solution of caustic soda.

"Sodium Salvarsan" is prepared by precipitating a solution of Salvarsan in caustic soda with alcohol, in the presence of a substance capable of stabilizing the product. It is said to be useful in the treatment of syphilis. Various other metallic derivatives and their methods of preparation have since been described. One of these is known as " Luargol," and is a compound of Salvarsan with silver bromide and antimony. It appears to have the composition $(C_{12}H_{12}O_2N_2As_2)_2$, AgBr, $SbO(H_2SO_4)_2$, and is said to be stable and to be less toxic and more effective therapeutically than the parent substance. Similar claims have been made for compounds of Salvarsan containing copper and gold.

In addition to the derivatives of Salvarsan, various more complex derivatives of arsenobenzene have been prepared and investigated. For example, 3-4-5-3'-4'-5'-hexamino-arsenobenzene is said to be a powerful spirillocide.

Eesorcinol readily reacts with arsenic acid giving 2-4-dihydroxyphenyl arsinic acid, from which a series of derivatives of arsenobenzene have been obtained by reduction.

From α-aminoanthraquinone, the corresponding arsinic acid has been prepared by Bart's reaction. This on reduction with sodium hydrosulphite gives l-l'-arsenoanthranol

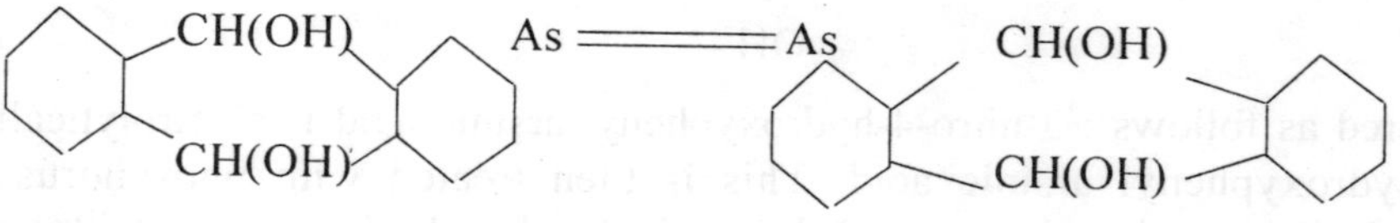

CH(OH) As ═══ As CH(OH)

CH(OH) CH(OH)

which is readily oxidized by air to anthraquinone-1-arsenoxide

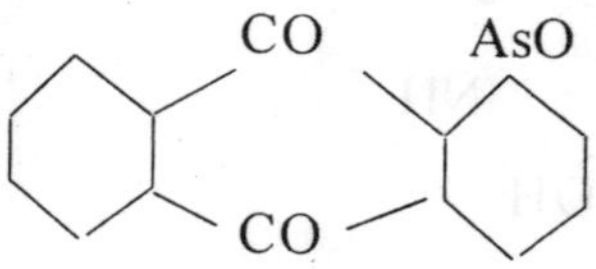

CO AsO

CO

These compounds are toxic, and they are easily decomposed into anthraquinone and arsenic acid.

A very large number of other derivatives of phenylarsinic acid and of arsenobenzene have been described, the abovementioned having been given as typical examples.

In addition to the symmetrical derivatives of arsenobenzene, R—As = As—R, already mentioned, mixed or unsym metrical derivatives of the type R—As = As—R can be obtained by the reduction of an equimolecular mixture of arsinic acids, $EAsO_3H_2$ and $R'AsO_3H_2$. Another method of obtaining unsymmetrical arsenobenzenes is as follows : The arsinic acid is first reduced to the corresponding arsine by means of metal and mineral acid

$$R.AsO_3H_2 \rightarrow R.\ AsH_2,$$

and this can then be condensed with arylarsenoxides or halides

$$R.AsH_2 \rightarrow OAsR' = R.As = As.R + H_2O$$

This method is also applicable to the production of mixed arsenostibino, and arseno-bismutho compounds.

$$R.AsH_2 \rightarrow R'SbCl_2 = R.As = Sb.R' + 2HCl$$

As can be seen from the preparation of Salvarsan, the use of sodium hydrosuiphite is not applicable to the formation of arseno compounds containing a nitro group as this is also re duced by the hydrosuiphite. It has been found, however, that hypophosphorus acid is a specific agent for reducing the arsinic acid group to the arseno group, and by this means arsinic acids containing a nitro group or an azo group can be reduced to the corresponding arseno compounds.

Organic Antimony Compounds

The trypanocidal action of antimony, compared with that of arsenic, has been discussed in the first section of this chapter. Unfortunately, the aromatic antimony compounds compare unfavourably with those of arsenic with regard both to ease of preparation and stability. For example, *p*-aminophenyl-stibinic acid, the antimony analogue of arsanilic acid, cannot be obtained by the interaction of antimony pentoxide and aniline, and the statement that it has been obtained from aniline, and antimony trichloride by a method analogous to Michaelis's synthesis of *p*-dimethylaminophenyl arsinic acid is quite incorrect.

Aryl-stibinic acids can, however, be obtained by the action of sodium antimonite on diazo solutions in a way somewhat analogous to Bart's reaction, and *p*-aminophenylstibinic acid has been prepared by this method, but the preparation is difficult and the yields are low. The compound, moreover, is amorphous and difficult to purify. Aryl-stibinic acids can also be obtained by the action of alkali on compounds obtained from antimony trichloride and diazo compounds.

Aryl-stibinic acids can be nitrated in the usual way, and *p*-chlorophenyl-stibinic acid on nitration gives 3-nitro-4-chloro phenyl-stibinic acid, which, on boiling with alkali, gives 3-nitro 4-hydroxyphenyl-stibinic acid. This, on reduction with sodium hydrosulphite, gives 3-3'-diamino-4-4'-dihydroxy-stibinobenzene, the antimony analogue of Salvarsan.

A large number of other aromatic antimony compounds, and mixed arseno-stibino compounds have been prepared.

None of the compounds which have been tried up to the present fulfil all the conditions necessary for a really efficient trypanocide. These conditions may be summed up as follows :

(1) The compound must be non-irritant, and capable of remaining in perfect solution at the temperature and alkalinity of the tissues.

(2) It must act quickly on the trypanosomes before they can acquire a tolerance to the drug.

(3) When the trypanosomes have been expelled from the blood by a single full therapeutic dose, there must be no recurrence in the majority of cases within some fixed time, which will depend to some extent on the particular host experimented with, and on the strain of parasites used.

Although the arsenic compounds which have been described do not fulfil the second and third of these conditions, yet the majority of them fulfil the first of these quite admirably, but in the case of antimony difficulty has been experienced in obtaining derivatives to fulfil even this condition.

Of the various compounds which have been tried therapeutically, those of a similar nature to ordinary tartar emetic (potassium antimonyl-tartrate) are amongst those giving the best results. Thomson and Cushny have experimented with many compounds of this type derived from various hydroxy acids, and the best results were obtained with compounds prepared from tartaric and malic acids

$$\begin{array}{c}HO\text{—}CH\text{—}COOH\\ |\\ HO\text{—}CH\text{—}COOH\end{array}\qquad\begin{array}{c}HO\text{—}CH\text{—}COOH\\ |\\ CH_2\text{—}COOH\end{array}$$

Tartaric acid *Malic acid*

The sodium and potassium antimonyl-tartrates seemed to be very nearly equal in their efficiency, but the ethyl ester of antimonyl-tartaric acid appeared to have some advantage over these alkali salts. Potassium ammonium antimonyl-tartrate is known as Antiluetin.

Chapter 6

Synthetic Antibiotics

Antibiotics are products of secondary metabolism, which inhibit growth processes of other organisms even when used at low concentrations. Growth inhibition of one microorganism by another in mixed culture has been known for a long time. The most famous example is the growth inhibition, which was observed by Alexander Fleming in 1929, when staphlococcal growth on a petriplate was inhibited by a contaminating Penicillium notatum culture. The Penicillium contaminant produced the antibiotic penicillin. During World War II, the demand for chemotherapeutic agents to treat wound infections led to the development of a production process for penicillin and the beginning of the era of antibiotic research. This continues to be the most important area of industrial microbiology today. Intensive screening programs in all industrial countries continue to increase the number of described antibiotics: 513 antibiotics were known in 1961, 7650 in 1972, 7650 in 1985, and currently around 8000. In addition, around 3000 antibiotically active substances have been detected in lichens, algae, higher animals, and plants. Each year, about 300 new antibiotically active materials are detected, of which 30-35% are secondary components from fermentations with known antibiotics.Of the large number of known antibiotics of microbial origin, only 123 are currently produced by fermentation. In addition, more than 50 antibiotics are produced as semisynthetic compounds, and three antibiotics, chloramphenicol, phosphonomycin, and pyrrolnitrin, are produced completely synthetically.

The significance of antibiotic production for the producing strain is unclear. Antibiotic production could be of ecological significance for the life of the organism in nature, but solid research to support this hypothesis is very limited. As secondary metabolites, antibiotics could serve regulatory roles during differentiation, perhaps acting as temporary inhibitory agents. To date, most new antibiotics have been detected strictly by empirical screening methods, with little attention to their possible roles for the producing strains.

Antibiotics are produced by bacteria, actinomycetes and fungi. In the fungi, only the antibiotics produced by the Aspergillaceae and Moniliales are of practical importance. The compounds isolated from basidiomycetes have not had any practical use. Only 10 of the known fungal antibiotics are produced commercially and only the penicillins, cephalosporin C, griseofulvin, and fusidic acid are clinically important. In the bacteria, there are many taxonomic groups, which produce antibiotics. The greatest variety in structure and number of antibiotics is found in the actinomycetes, especially in the genus Streptomyces. Another important group of substances are the peptide antibiotics produced by bacteria of the genus Bacillus. A number of new compounds have also recently been isolated in other taxonomic groups.The significance of antibiotic production for producer strains is still unclear. The status of the controversy can be found in the references of Zahner and Barabas cited in the bibliography.

CLASSIFICATION OF ANTIBIOTICS

Although there are several classification schemes for antibiotics, based on bacterial spectrum (broad, narrow) or route of administration (injectable, oral, topical), or type of activity (bactericidal, bacteriostatic), the most useful is based on chemical structure. Antibiotics within a structural class will generally have similar patterns of effectiveness, toxicity, and allergic potential.

Most commonly used types of antibiotics are:

— *Penicillins*. The penicillins are the oldest class of antibiotics. Penicillins have a common chemical structure which they share with the cephalopsorins. Penicillins are generally bactericidal, inhibiting formation of the cell wall. There are four types of penicillins:

  — The natural pencillins are based on the original penicillin-G structure. Penicillin-G types are effective against grampositive strains of streptococci, staphylococci, and some gramnegative bacteria such as meningococcus.

  — Penicillinaseresistant penicillins are active even in the presence of the bacterial enzyme that inactivates most natural penicillins.

  — Extended spectrum penicillins which are effective against a wider range of bacteria.

  — Aminopenicillins such as ampicillin and amoxicillin have an extended spectrum of action compared with the natural penicillins.

  Penicillins are among the least toxic drugs known. The most common side effect of penicillin is diarrhea. Nausea, vomiting, and upset stomach are also common. In rare cases penicillins can cause immediate and delayed allergic reactions - specifically, skin rashes, fever, and anaphylactic shock. Penicillins are classed as category B during pregnancy.

— *Cephalosporins*. Cephalosporins have a mechanism of action identical to that of the penicillins. However, the basic chemical structure of the penicillins and cephalosporins differs in other respects, resulting in some difference in the spectrum of antibacterial activity. Like the penicillins, cephalosporins interfere with synthesis of the bacterial cell wall and so are bactericidal. Cephalosporins are among the most diverse classes of antibiotics, they are grouped into generations by their antimicrobial properties. Each generation has a broader spectrum of activity than the one before.

  The first generation cephalosporins include: cephalothin, cefazolin, cephapirin, cephradine, cephalexin, cefadroxil. Their spectrums of activity are quite similar. They possess generally excellent coverage against most grampositive pathogens and variable to poor coverage against most gram negative pathogens.

The second generation cephalosporins include: cefaclor, cefamandole, cefonicid, ceforanide, cefuroxime. In addition to the grampositive spectrum of the first generation cephalosporins, these agents have expanded gramnegative spectrum. Cefoxitin and cefotetan also have good activity against Bacteroides fragilis.

The third generation cephalosporins have much expanded gramnegative activity. However, some members of this group have decreased activity against grampositive organisms. The third generation cephalosporins include: cefcapene, cefdaloxime, cefditoren, cefetamet, cefixime, cefmenoxime, cefodizime, cefoperazone, cefotaxime, cefpimizole, cefpodoxime, ceftibuten, ceftriaxone. They have the advantage of convenient dosing schedules, but they are expensive.

The fourth generation cephalosporins are extended spectrum agents with similar activity against grampositive organisms as first-generation cephalosporins. They also have a greater resistance to betalactamases (bacterial enzymes that may destroy antibiotic before it can do its work) than the third generation cephalosporins. Many fourth generation cephalosporins can cross blood brain barrier and are effective in meningitis. The fourth generation cephalosporins include: cefclidine, cefepime, cefluprenam, cefozopran, cefpirome, cefquinome.

Cephalosporins generally cause few side effects. Common side effects associated these drugs include: diarrhoea, nausea, mild stomach cramps or upset. Approximately 5?10% of patients with allergic hypersensitivity to penicillins will also have cross-reactivity with cephalosporins. Thus, cephalosporin antibiotics are contraindicated in people with a history of allergic reactions (urticaria, anaphylaxis, interstitial nephritis, etc) to penicillins or cephalosporins. Cephalosporin antibiotics are classed as pregnancy category B.

— *Fluoroquinolones*. Fluoroquinolones are the newest class of antibiotics. Their generic name often contains the root floxacin. They are synthetic antibiotics that belong to the family of antibiotics called quinolones. The older quinolones are not well absorbed and are used to treat mostly urinary tract infections. The newer fluoroquinolones are broad-spectrum bacteriocidal drugs that are chemically unrelated to the penicillins or the cephaloprosins. Because of their excellent absorption fluoroquinolones can be administered not only by intravenous but orally as well.Commonly used fluoroquinolones include ciprofloxacin, levofloxacin, lomefloxacin, norfloxacin, sparfloxacin, clinafloxacin, gatifloxacin, ofloxacin, trovafloxacin.

Fluoroquinolones are well tolerated and relatively safe. The most common side effects include nausea, vomiting, diarrhea, abdominal pain. Other more serious but less common side effects are central nervous system effects (headache, confusion and dizziness), phototoxicity (more common with lomefloxacin and sparfloxacin). All drugs in this class have been associated with convulsions. Fluoroquinolones are classed as pregnancy category C.

— *Tetracyclines*. Tetracyclines got their name because they share a chemical structure that has four rings. They are derived from a species of Streptomyces bacteria. Tetracycline antibiotics are broad-spectrum bacteriostatic agents, that inhibit bacterial protein synthesis.

Tetracyclines may be effective against a wide variety of microorganisms, including rickettsia and amebic parasites.Tetracyclines are used in the treatment of infections of the respiratory tract, sinuses, middle ear, urinary tract, skin, intestines. Tetracyclines also are used to treat Gonorrhoea. Their most common current use is in the treatment of moderately severe acne and rosacea.The most commonly prescribed tetracycline antibiotics are: tetracycline, doxycycline, minocycline, oxytetracycline.Drugs in the tetracycline class become toxic over time. Expired drugs can cause a dangerous syndrome resulting in damage to the kidneys.

Common side effects associated with tetracyclines include cramps or burning of the stomach, diarrhea, sore mouth or tongue. Tetracyclines can cause skin photosensitivity, which increases the risk of sunburn under exposure to UV light. This may be of particular importance for those intending to take on holidays longterm doxycyline as a malaria prophylaxis. Rarely, tetracyclines may cause allergic reactions. Very rarely severe headache and vision problems may be signs of dangerous secondary intracranial hypertension.Tetracycline antibiotics should not be used in children under the age of 8, and specifically during periods of tooth development. Tetracyclines are classed as pregnancy category D. Use during pregnancy may cause alterations in bone development.

— *Macrolides.* The macrolide antibiotics are derived from Streptomyces bacteria, and got their name because they all have a macrocyclic lactone chemical structure. The macrolides are bacteriostatic, binding with bacterial ribosomes to inhibit protein synthesis. Erythromycin, the prototype of this class, has a spectrum and use similar to penicillin. Macrolide antibiotics are used to treat respiratory tract infections (such as pharyngitis, sinusitis, and bronchitis), genital, gastrointestinal tract, and skin infections.The most commonly prescribed macrolide antibiotics are: erythromycin, clarithromycin, azithromycin, roxithromycin, troleandomycin.Side effects associated with macrolides include nausea, vomiting, and diarrhea; infrequently, there may be temporary auditory impairment. Azithromycin has been rarely associated with allergic reactions, including angioedema, anaphylaxis, and dermatologic reactions. Oral erythromycin may be highly irritating to the stomach and when given by injection may cause severe phlebitis. Macrolide antibiotics should be used with caution in patients with liver dysfunction. Pregnancy category B: Azithromycin, erythromycin. Pregnancy category C: Clarithromycin, dirithromycin, troleandomycin.

Applications of Antibiotics

Chemotherapeutic antibiotics can be either broad-spectrum antibiotics, active against many organisms, or narrow-spectrum antibiotics, active against only a restricted range of organisms. Most antibiotics are manufactured as antimicrobial agents for chemotherapy, but some have other applications, as outlined below.

Antitumor Antibiotics

Such antibiotics are clinically used as cytostatic agents. Although generally toxic, with careful control of the dose certain of these antibiotics are effective in the treatment of certain kinds of tumors.

Antibiotics for Plant Pathology

Antibiotics may be more useful than synthetic chemicals in the control of plant diseases for the following reasons: they may be applied selectively in low concentrations, they are only slightly toxic to warm-blooded animals and beneficial insects, and they are normally easily broken down by soil microorganisms. Initially, medically useful antibiotics were also used on plants (especially streptomycin to combat plant disease caused by Pseudomonas sp. and Xanthomonas oryzae). Now antibiotics have been developed which are used exclusively for plant application.

Antibiotics as Food Preservatives

Government regulations in each country directly control the, use of antibiotics as food preservatives. The following compounds are available: pimaricin, a fungicide applied to food surfaces; tylosin (effective against Bacillus spores) and nisin (effective against clostridia), both used in the canning industry; chlortetracycline, used to maintain freshness in fish, meat and poultry by incorporation into ice (5 ppm) or by addition to an immersion bath (10 ppm). Of these, pimaricin and nisin are currently used commercially.

Antibiotics used as Animal Growth Promoters

Animal feed is more efficiently processed in the animal's digestive system if an antibiotic additive is used in subtherapeutic concentrations, (1-10 mg/kg feed). Weight gain may also be accelerated. The cause of this growth promotion 'may' be traced to changes in the microflora of the gastrointestinal tract. This is further confirmed by the fact that the growth of animals raised in germ-free conditions is not stimulated by antibiotics. Originally, large quantities of therapeutically useful antibiotics were added to feed. Because such extensive use of common antibiotics encouraged the risk of rapid development of antibiotic resistance, government regulations have been established to eliminate the parallel use of antibiotics in human medicine and in animal feed. In veterinary medicine, the use of antibiotics is similarly regulated. Antibiotics, which are used solely for veterinary work, include hygromycin B, thiostrepton, tylosin, and the coccidiostatic, agents monensin, lasalocide, and salinomycin.

Antibiotics as Tools in Biochemistry

The use of antibiotics as selective inhibitors has made a vital contribution to the understanding of certain cell functions, such as DNA replication, transcription, translation, and cell wall synthesis. The uses of antibiotics in these areas are discussed in textbooks of biochemistry and microbiology. In addition, some antibiotics have been added to animal cell cultures to control contamination.

Economic Significance

Worldwide antibiotic production is over 100,000 tons per year and estimated gross sales for 1980 were $4.2 billion. The annual gross sales in the United States alone are $1 billion, with cephalosporin in the leading position, followed by ampicillin and the tetracyclines. Feed-additive antibiotics are believed to have a world market of $100 million annually.

ANTIBIOTIC RESEARCH

Before 1960, about 5% of the newly isolated antibiotics were therapeutically useful. In the following years new antibiotics were discovered at an approximately constant rate, but the percentage of the new antibiotics, which actually came on the market decreased from 2.6% in 1961-1965 to 1 % in 1966-1971. This is primarily because of severe cost increases in development and clinical testing, so that manufacturers produce only those compounds, which clearly show promising therapeutic progress. On the average, it takes around 8-10 years to develop a new antibiotic, at an average cost of $10,000,000 to $20,000,000. Considering the very large number of known compounds, it may seem questionable whether the search for new antibiotics should continue. The reasons for continued research are:

In many cases the properties of natural antibiotics are not optimal for therapeutic application. The following improvements are needed greater activity with unchanged or diminished toxicity, decreased side effects, broader antimicrobial range, greater selectivity against certain pathogens, improved pharmacological properties.

Numerous tests have been made of new and semisynthetic substances, but not significant breakthroughs have been made. Since the beginning of chemotherapy, the number of resistant strains has increased. Multiple- and cross-resistance can occur, i.e., if resistance develops to one antibiotic it may simultaneously develop to others having the same mode of operation or uptake mechanism. Careless use of antibiotics has been responsible for much increase in resistance, but even with careful use in chemotherapy, resistance still develops, albeit at a slower rate. Currently, the only alternative for overcoming the resistance problem is the discovery of new antibiotics.

B–LACTAM ANTIBIOTICS

The α–lactam antibiotics can be divided into five distinct classes. The penicillins and the cephalosporins belong to the most effective of all therapeutic agents for the control of infectious diseases. In addition to the development of numerous semisynthetic β–lactam based on the known β–lactam rings, antibiotics with completely new β–lactam ring systems have been isolated in the past few years using new specific and sensitive screening methods.

PENICILLINS

Fleming described penicillin in 1929. A research group at Oxford under Florey and Chain isolated it from surface cultures of Penicillium notatum in 1940 and the first clinical application of penicillin was in 1941. Many fungi, particularly Penicillium and Aspergillus species, produce penicillins. Natural penicillins are effective against numerous grampositive bacteria. They are labile in acid and may be inactivated by splitting the β–lactam ring with penicillin- β–lactamases. Because of its low toxicity, large doses of penicillin can be used; only a small percentage of patients develop allergies (0.5-2%). β–Lactam antibiotics are specific inhibitors of bacterial cell wall (peptidoglycan) synthesis. They combine specifically with the socalled penicillin-binding protein (PBP) of the bacterial cell and, by inhibiting the enzyme activity of this protein, bring about :ell death. In E. coli it has been shown that PBP1a and 1b are transpeptidases involved in the crosslinking of the peptidoglycan, PBP3 plays a role in septum formation, and PBP4 and 5 are carboxypeptidases. A further point of attack of the β–lactam antibiotics seems to involve phospholipid synthesis.

Chemical structure

The basic structure of the penicillins is 6-aminopenicmanic acid (6-APA), which consists of a thiazolidine ring with a condensed β–lactam ring. The 6-APA carries a variable acyl moiety in position 6.

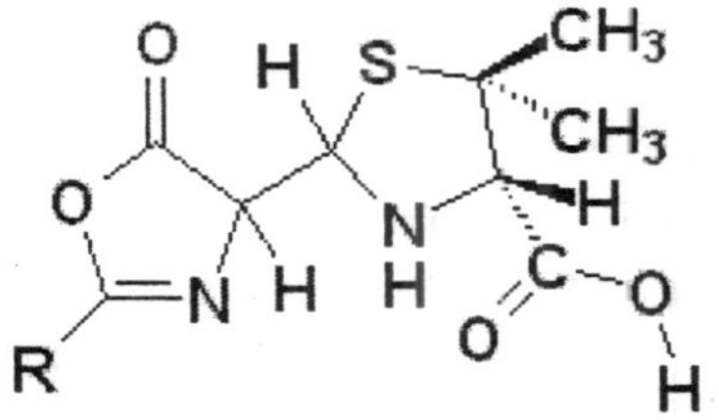

Structure of Penicillin

If the penicillin fermentation is carried out without addition of side-chain precursors, the natural penicillins are produced. From this mixture, only benzylpenicillin is therapeutically useful; the other compounds must be removed at the product recovery stage. Adding a sidechain precursor can better control the fermentation, so that only one desired penicillin is produced. Over 100 biosynthetic penicillins have been produced in this way. In commercial processes, however, only penicillin G, penicillin V, and very limited amounts of penicillin a have been produced. In order to produce semisynthetic penicillins, penicillin G (sometimes penicillin V is used instead) is chemically or enzymatically split to form 6-AP A, which is chemically recycled to make yet another penicillin derivative..

Due to their improved characteristics (acid stability, resistanceto plasmid or chromosomally coded β-lactamases, expanded antimicrobial effectiveness), semisynthetic penicillins have come to be extensively used in therapy. Because of their broadened action spectrum, these semisynthetic penicillins are sometimes compared with the so- called third generation cephalosporins. About 38% of the penicillins produced commercially are used in human medicine, 12% in veterinary medicine, and 43% are used as starting materials for the production of semisynthetic penicillins. In the early 1980's, worldwide production of penicillins amounted to 12,000 tons, an exceedingly large amount considering how highly active these agents are.

Biosynthesis and Regulation

The β–lactamthiα–zolidine ring ofpenicillin is constructed from L--cysteineand L-valine. Biosynthesisoccurs in a nonribosomal process by means of a dipeptide composed of L-α-aminoadipic acid (L-α-AAA) and L-cysteine or a breakdown product of cystathionine.

Subsequently, L-valine is connected via an epimerization reaction, resuiting in the formation of the tripeptide d-(L-α–aminoadipyl)-cysteinyl-D-valine. The first product of the cyclization of the tripeptide which can be isolated is isopenicillin N, but the biochemical reactions leading to this intermediate are not understood. Benzylpenicillin is produced in the exchange of L-α–AAA with activated phenylacetic acid . 6-AP A, which is not an intermediary product of biosynthesis, is excreted in the absence of a side chain precursor. Several regulatory mechanisms are known in penicillin biosynthesis. The amino acid lysine is synthesized from a pathway that involves L-α–aminoadipic acid, so that penicillin and lysine share a common branched biosynthetic pathway. Lysine inhibits penicillin synthesis because it is a feedback inhibitor of homocitrate synthase, an enzyme involved in L-α–AAA synthesis. If L-α–AAA is deficient, penicillin cannot be synthesized.However, feedback regulation by lysine does not seem to be a rate-limiting step in penicillin biosynthesis.

Penicillin biosynthesis is affected by phosphate concentration and also shows a distinct catabolite repression by glucose, in addition to a regulation by concentration of ammonium ion, the latter by an unexplained mechanism. Because of the glucose repression, penicillin fermentations were originally done only with the slowly metabolizable sugar lactose.

Strain Development

The penicillin production of Fleming's isolate was about 2 International units/ml; today's processes yield a penicillin titer of about 85/000 units/ml. This is an increase from 0.0012 g/l to about 50 g/l and well illustrates the value and power of a strainselection program.

The initial culture improvement came in 1943 with the isolation of Penicillium chrysogenum strain NRRL 1951. This organism was better suited for submerged production than the original P. notatum strain. Through further mutagenesis strain Wis Q 176 was isolated, the original strain in the famous Wisconsin culture line. Wis Q 17'b was adopted by most penicillin manufacturers and was used as the original strain for the various commercial strain improvement programs, about which little has been published. Newer data have been released only by Pan labs Inc., a corporation which since 1973 has specialized in commercial strain development.

Yield increases have been the main objective of strain development, but other factors, which have an effect on fermentation and efficiency of product recovery, have also been optimized. Major advances in processing have resulted through empirical screening of mutants. Until the mid1960's, the most frequently used mutagens were X-rays, methylbis- (β–chloroethyl)amine (nitrogen mustard), and short-wave ultraviolet radiation. More recently, nitrosoguanidine, alkylating agents, and nitrite have been used as mutagens.

In the early 1970's, strain improvement by using mere mutation had reached its limit. The discovery of a parasexual cycle in P. chrysogenum provided a means of utilizing genetic recombination for strain development. Heterozygous diploids were described which had penicillin titers above those of the haploid parent strains. Several such diploids have found use as production strains. However, the highest percentage of strains with increased penicillin production arose from haploid segregants of crosses between mutants of different strain lines. The protoplast fusion technique has opened up a new approach to the development of high yielding strains. Yield increases of 8% have been obtained over those obtained by mutation and selection. In addition, some of the resulting strains have had better growth characteristics, such as more stable sporulation and improved growth for inoculum production. Some of the genes involved in penicillin synthesis in P. chrysogenum have been identified by analysis of mutants blocked in penicillin synthesis (npe). Twelve npe mutants have been mapped into 5 loci (npe V, W, X, Y, and Z) by means of complementation experiments with heterozygous diploids. Cosynthesis studies showed a defect in the LLD-tripeptide synthesis for npe X, Y, and Z, a block between the trireptide and isopenicillin N in npe W, and a block between isopenicillin N and penicillin G in npe V. The function of npe Y was elegantly demonstrated as follows: Protoplasts of the various mutants were fused with liposomes into which the tripeptide had been preloaded, and penicillin synthesis was obtained.

The use of recombinant DNA technology to increase the formation of rate-limiting enzymes via gene amplification or improved transcription has not yet been possible in P. chrysogenum, due to the absence of precise biosynthetic data and the absence of good host-vector systems. However, a gene bank for P. chrysogenum has been constructed, and a transformation system has been developed.

Production Methods

Penicillin G and V are produced using submerged processes in 40,000-200,000 liter fermenters. Due to difficulties with the O_2 supply, larger tanks cannot be employed. Penicillin fermentation is an aerobic process with a volumetric oxygen absorption rate of 0.4-0.8 mM/1.min. The required aeration rate is between 0.5-1.0 vvm depending on the strain, on the bio-reactor, and on the impeller system; various turbine impellers are used for mixing (120-150 rpm). Some manufacturers use Waldh of fermenters or airlift fermenters, but this is only possible in mutants which generate low viscosity. Depending on the production strain used, the optimal temperature range is between 25-27°C.

The inoculum is started using lyophilized spores. Because of the great variability of highyielding strains, careful strain maintenance is necessary. Spore concentration (optimal 5 x 10^3/ ml) and pellet formation are crucial for the subsequent yield. If an optimal penicillin formation rate is to be achieved, pellets must grow not as compact balls, but in a loose form. After several stages of growth the production culture is ready. In the typical penicillin fermentation , there is a 'growth phase of about 40 hours duration, with a doubling time of 6 hours, during which time the greatest part of the cell mass is formed. The oxygen supply in the growing culture is critical, since the increasing viscosity hinders oxygen transfer. This can be resolved by engineering changes in the fermenter or by the use of mutants, which develop reduced viscosity. After the growth phase the culture proceeds to the actual penicillin production phase. In cases of high penicillin production, growth is sharply reduced. By feeding with, various culture medium components, the pr9duction phase can be extended to 120-180 hours.

The medium of a typical fed-batch culture may vary depending on the strain and usually consists of: corn steep liquor (4-5%, dry weight), which in present processes may be replaced by other nitrogen sources (e.g., a commercial nitrogen source called Pharmamedia); an additional nitrogen source, such as soy meal, yeast extract, or whey; a carbon source (such as lactose); and various buffers. The pH is kept constant at 6.5. Phenylacetic acid or phenoxyacetic acid is fed continuously as a precursor. Processes with glucose or molasses feeding are also successful: feeding rates of 1.0-2.5 kg.m-3h-1 with a glucose concentration of 500 kg.m^{-3}. About 65% of the metabolized carbon source is used for maintenance energy, 25% for growth, and only 10% for penicillin production. When it is considered that 50% of the production cost derives from the medium ingredients, it is obvious that control of carbon metabolism offers an important point of attack for further optimization of the production process. Yield increases of 25% have been reported by adding glucose and acetic acid. Critical parameters in these fedbatch systems are the rate of sugar utilization and the oxygen supply.

Production with continuous penicillin fermentation has been attempted, but has been difficult due to the instability of the production strains. A □batch fill and draw□ system has been suggested as an alternative. In this procedure, 20-40% of the fermentation contents is drawn off and replaced with fresh nutrient solution; this process may be repeated up to 10 times without yield reductions.Extensive research is being carried out to produce penicillin with immobilized cells. In one laboratory-scale study, the advantage of this approach over the use of a fedbatch system was demonstrated, but this technique has not as yet been introduced commercially.

Penicillin is excreted into the medium and less than 1 % remains mycelium-bound. After separation of the mycelium, product recovery is accomplished by means of a two-stage continuous countercurrent extraction of the fermenter broth with amyl or butyl acetate at 0-3°C and pH 2.5-3.0. The yield is around 90%.

CEPHALOSPORINS

Cephalosporins are β–lactam antibiotics containing a dihydrothiazine ring with D-α–aminoadipic acid as acyl moiety. Cephalosporin C was discovered in culture filtrates of Cephalosporium acremonium in 1953. The strain isolated by Brotzu in 1945 was later classified as Acremonium chrysogenum and produces several antibiotics: cephalosporin C, penicillin N (α 6-APA derivative with D-α–AAA as side chain), and the steroid cephalosporins P1 - P5. The cephalosporin antibiotics are also produced by other fungi, such as Emericellopsis and Paecilomyces. In 1971, in a screening program designed to discover β–Lactam antibiotics, the first cephamycins (7-methoxyephalosporins) were found, which are produced from various Streptomyces species, such as S. lipanii, S. clavuligerus, or Nocardia lactamdurans. Czephalosporins are valued not only because of their low toxicity but also because they are 'broad-spectrum antibiotics, comparable in action to ampicillin.

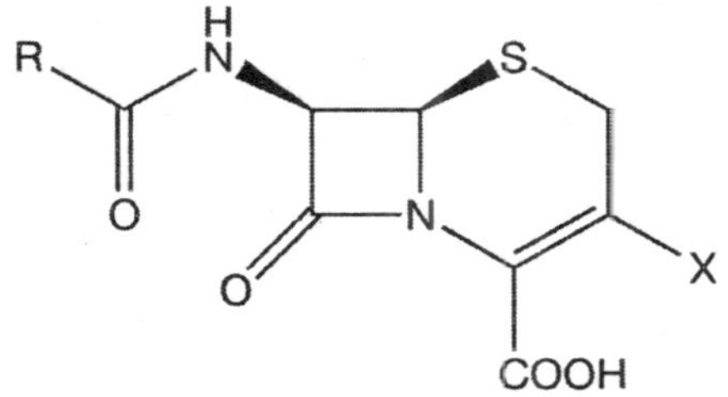

Structure of Cephalosporin

With about 29% of the antibiotic market, the cephalosporins are the single most important group of antibiotics. Therapeutically only semisynthetic derivatives of cephalosporins and cephamycin are used. For oral use, the first generation cephalosporins have been modified so that they have improved stability against -β–lactamases. These second generation antibiotics were then further modified so that they were more active against gramnegative bacteria. Further modifications have led to the commercial development of third generation antibiotics with excellent α–lactamase stability and a broadened action spectrum. Thirteen therapeutically important semisynmetic cephalosporins are commercially produced. These have been synthesized by chemical splitting to form 7-aminocephalosporanic acid (7-ACA) with subsequent chemical acylation, as well as by modifications on the C-3 site. Cephalosporins are as economically significant as penicillins.

Biosynthesis and Regulation

As in the case of benzylpenicillin, the first stages of cephalosporin Diosynthesis proceed from δ- (α–aminoadipyl)-L-cysteinyl- D-valine to isopenicillin X. In the next stage, penicillin N is produced by transformation of the L-α–AAA side chain into the D-form, via the action of a very labile racemase. After ring expansion to deacetoxycephalosporin C by the so-called expandase reaction hydroxylation via a dioxygenase to deacetyl cephalosporin C occurs. The acetylation of cephalosporin C by an acetyl-CoA dependent transferase is the end point of the biosynthetic pathway in fungi. In streptomycetes, further uansformations occur cephalosporin C or the carbamoyl derivative of deacetylcephalosporin C is converted in a two-step reaction with molecular oxygen and S-adenosylmethionine to 7- methoxycephalosporin or cephamycin C. In contrast to penicillin, the D-α–AAA moiety cannot be changed with precursor feeding.

Cephalosporins production is affected by phosphate regulation as well as by nitrogen and carbohydrate catabolite regulation.

Rapidly metabolizable carbon sources, such as glucose, maltose, or glycerol, reduce the formation of cephalosporin. The repression of the Expandase seems to be the most significant effect. The addition of lysine in low concentrations promotes cephalosporin production, whereas the inhibition caused by higher amounts of lysine can be overcome by addition of DL-α–AAA. Methionine stimulates cephalosporin synthesis in C. acremonium, but not in the streptomycetes.

In the biosynthesis of cephamycins by streptomycetes, it should be noted that α–AAA is not a precursor of L-lysine because streptomycetes synthesize lysine via the diaminopimelic acid pathway. L-lysine is converted by an L-lysine αketoglutarate-6-aminotransferase into α–AAA via L-J - piperidine-6-carboxylate as intermediate.

Strain Development

Strain development has been conducted with the original Brotzu strain, C. acremonium CMI 49 137, using mutation, selection, and parasexual techniques. Mutants have also been isolated with defective sulfur metabolism. Some of these regulatory mutants, which were isolated because they were resistant to sulfur analogs, exhibit an increased cephalosporin production or utilize inorganic sulfate as well as methionine. Parasexual recombination with mycelium of C. acremonium is difficult because the compartments of the hyphae are almost always mononuclear. By using protoplasts of different strains in fusion experiments, haploid recombinants have been isolated which showed a 40% increase in cephalosporin C production, as well as increased growth and improved sporulation. Protoplast fusion techniques have been combined with parasexual reproduction to develop an overall map of the biosynthetic genes for cephalosporin C. The gene for isopenicillin K synthetase from C. acremonium has been cloned in E. coli and an effective transformation system for C. acremonium has been developed, making it likely that the key steps in cephalosporin production will become subject to analysis and modification. The cephalosporin C titer of contemporary highyielding strains, in which the proportion of penicillin N is reduced, is about 20 g/l.

Production Method

The fermentation of cephalosporin is similar to that of penicillin. Complex media with corn steep liquor, meat meal, and sucrose, glucose and ammonium acetate are used. Fermentations are carried out at as fedbatch processes with semicontinuous addition of nutrients at pH 6.0-7.0 and at temperatures between 24-28°C.) In the main growth phase, a high aeration rate is necessary, but during the production phase (48-160 h), O_2 consumption decreases sharply. The chemical synthesis of cephalosporin by ring-expansion of penicillin has been developed in recent years. This process has become especially attractive because of the low price of the penicillin starting material. An example of how this approach has been commercialized is the use of penicillin V to produce oraspor, an orally active cephalosporin.

New β–Lactam Ring Systems

New β–Lactam ring systems have been discovered since the 1970's by the introduction of screening methods such as using as test organisms β–Lactam hypersensitive strains or using enzymatic tests to detect β–lactam inhibitors. Because they may be active as either Droad-spectrum antibiotics or β– lactamase inhibitors, these new compounds, especially their semisynthetic derivatives, can be expected to find reconsiderable use in chemotherapy.Clavulanic acid, a β–lactamoxazolidine ring system, is produced along with cephalosporins by Streptomyces clavuligerus. Clavulanic acid only slightly effective antibacterially but inhibits a wide range of β–lactamases in concentrations under 0.1,i.g/ml. The inhibition is irreversible; in combination with β–lactamasesensitive penicillins and cephalosporins, this compound causes distinct increase in activity of these antibiotics, even against bacteria, which are normally resistant to β–lactam antibiotics. Clavulanic acid has been marketed in combination with amoxycillin. Thienamycins and olivanic acids have a β–lactampyrroline ring system in common. Thienamycin, the first in a series of carbapenems with very similar structures, was isolated in 1976 from Streptomyces cattleya. Thienamycin acts as a β–lactamase inhibitor and as a broad-spectrum antibiotic against Grampositive and Gram-negative bacteria, including Pseudomonas aeruginosa, but it is extremely unstable.

The semisynthetic N-formimidoylthienamycin, resistant to dehydropeptidases, has good stability. It is the β–lactam with the broadest spectrum and is undergoing clinical trials. The olivanic acids and the closely related epithienamycins show a lower activity against Pseudomonas aeruginosa and Staphylococcus aureus.

Nocardicins are monocyclic β–lactams. From a mixture of seven components, nocardicin A has been isolated as the most active compound. It is weakly active against gram-negative bacteria, but research is being carried out to improve its activity by formation of chemical derivatives. Monobactams, which are also monocyclic β-–lactams, have beenisolated from bacterial cultures.However, they are only weakly antibacterial. Compounds with a 3α–methoxy group are resistant to β–lactamases, whereas the rest of the monobactams show varying degrees of sensitivity to J3-lactamases. Azthreonam, a semisynthetic monobactam made from threonine, has low toxicity and excellent activity against gramnegative pathogens; it is in clinical trials.

Amino Acid and Peptide Antibiotics

This group of antibiotics includes amino acid derivatives such as cycloserine and azaserine, most of the β–actam antibiotics, chromopeptide antibiotics, depsipeptide antibiotics, and linear or cyclic peptide antibiotics. Over 400 peptide antibiotics are currently known.

D-cycloserine

Cycloserine, a D-4-amino-3-isoxazolidone, may be synthetically produced or may be isolated from cultures of Streptomyces or-chidaceus, S. lavendulae, S. garyphalus, or S. roseochromogenes. D-CyCloserine, a D-alanine an-alog, inhibits cell wall synthesis by inhibiting alanine racemase, the enzyme responsible for the production of the D-alanine moiety of the peptidoglycan. Cycloserine is very effective againS mycobacteria, particularly M. tuberculosis, and is used in the treatment of tuberculosis in commnation with isonicotinic acid hydrazide (INK) and rifampicin or streptomycin.

Chromopeptide Antibiotics

This group includes the actinomycins, which are produced as a mixture of very similar substances by different streptomycetes especially Streptomyces antibioticus and S. chrysomallus. The following actinomycins have been described: A, B, CI, C_2, i-C_2, C_3, D, F_1, Z_O-Z_5. A actinomycins which have thus far been isolate have the same phenoxazone-chromophore (actinocin), which is linked to two pentapeptide lactones. These two lactones have varied amino acid sequences, and by adding different amino acids to the fermentation broth, 30 distinct biosynthetic actinomycins have been produced. The production of actinomycin from 2 molecules of 3-hydroxy-4-methylanthranilic acid pentapeptide lactone is catalyzed by phenoxα–zine synthase. This enzyme is produced only after the trClnsition into the idiophase and shows pronounced catabolite regulation by glucose. The phenoxazone rings of actinomycins intercalate at a GC pair in 5'- T -G-C-A-3' palindromes of DNA, thus blocking DNAdependent RNA-polymerases. Actinomycins are very toxic causing liver and kidney damage, but are beneficial in tumor treatment. The most important actinomycin used in cancer chemotherapy is dactinomycin.

Depsipeptide Antibiotics

In depsipeptides, the subunits are usually linked with alternating ester and acid amide bonds. Valinomycin, a cyclododecadepsipeptide produced by Streptomyces fulvissimus, has no economic significance but finds wide use in bioemical reasearch as an agent for uncoupling oxidative phosphorylation. It is also useful as potassium carrier because of its selective binding of K+ ions. Antibiotics of the virginiamycin family also have a depsipeptide structure. They consist of a mixture of structurally different macrocyclic lactone rings. Antibiotics of the virginiamycin group are chiefly effective against Gram-positive bacteria and are widely used as growth promoters for poultry, swine, and calves. Seventeen commercially are commercially produced the most important are the mikamycins and virginiamycins, which are produced by Streptomyces mitakaensis Or S. virginiae.

Linear and Cyclic Peptide Antibiotics

The majority of the antibiotics formed by species of the genus Bacillus are peptides; other peptide antibiotics are produced by streptomycetes. The molecular weights of peptide antibiotics are between 270 (bacilysin) and 4500 (licheniformin). Some peptide antibiotics are linear but most are cyclic compounds. They may consists of either amino adds alone or they may contain other components in addition to the amino acids. The peptide antibiotics produced by bacilli are formed at the on-set of sporulation and seem to play a regulatory role in the sporulation process. Therapeutic application of peptide antibiotics is limited, due to their toxicity. Gramicidin S, linear gramicidins, tyrocidin, and bacitracin are used topically. Polymyxins are used for infections caused by Gram-negative pathogens, and viomycin and capreomycin are used in tuberculosis therapy.

Bacitracin

Bacitracin is the most economically important of the peptide antibiotics. In addition to its use as a topical antibiotic, it is also used as a growth promoter in animal feeds. As a feed supplement, it can be used as the zinc or manganese salt, as the lignin-bacitracin complex, or as bacitracin methylene disalicylate.Total production of baci-tracin worldwide was estimated at 500 tons in 1976.

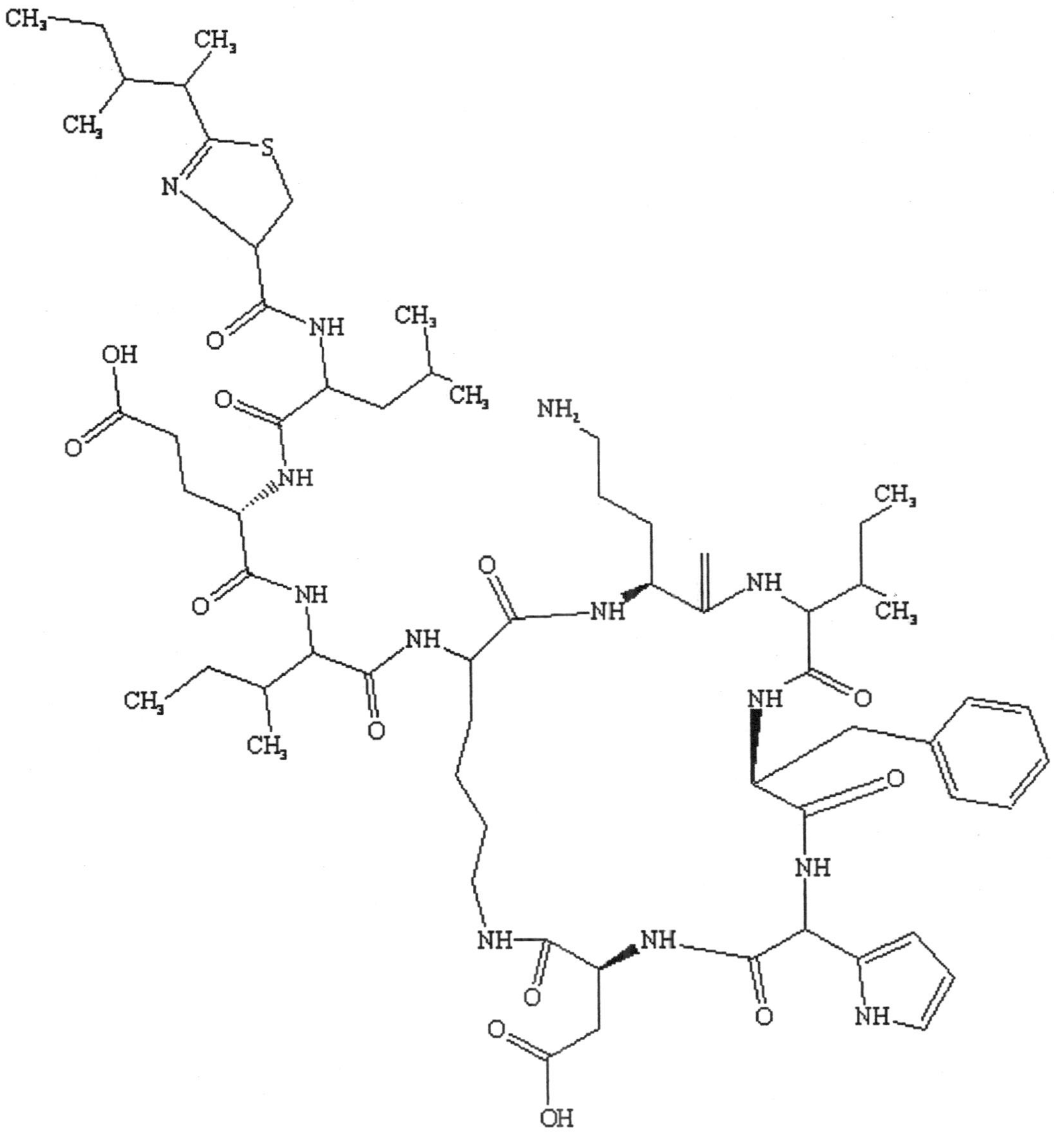

Structure of bacitracin

Various strains of B.licheniformis produce a mixture of bacitracin A, A', B, C, D, E, F, F_1, F_2, F_3 and G. Bacitracin A is the main component of the mixture (about 70%) and has the greatest biological activity. It is a dodecapeptide consisting of 8 L- and 4 D-amino acid and contains both α–cyclic hexapeptide and a thiazoline ring structure.

Bacitracin inhibits the formation of the bacterial cell wall at the level of peptidoglycan biosynthesis. It prevents the dephosphorylation of C55-isoprenylpyrophosphaa to C55-isoprenylphosphate.

Peptide formation can occur by either ribosomal or nonribosomal mechanisms. The biosynthesis of high molecular-weight antibiotics such as nisin and neocar-rinostatin involves ribosomes, mRNA, and t RNA.However, the biosynthesis of most peptide antibiotics does not involve this proteinsynthesizing machinery, but takes place instead on multienzyme complex. This nonribosomal peptide synthesis process, which has also been called the □thiotemplate□ mechanism, has been described for gramicidin 5, tyrocidin, and the linear gramicidins, as well as for bacitracin. Bacitracin synthetase consists of subunits A, B, and C, with molecular weights of 200,000, and 380,000, respectively.

The first step, the activation of amino acids in the presence of A TP and Mg 2+ to aminoacyl adenylates, takes place at and activation site for each amino acid on the enzyme complex. In a transfer reaction, the activated amino acids are then bound to specific thiol groups in thioester linkages. The amino acid sequence of the subsequent peptide is determined by the arrangement of the thiol groups on the enzyme. The racemization to the D-enantiomers probably takes place at this level. The transfer of the growing peptide chain (growth from the N-terminal end toward the C-terminal end) to the next amino acid and to the next en-zyme subunit is catalyzed by 4' - phosphopan-tetheine, a cofactor of each subunit. The cycli-zation of the peptide chain takes place in the multienzyme complex, but the formation of the thiazoline ring is not yet fully understood.

Production Methods

Bacitracin was initially obtained from surface cultures; today its production occurs in aerobic submerged culture, like that of all other antibiotic processes. The yields of this process were increased from about 13 mg/l to about 9 g/l through strain development and medium optimization. A continuous process has also been developed for bacitracin production using immobilized cells.

Chelate-forming Peptide Antibiotics

Siderochromes and bleomycins are peptide antibiotics, which form chelates with metals. Siderochromes are iron-containing or iron-binding natural substances containing hydroxamic acid groups, and bring about the formation of ferrihydroxamate complexes. Albomycins and ferrimycins have been isolated as antibiotically active siderochromes sideromycins), but there is no known industrial application for these substances. Sideramines are siderochromes which act as growth stimulants, and ferrioxamine B, a sideramine from Streptomyces pilosus ETH 21748, is commercially produced. It is used in human therapy as the desferri compound for the treatment of diseases involving iron storage and also in agriculture in such applications as the elimination of soil iron deficiencies in peach orchards. Bleomycins are isolated as Cu(II) complexes from the culture broth of Streptomyces verticillus. The glycopeptide exhibits some antitumor action and is marketed as a mixture of bleomycin A_2 and B_2.

Carbohydrate Antibiotics

Carbohydrate antibiotics include the therapeutic important aminoglycosides, the orthosomycins, and various sugar derivatives.

Glycosides and Sugar Derivatives

Several commercially important antibiotics from heterogeneous group are described below:

Nojirimycin, shows anti-bacterial activity against Sarcina lutea and Xanthomonas oryzae, and is of interest because of its high inhibitory activity against α- and, β-glucosidases and amylases.

Vancomycin, a proadspectrum antibiotic, consists of the amino sugar vancosamine, two β-hydroxychlortyrosine moieties, three phenylglyine derivatives, N-methyl leucine, and aspartic acid. It is effective against Grampositive organisms, particularly staphylococci. However, it has adverse side effects in humans and thus is used only to treat infections caused by pathogens which are resistant to other antibiotics or with patients who are hypersensitive to penicillin. No development of resistance to vancomycin has yet been observed in therapy.

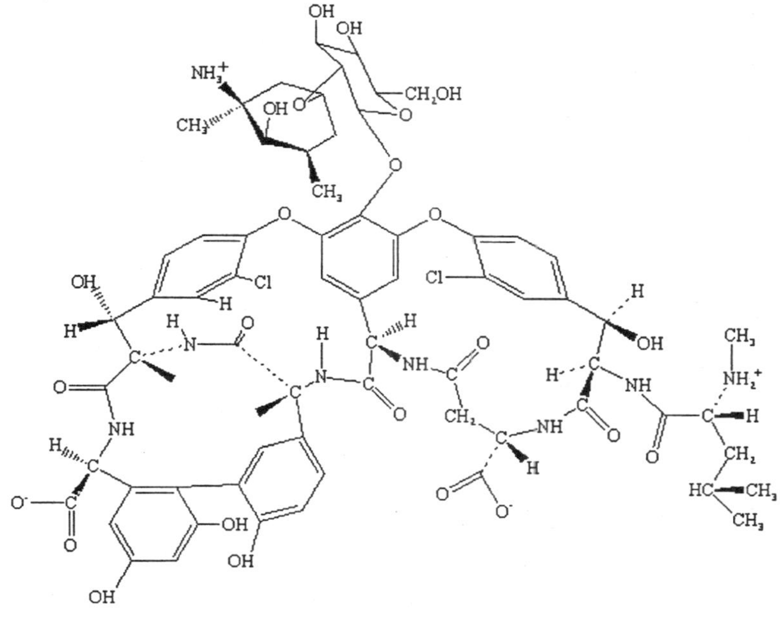

Structure of vancomycin

Lincomycin is used in the treatment of staphylococcus, streptococcus, and pneumococcus infections; only minor side effects are observed, but during therapy, resistance acquired rapidly. Clindamycin (7-chlor-7-deoxy lincomycin), a semisynthetic derivative with improved pharmacological properties, is principle marketed today. In protein synthesis, the bonding of aminoacyl-tRNA to the A-binding site at the ribosome is inhibited by lincomycin, probα–bly through interaction with the ribosomal protein L6 of the 50S subunit.

Moenomycin is a phosphoglycolipid which shows activity against Gram-positive bacteria as well as Gram-negative bacteria containing a resistance plasmid. Moenomycin A and the related antibiotics macarbomycin and quebemycin are marketed for use as nutritional feed supplements. Phosphoglycolipids inhibit bacterial cell wall synthesis by affecting the transglycosylation reaction in peptidoglycan biosynthesis.

The orthosomycins are a class of antibiotics newly discovered in the late 1970's. They are characterized by an oligosaccharide structure and two noncarbohydrate ester groups. The following antibiotics have been described the avilamycins produced by 5treptomyces viridochromogenes), the curamycins S. curacoi) the everninomicins (Micromonospora carbonaeae), and' flambamycin (S. hygroscopicus).The avilamycins are being tested for use as animal feed supplements.

Aminoglycoside Antibiotics

Aminoglycosides are oligosaccharide antibiotics and consist of an aminocyclohexanol moiety (e.g.. deoxystreptamine, streptidine), which is glyrosidically linked to other amino sugars. Aminoglycosides are primarily used against Gram-negative bacteria in a wide range of applications. Although the semisynthetic cephalosporins are also now widely used for Gram-negative infections, in severe infections the aminoglycosides are often the antibiotics of choice. Although streptomycin and dihydrostreptomycin are broadspectrum antibiotics, they are primarily used to treat tuberculosis. They are considered reserve antibiotics, mainly due to the fact that resistance can develop rapidly and in a single step by a chromosomal mutation.In 1976 the world market for aminoglycosides was estimated at $500 million, with a yearly inrease of 10-15%.

All of the aminoglycoside antibiotics cause kidney damage and deafness (nephro- and ototoxicity) as side effects. In addition to resistance development by chromosomal mutation, plasmiddetermined resistance due to the formation of inactivating enzymes is known. Compounds with reduced toxicity and/or increased effectiveness against resistant organisms are being sought via screening, mutasynthesis, and formation of chemical derivatives. As a result of a screening program, the new group of fortimicins was discovered. This is a group of pseudodisaccharides with a broad spectrum of activity, particularly against aminoglycoside-resistant bacteria. The kanamycin derivatives ami-kacin and dibekacin, and netilmicin (l-N-ethylsisomicin) are semisynthetic compounds with partial resistance to inactivating enzymes. These compounds have already been introduced for clinical use. The compound 5-episisomicin, isolated by mutasynthesis, has been tested clinically because of its action against sisomicin-resistant organisms. However, for cost reasons it is being produced chemically rather than by fermentation. Besides the antibiotics noted above which are useful for human therapy, kasugamycin and validamycin are used as important antibiotics in rice cultivation.

Mode of Action

The mode of action of amino-glycosides is on protein synthesis in sensitive bacteria. Streptomycin, the aminoglycoside which has been studied the most, binds to protein S12 of the 30S ribosome and causes mis-reading of the code and hence'inhibition of protein synthesis. S12 is a component of the A site at which the aminoacyl-tRNA and fMet-tRNA =d. 2-Desocystreptamine-containing antibiotics inhibit the translocation process of the peptidyl-tRNA. In addition, aminoglycoside antibiotics cause damage to the bacterial cell membrane, thus bringing about loss from the cell of low-molecular weight constituents.

Biosynthesis of Aminoglycosides

Glucose is the precursor of all the subunits of those aminoglycoside antibiotics which have been examined. Although a number of enzymatic reactions have been uncovered, the exact biosynthesis of any aminoglyoside antibiotic is not yet fully understood. Over 30 separate enzymatic steps are known. One of the principal components of streptomycin, streptidine, is synthesized from glucose-6-phosphate via myoinositol. A number of enzymatic steps, involving oxidation, amination, phosphorylation, carbamidinylation, and dephosphorylation, lead to the formation of streptidine-6-P. The second dephosphorylation step takes place during the final biosynthetic step to streptomycin synthesis. Although the pathway for streptose biosynthesis is known, that for N-methyl-L-glucosamine is still unclear. The combination of the three subunits involves two steps. Only in the last step is the phosphate moiety removed, leading to the formation of the biologically active streptomycin from the biologically inactive streptomycin phosphate.

Strain Development

Strain-development programs have been conducted to increaseyields for all industrially produced aminoglycoside antibiotics, since yields of wild strains range only from 10-200 mg/ml, which is not sufficient for a costeffective process. The following yields have been published: kanamycin 2000, mg/ml; tobramycin 1500 mg/ml; neomycin 2000 mg/ml; gentamicin 1400 mg/ml. Streptomycin yields have been increased from 100-200 mg/ml to about 15 mg/ml by strain-improvement techniques. The current yields of production processes are probably far more than the published values in many cases.

Mutation and selection are the principal methods which have been used in strain development) there have also been some cases in which recombination (conjugation and protoplast fusion) has been employed. The strain development of kasugamycin is a typical example. Using the agar disk method in one year 650,000 isolates were screened after mutagenesis, resulting in a yield increase from 200 mg/ml to about 7000 mg/ml. Conjugation of auxotrophs produced hybrid strains, whose offspring in one case produced up to 26% more kasugamycin (9300 mg/l) than the initial prototroph strain. Cell fusion and recombinant DNA methods have been used in several cases to produce higheryielding strains. Protoplast fusion has been used to increase yield of sisomicin producers and amplification of a resistance gene, 6'-N-acetyltransferase, has increased production of kanamycin and neomycin.

Production methods

Aminoglycoside antibiotics are produced in fermenters with volumes up to 150,000 1. Optimal oxygen supply is required (aeation rate 0.5-1.0 vvm) and temperatures are held between 28-30°C with the pH in the neutral range. The length of fermentation is between 4-7 days, depending on the strain.

Glucose in combination with starch or dextrin serves as the carbon source. Originally, the antibiotic butirosin could be produced only with glycerol as carbon source, due to catabolite repression by glucose, but the isolation of mutants not subject to catabolite regulation has made it possible to use glucose. In several proesses, continuous feeding of glucose has been successfully used. Soy meal is an ideal nitrogen source for aminoglycoside production because of its slow catabolism. NaCl (1-3 g/l) must be added to the streptomycin fermentation. The production of gentamicin, sisomicin, and fortimicin A is significantly accelerated by addition of Co|, which is active in the methylation steps during biosynthesis. Addition of Zn^{2+} increases the yield of nebramycin and addition of Mg^{2+} stimulates kanamycin and neomycin production. stimulates kanamycin and neomycin production. Aminoglycoside processes are typical secondary metabolite fermentations with trophoand idiophases for the butirosin fermentation.

Most of the aminoglycoside antibiotics are exreted and are present in the culture supernatant, from which they are removed by adsorption to ion-exchange columns. Cell-bound aminoglycosides, such as gentamicin, sisomicin, and fortimicin, must first be released by acidification to pH 2-2.5 with H_2SO_4.

MACROCYCLIC LACTONE ANTIBIOTICS

The macrolide group of antibiotics consists of compounds with large lactone rings. In addition to those antibiotics conventionally called mac-rolides, this group includes polyenes, macrotetrolides, and a heterogeneous group of atypical macrolides.

Macrolide Antibiotics

Mactolides are hydrophobic, usually basic compounds. They consist of 12-,14-,16-, or 17-membered lactone rings with 1-3 sugars glycosidically linked with the aglycone (lactone ring) and with each other. The sugars are amino sugarand/or 6-deoxyhexoses. Macrolides are very effective against ,Grampositive bacteria, particularly staphylococci, and mycoplasmas and were frequently used against penicillin-resistant organisms, but because of the development of the newer β–lactam antibiotics, the macrolides are no longer of as great medical significance. The aglyclone and sugar moieties are both necessary foe antibacterial activity. Therapy with macrolides results in a rapid buildup of resistance and there is also cross-resistance between the individual antibiotics. Because of the low toxicity, efforts are being made to isolate new compounds of this class, to chemically modify known macrolides, and to modify known macrolides through biotransformation. All macrolide antibiotics are inhibitors of protein synthesis, binding to the 50 S subunit of bacterial ribosomes. Erythromycin is known to inhibit the elongation-factor G-dependent re-lease of deacylated tRNA from the P site of the ribosome.

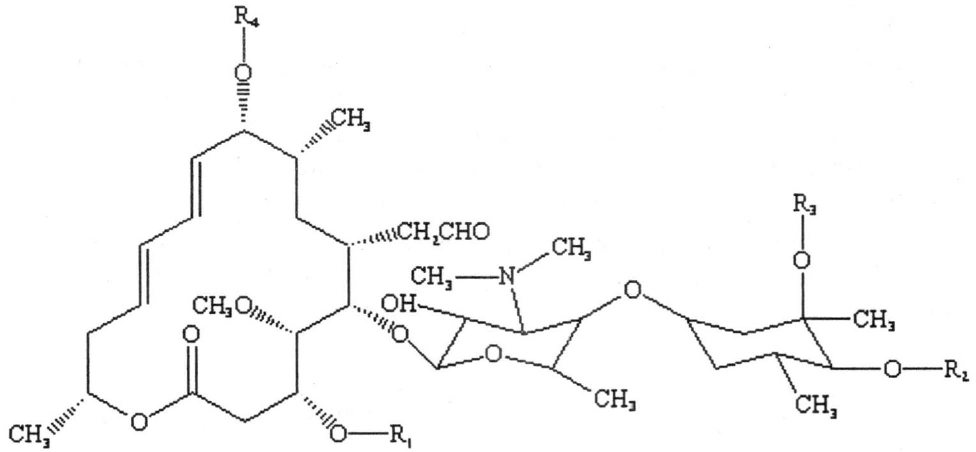

Chemical structure of a macrolide with a 16-membered ring

Biosynthesis and Regulation

The aglycone moiety of macrolides is built either of propionate subunits or of a combination of acetate and propionate subunits, sometimes with the participation of butyrate or related compounds. One of the first intensively studied macrolides was erythromycin; the biosynthesis of this antibiotic was studied by labeling experiments and from the behavior of mutants of Streptomyces erythreus blocked in antibiotic biosynthesis. Analogous to fatty acid synthesis, erythronolide production proceeds from propionyl-CoA as a starting material, onto which 6 molecules of 2-methylm-α–lonyl-CoA condense. This polyketide synthesis takes place on a multi-enzyme complex called lactone synthase, which is very similar to the fatty acid synthetase of S. erythreus. After cyclization to the aglycone, the first free compound, 6-deoxyerythronolide B, can be isolated.

A number of observations have been made about the regulation of macrolide biosynthesis. Biosynthesis of the macrolide tylosin is stimu-lated by longhain fatty acids; glucose inhibits the metabolism of fatty acids and tylosin production, probably by inducing an acetyl-CoA deficiency. Phosphate regulation is also observed with many macrolide antibiotics, as well as nitrogen (NH^{4+}) catabolite regulation.

Erythromycin biosynthesis is regulated at various levels. For instance, erythronolide B inhibits its own production, the synthesis of deoxyerythronolide B, and presumably also the activity of the lactone synthase. End-product inhibition of the transmethylase by erythromycin A has been observed. This enzyme catalyzes the methylation of mycarose to cladinose as a last -step of the biosynthetic pathway. In highyielding strains used for industrial production, this regulation is eliminated. When propanol (0.2-0.5%) is added to the fermentation broth of S. erythreus, it induces the synthesis of acetyl-CoA-carboxylase and causes a 100% increase in erythromycin production.

Production Method

Macrolide antibiotics are produced in aerobic submerged fermentations. Yields in largescale industrial processes are around 20 g/l. A typical medium for S. erythrea contains glucose 50 g, soy meal 30 g. Complex culture medium ingredients may as be used, such as starch, corn steep liquor, yeast extract, and oils (e.g., soy bean oil). Fermentation temperatures for most macrolides are 25-28°C and for erythromycin 33°C. The length of the fermentation is 3-7 days.

The macrolide tylosin can be produced in either batch or continuous fermentation. The specific production rate is increased to 1.1 mg/g biomass.h if glycerol is used instead of glucose and if the feeding rate of sodium glutamate is optimized at a low specific growth rate of 0.03/h.

POLYENE MACROLIDE ANTIBIOTICS

Polyenes consist of 26- to 38-membered lactone rings with 3-7 conjugated double bonds. Depending on the length of the chromophore, pdrenes are classified as trienes, tetraenes, pa:r taenes, hexaenes, or heptaenes. Many polyenes; also contain a glycosidically-bound amino sugar (mycosamine or perosamine). Some heptaenes carry p-aminoacetophenone or N-methyl-paminoacetophenone as an aromatic moiety on the carbon atom which is involved in lactone formation.

More than 90 polyenes have been characterized primarily from streptomycetes. These antibiotics are active against fungi and yeast but they are ineffective against bacteria. Polyenes affect membrane sterols of sensitive eucaryotes, causing a change in membrane permeability. Some polyenes are relatively toxic (nephro- or hepatotoxic) but they are widely used commercially for topical application because no better antifungal antibiotics are available. Polyenes are not suitable for plant-disease control because they are sensitive to light and heat.

Biosynthesis and regulation

Like macrolides, po-lyene antibiotics are produced from C and C units via the polyketide biosynthetic pathway. Among the aromatic polyenes, paminobenzoyl-CoA, the precursor of p-aminoacetophenone and N-methyl-p-aminoacetophenone, seems to function as the starting point for polyketide polymerization. The chromophore consists solely of acetate units. The amino sugar is derived from glucose, with thymidine diphosphate-4-keto-6-deoxy-D-glucose as an intermediate. Polyene biosynthesis shows phosphate regulation, catabolite repression, and feedback inhibition.

Production methods

Glucose is usually the best carbon source for polyene production. Most fermentations using glucose are continuous, beause catabolite repression occurs at the required glucose concentration (for candicidin 7-9.6%). The candicidin production phase can be extended to 300 h by glucose feeding. The addition of acetate, propionate, malonate, or n-propanol stimulates polyene production, and soy meal or soy peptone can be used as nitrogen sources.

For maximal polyene production, carefu' control of aeration is essential. During the growth phase, the required oxygen transfer rate is around 0.8 mmole 02/l.min, but during the production phase, O_2 concentration should not exceed 40% of saturation.

Since polyenes are not- watersoluble, they accumulate as crystals or microparticles at the surface of the cell or in the medium, from which they can be extracted with appropriate solvents.

MACROTETROLIDES

Macrotetrolides are cyclic polylactones which function as ionophores. These sugarfree compounds are synthesized from four tetrhydrofuranyl hydroxy acids. Tetranactin, produced by S. flaveolus, is used as an insecticide.

This is a heterogeneous group, classified together because all structures contain the lactone ring. Examples include avermectin, milbemycin, venturicidin, bafilomycin, and irumamycin. Avermectin is produced by S. avermitilis. It has found use in veterinary medicine because of its high activity against nematodes and arthropods. Irumamycin is undergoing field trials in agriculture as a fungicide.

ANSAMYCINS

Ansamycins are macrolactam antibiotics with an aromatic chromophore to which an aliphatic polyketide derivative is bound at both ends, via amide, ether, and/or carbon-carbon bonds, the so-called ansa chain. Ansamyins are distinguished by their chromophore. Those compounds with a benzene ring, such as geldanamycin or ansamitocin, act mainly on euaryotes, whereas those with a naphthalene structure are antibacterially effective. In this latter class are the rifamycins, halomycins, and streptovaricins.

Rifamycins are important in chemotherapy. They are very effective against Gram-positive bacteria and mycobacteria, but less effective against Gram-negative bacteria; they show no cross resistance with other antibiotics. They are produced by Nocardia mediterranei. Approximately 20 different rifamycins have been isolated from culture solutions of Nocardia mutants, and genetic studies have shown that the genes for the last steps in the biosynthetic pathway lie in a single cluster. The wild strain produces rifamycin B as a major component and smaller amounts of the so-called rifamycin complex (A, C, D, E). The production of this complex can be largely suppressed by the addition of sodium di-ethylbarbiturate. Rifamycin B is completely inactive but during its breakdown or via chemical modification, the biologically active rifamycin O, S, and SV are formed. Rifamycin S and SV are biosynthetic precursors to rifamycin B. Mutants have been isolated which excrete rifamycin SV due to a block in the subsequent steps.The ansa chain of rifamycin is formed from two malonyl-CoA and eight methylmalonyl-CoA and methionine. The polyketide condensation -process proceeds from an aromatic C7 unit. The biosynthesis of the naphthohydroquinone hromophore branches off from the aromatic biosynthetic pathway between 3-deoxy-D-arabinoheptulose-7-P and shikimate. Rifampicin, a semisynthetic compound derived from rifamycin SV, is widely used in the treatment of tuberculosis. It is a specific inhibitor -of bacterial DNA-dependent RNA polymerase. It binds with the bsubunit of the polymerase and RNA polymerases of higher organisms are not -affected.

Tetracyclines

The tetracyclines are important antibiotics with -widespread medical use. Chlortetracycline, the -first antibiotic of this group, was isolated from -cultures of Streptomyces aureofaciens in 1945. Tetracycline itself was first prepared by dehalogenation chlortetracycline but was subsequent found to be produced by cultures of S. viriifaciens as well. Presently about 20 streptomycetes are known which produce a mixture of different tetracyclines.

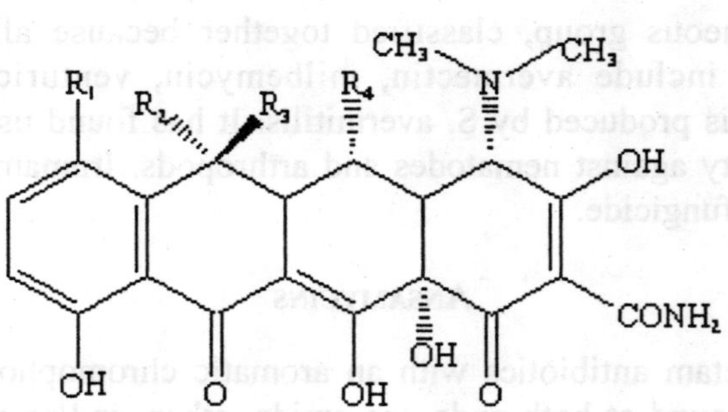

Chemical structure of a tetracycline

Tetracyclines are broad-spectrum antibiotics, which are effective against Gram-positive and Gram-negative bacteria, as well as rickettsias, mycoplasmas, leptospiras, spirochetes, and chlamydias. There is a marked cross-resistance between different tetracyclines. After cephalosporin and penicillin, the tetracyclines are the most widely used antibiotics. Chlortetracycline and oxytetracycline are commonly used in human and veterinary medicine. In many countries they are also widely used as nutritional supplements in poultry and swine production and in some countries they are also used to preserve fish, meat and poultry. Tetracyclines act as inhibitors of protein synthesis. Their site of action is the 30S ribosome, where binding of aminoacyl-tRNA to the ribosomal A-site is inhibited.

Biosynthesis and Regulation

Chlortetracycline and oxytetracycline are the tetracyclines most frequently produced by the streptomycetes, tetracycline itself usually being only a minor com-ponent. However, S. aureofaciens mutants with a block in the chlorination reaction excrete tetrα–cycline as the major product.

The biosynthesis can be subdivided into three sections:

— Glucose is converted into acetyl-CoA.

— Malonyl-CoA is produced in the transformation of acetyl-CoA by means of acetyl-CoA-carboxylase. After transamination to malonamoyl-CoA, which is bound to an enzyme complex (anthracene synthase), this compound condenses with 8 molecules of malonyl-CoP. The polyketide presumably formed from this condensation has not been isolated. An alternative pathway to malonyl-CoA is via the production of oxalacetate catalyzed by phosphoenol pyruvate carboxylase, oxalacetate being converted to malonyl-CoA by oxidative decarboxylation. Cyclization into the tricyclic intermediate also takes place on the enzyme anthracene synthase.

— The step-by-step transformation of the assumed tricyclic intermediate, which has not yet been isolated, into the variety of known tetracyclines.

The scheme of chlortetracycline biosynthesis, still mainly hypothetical, has been deduced from studies of mutants blocked in tetracycline biosynthesis and from cosynthesis studies with S. aureofaciens. There are 72 intermediate products; 27 substances have been characterized but 45 have not yet been isolated and the chlorination reaction itself is not yet understood. There is a clear correlation between tetracy-cline production and carbohydrate catabolism. Highyielding strains have a lower glycolysis rate than strains with low tetracycline production. The addition of 0.5-3 mg/ml benzylthiocyanate, an inhibitor of sugar uptake, causes a 50% increase in chlortetracycline production; simultaneously, the activity of the pentose-phosphate cycle increases. In the case of tetracycline biosynthesis, phosphate regulation is well understood. The addition of Pi to the producing culture decreases the rate of tetracyline formation by inhibition of ATC oxygenase. The rate of tetracycline synthesis is directly proportional to the activity of this enzyme, which is only detectable in the culture after phosphate has been completely depleted.

Strain Development

Studies on the genetics of S.aureofaciens have shown that over 300 genes are involved in the biosynthesis of chlortetracycline. Considering also the dearth of knowledge of biosynthesis and regulation in this pathway, it is understandable that a rational approach to strain improvement has not been possible. So far, all developments of higheryielding strains have come from empirical mutagenesis studies. Ultraviolet radiation, alone or in combination with other mutagens, has brought about the greatest number of higheryielding strains using S. aureofaciens and S. rimosus.

Strains with increased tetracycline production have been isolated by selecting revertants of mutants blocked in antibiotic synthesis and from prototrophic revertants of auxotrophs. They may also be isolated from mutants showing increased resistance to their own product. The highest published yield for tetracycline is 25 g/ liter. Hybridization has not resulted in strains with increased product formation, but in combination with mutation it has caused an increase in strain variability. Protoplast fusion has been used to increase yields.

Production Methods

The production of tetracyclines is carried out in stirred fermenters with volumes up to 150,000

If glucose is used, continuous feeding is necessary; starch can be used as an additional carbon source. Because of phosphate repression, tetrα–cycline fermentations are run with phosphate limitation. For maximal yield, an optimal oxygen supply is important, particularly in the first hours (6-12 h) of the fermentation. Aeration rates of 1 vvm are customary; yield increases result when oxygen-enriched air is used.

There are several routes to tetracycline:

— By a chemical process using chlortetracycline-.

— By fermentation in chloride-free culture medium, which can be made by pretreatment with ion exchangers.

— By adding chlorination inhibitors to the medium, such as mercaptobenzothiazole, 2-, thiouracil, or thiourea. Bromide inhibits chlorination in some cultures, but causes formation of 7-bromtetracycline in other strains.

— With mutants, which are blocked in the chlorination reaction.

Anthracyclines

Anthracyclines are glycosides which consist of 7,8,9,10-tetrahydro-5,12-naphthacenequinone as an aglycone combined with 1-3 sugar residues. Anthracyclines are very toxic compounds.They intercalate into DNA and inhibit DNA-dependent DNA polymerase in DNA replication. Daunomycin (daunorubicin) and adriamyin (doxorubicin), produced by Streptomyces peuetius, are used clinically in cancer treatment. Daunomycin acts against certain forms of leukemia; adriamycin, which has low toxicity, is effective in treating various tumors. However, these compounds have significant side effects and research is under way, using mutasynthesis and protoplast fusion as well as by the formation of chemical derivatives, to produce compounds with decreased problems.

Nucleoside Antibiotics

About 200 antibiotics with nucleoside-like structures are known. Because of the diversity of structures, marked differences in biological activity are seen. Aminoacyl nucleosides used as fungicide antibiotics in plant protection include blasticidin S from Streptomyces griseochromogenes, mildiomycin, from Streptoverticillium rimofaciens, and the polyoxins from S. cacaoi var. asoensis. Polyoxins inhibit the incorporation of N-acetyl-D-glucosamine into the chitin moiety of fungal cell walls by competitive inhibition of chitin synthase. The dapiramines as well as nikkomycin, which also act as chitinase inhibitors, are undergoing field tests as possible herbicides. Several compounds with cytostatic or antiviral activity are under development. for example, neplanosin A from cultures of Ampullariella regularis. Puromycin from S. alboniger has found wide use as a tool in studying ribosomal function in protein biosynthesis but has no clinical significance. Puromycin resembles the aminoacyl end of transfer RNA. In both procaryotes and eucaryotes it binds to the A-site of the ribosome and is transferred to the growing polypeptide chain, thus causing a break in the po-lymerization process.

Aromatic Antibiotics

This is a heterogenous group of antibiotics with aromatic rings in the molecule. The biosynthesis of individual compounds differs considerably. Three antibiotics of this group are produced commercially.

Chloramphenicol

Chloramphenicol is produced by Streptomyces venezuelae, S. phaeochromogenes var. chloromyceticus, S. omiyaensis, and other streptomycetes. Chloramphenicol is a broad-spectrum antibiotic which acts on Grampositive and Gram-negative bacteria, actinomycetes, rickettsias, and chlamydias. It causes significant side effects, particularly bonemarrow damage, but risk of toxicity can be reduced if therapy is conducted carefully. Because of its potential toxicity, chloramphenicol has not been as widely used as would be expected from its antimicrobial spectrum and it is presently considered a reserve antibiotic. Nevertheless, chloramphenicol is indispensible in the treatment of persistent Salmonella infections.

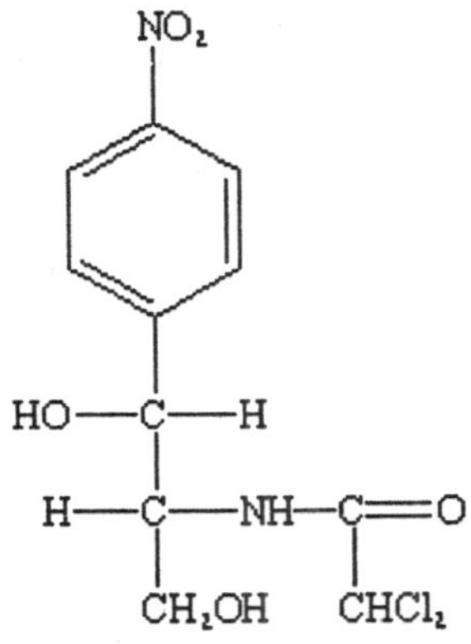

```
          NO2
           |
       [benzene ring]
           |
    HO—————C—————H
           |
     H—————C—————NH—C==O
           |        |
         CH2OH    CHCl2
```

Chemical structure of a chloramphenicol

Chloramphenicol binds specifically to the 50S subunit of 70S ribosomes and blocks the peptidyl transferase reaction, causing a premature chain break. In molecular biology, chloramphenicol is used as a research tool to inhibit the translation process in procaryotes without affecting the synthesis of nucleic acids. The pathway for chloramphenicol biosynthesis is via the aromatic pathway from chorismic acid. However, chloramphenicol has been produced by chemical synthesis since about 1950 and even improved fermentation processes have not succeeded in replacing this chemical process.

Griseofulvin

Griseofulvin, a benzofuran derivative, is produced by Penicillium griseofulvum, P.patulum, and other species of Penicillium. Griseofulvin is a fungistatic antibiotic but acts only on fungi with chitinous cell walls. In humans, griseofulvin is administered orally for the treatment of fungal skin infections (dermatomycoses). Although the antibiotic has been used in plant pathology in the treatment of leaf diseases due to Botrytis, Alternaria solani, and species of mildewcausing fungi, it is currently not in wide use here because of the high cost (due to the high production costs). Griseofulvin affects the morphogenesis of many fungi and in concentrations of 10-1000 ppm causes a characteristic curling effect of the hyphae. The exact mechanism of action has not yet been determined, but mitosis and chitin biosynthesis are presumed to be affected.

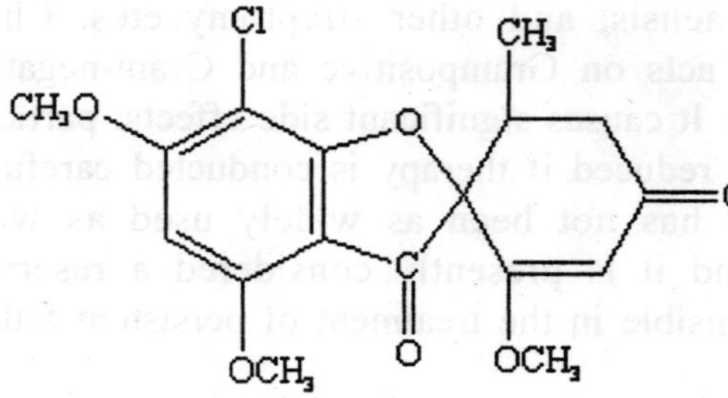

Chemical structure of Griseofulvin

The biosynthesis of griseofulvin occurs on a rultienzyme complex. The antibiotic is produced from acetate, 6 molecules of malonate, and 2 methyl groups derived from methionine. An intermediate stage in the biosynthetic process is a 14-member polyketide, from which the antibiotic is produced via cyclization, chlorination, and methylation reactions. The chemical synthesis of griseofulvin has been achieved, but production by fermentation with P. patulum is more costeffective.

Griseofulvin production is carried out as an aerobic submerged process in a glucoserich medium. In some processes, glucose feeding is used. Starch, lactose, or sucrose can also be used as carbon sources and usually $NaNO_3$ is used as a source. The addition of potassium chloride (0.1-0.3%) stimulates production. The fermentation takes place at 25°C for 7-9 days within an optimal pH range between 6.8-7.2. High aeration (1 vvm °2) is required when media with high carbon or nitrogen concentrations are used. The entire culture must be processed for recovery, since griseofulvin is only partially mycelium-bound.

Novobiocin

Novobiocin is a coumarin glycoside, which can be isolated from culture filtrates of Streptomyces spheroides, S. niveus, S. griseus, or S. griseoflavus. Novobiocin is effective against Grampositive bacteria, particularly staphylococci, against Gram-negative meningoocci and gonococci, and also against Haemophilus and some Proteus strains. Novobiocin shows no cross resistance with other antibiotics, but organisms do develop rapid resistance, a marked disadvantage for its medical use. Novobicin acts as a specific inhibitor of DNA gyrase during DNA replication.The biosynthesis of the coumarin moiety is via aromatic metabolism.

L- Tyrosine is incorporated into ring A and B; ring C is derived from glucose. The fermentation is carried out in a medium with glucose or sucrose as a carbon source. Organic acids, such fumaric acid, can be added, and peptone and meat extract are favorable nitrogen sources. p-Aminosalicylic acid can be used as a precursor and addition of tyrosine and phenylalanine stimulate the formation of the antibiotic. Fermentations take place over about 90 h at 25-28°C and at a slightly alkaline pH range, with aeration rates of up to 1.5 vvm.

FUSIDIC ACID

Among the steroid antibiotics, only fusidic acid is therapeutically important. This compound, isolated from culture filtrates of Fusidium coccineum, inhibits the ribosomal trans-location process during protein synthesis in both procaryotes and eucaryotes.

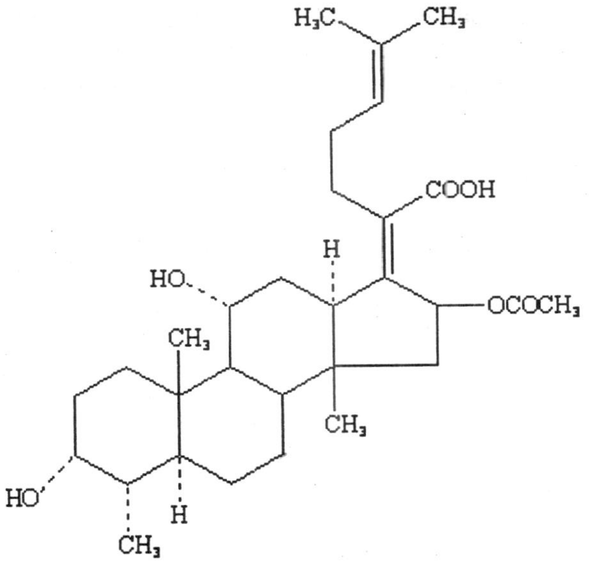

Chemical structure of fusidic acid

Fusidic acid finds special use for the treatment of staphylococcal infections, especially wound infections and those of the skin.

POLYETHER ANTIBIOTICS

Due to the numerous tetrahydropyran and tetrafuran moieties in these antibiotics, they are called polyethers. Over 150 different compounds are known. Their biosynthesis is derived from a polyketide derivative which is produced on a multienzyme complex. Monensin, commercially the most important compound of this group, consists of 5 acetate, 4 propionate and 3 butyrate units. Because of their ability to transport mono- or divalent cations through biological or artificial membranes, these antibiotics are classified with the ionophores.

Antibiotics in this group which are commercially prepared include monensin, produced by Streptomyces cinnamonensis, lasalocide, produced by S. lasaliensis, and salinomycin, produced by S. albus. Salinomycin yields of 60 g are attained on media containing a fatty acid. Polyethers are active against coccidia (protozoa),as well as Grampositive bacteria, mycobactei. and fungi. They are added to poultry feed (about 100 ppm) to prevent coccidioses. Monense which was first introduced in 1971, has alx:a 80% of the market; current annual gross world sales are estimated at $100 million.

MITOMYCINS

The mitomycins, produced by Streptomyces caes-pitosus, are benzoquinones. Erythrose and pyruvate have been detected as precursors of the methylbenzoquinone. Mitomycin C is very effective against tumors, but it is relatively toxic. Although this compound was first described in the late 1950's, it was not marketed until 1974. Mitomycin reacts with double-stranded DNA, producing cross-linkages which inhibit DNA replication. Although various derivatives have been produced with the aim of achieving reduced toxicity, success along these lines has not yet been achieved.

Chapter 7

Chemistry of Alkaloids

An alkaloid is, a naturally occurring amine produced by a plant, but amines produced by animals and fungi are also called alkaloids. Many alkaloids have pharmacological effects on humans and other animals. The name derives from the word alkaline; originally, the term was used to describe any nitrogen-containing base. Alkaloids are usually derivatives of amino acids, and many have a bitter taste. They are found as secondary metabolites in plants (e.g. potatoes and tomatoes), animals (e.g. shellfish) and fungi. Many alkaloids can be purified from crude extracts by acid-base extraction. Alkaloid is used in quinine While many alkaloids are poisonous, some are used medicinally as analgesics (pain relievers) or anaesthetics. particularly morphine and codeine, and for other uses.

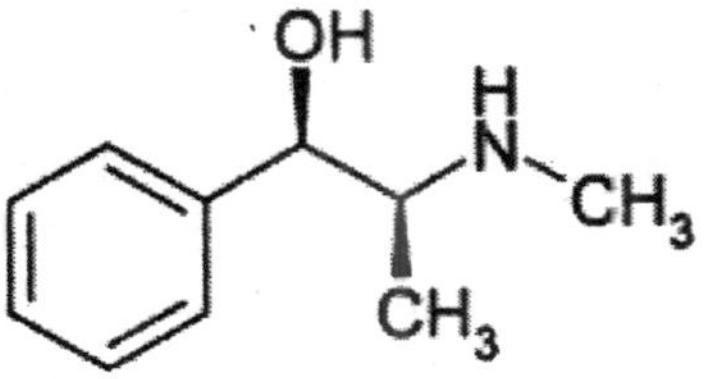

OH

H

N

CH_3

CH_3

Ephedrine

Alkaloids are usually classified by their common molecular feat, based on the metabolic pathway used to construct the molecule. When not much was known about the biosynthesis of alkaloids, they were grouped under the names of known compounds, even some non-nitrogenous ones (since those molecules' structures appear in the finished product; the opium alkaloids are sometimes called "phenanthrenes", for example), or by the plants or animals they were isolated from. When more is learned about a certain alkaloid, the grouping is changed to reflect the new knowledge, usually taking the name of a biologically-important amine that stands out in the synthesis process.

— *Pyridine group*: piperine, coniine, trigonelline, arecaidine, guvacine, pilocarpine, cytisine, nicotine, sparteine, pelletierine.

— *Pyrrolidine group*: hygrine, cuscohygrine, nicotine

- *Tropane group*: atropine, cocaine, ecgonine, scopolamine, catuabine
- *Quinoline group*: quinine, quinidine, dihydroquinine, dihydroquinidine, strychnine, brucine, veratrine, cevadine
- *Isoquinoline group*: The opium alkaloids (morphine, codeine, thebaine, Isopapa-dimethoxy-aniline, papaverine, narcotine, sanguinarine, narceine, hydrastine, berberine)
- *Phenethylamine group*: mescaline, ephedrine, dopamine, amphetamine
- *Indole group*:
  - Tryptamines: DMT, N-methyltryptamine, psilocybin, serotonin
  - Ergolines: the ergot alkaloids (ergine, ergotamine, lysergic acid, etc.)
  - Beta-carbolines: harmine, harmaline, yohimbine, reserpine, emetine
  - Rauwolfia alkaloids: Reserpine
- Purine group:
  - Xanthines: caffeine, theobromine, theophylline
- Terpenoid group:
  - Aconite alkaloids: aconitine
  - Steroids: solanine, samandaris (quaternary ammonium compounds): muscarine, choline, neurine
  - Vinca alkaloids: vinblastine, vincristine. They are antineoplastic and binds free tubulin dimers thereby disrupting balance between microtuble polymerization and delpolymerization resulting in arrest of cells in metaphase.

No precise and satisfactory definition of the term alkaloid can be given, and it has generally been used to denote nitrogenous basic substances of a cyclic structure found in plants, but it seems advisable to restrict its use to substances possessing a more or less marked physiological action. The alkaloids possess a special interest for the study of the relation between physiological action and chemical constitution, owing to the fact that many of them, even in small doses, produce very marked and definite physiological effects, and that a slight alteration in the chemical structure of the molecule often produces a decided change in this effect.

Mention has previously been made of the fact that the re duction of a nitrogenous ring generally produces a marked in crease in the toxicity and strength of the action of the drug, and sometimes completely alters its character. For example, pyridine has no marked toxic action and lowers blood-pressure, but piperidine is very toxic and raises blood-pressure, β–naphthylamine is not very poisonous, and contracts the pupil of the eye, but tetrahydro-β-naphthylamine is more poisonous and dilates the pupil. Many other examples of this type will be encountered from time to time. In all cases it is probably connected with the fact that reduction causes the substance more nearly to approach the aliphatic bodies in its properties, while in the case of pyridine and quinoline rings, it is connected with the change from tertiary nitrogen to the more reactive secondary form.

Although the substances containing reduced rings are more active physiologically than those containing pyridine or quino line rings, yet the activity is often apparently dependent upon the cyclic structure, as the open chain compounds are in general much less active than the corresponding alicyclic ones, and nearly all the active alkaloids are possessed of a cyclic structure. For example, δ-amino-valerianic acid and y-amino butyric acid are without any particular physiological action, but their anhydrides, the cyclic bases piperidone

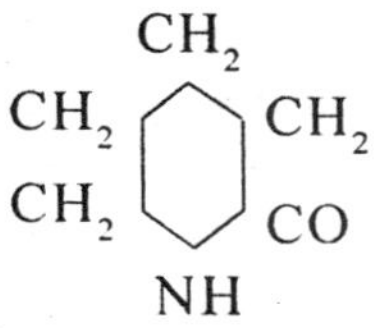

```
        CH2
  CH2 /    \ CH2
  CH2 \    / CO
        NH
```

and α-pyrrolidone

```
  H2C——CH2
   |    |
  H2C   CO
     \ /
     NH
```

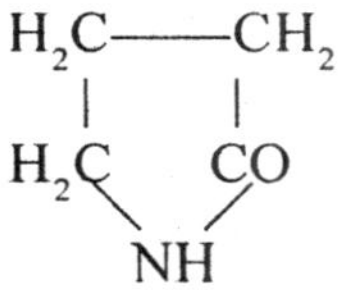

have a powerful action.In the same way pentamethylene diamine, $NH_2[CH_2]_5NH_2$, is not toxic, while piperidine is.

The size and position of the side chains attached to the ring have an important effect, and in some cases relationships have been traced between these factors and the physiological action. Thus, most of the alkaloids are derivatives either of pyridine itself, or of quinoline or isoquinoline, both of which latter contain a conjugated pyridine nucleus. Pyridine has only a slight physiological action, but more active bodies are generally obtained by reduction or by the entrance of aliphatic side chains. The introduction of the latter is accompanied by the appear ance of intoxicating action, which increases with the length and the number of the side chains. The toxic action of piperi dine itself is not very strong, but it is increased in α-methyl piperidine (pipecoline), and still more in a-ethyl piperidine and a-propyl piperidine (coniine). The toxicity of these substances

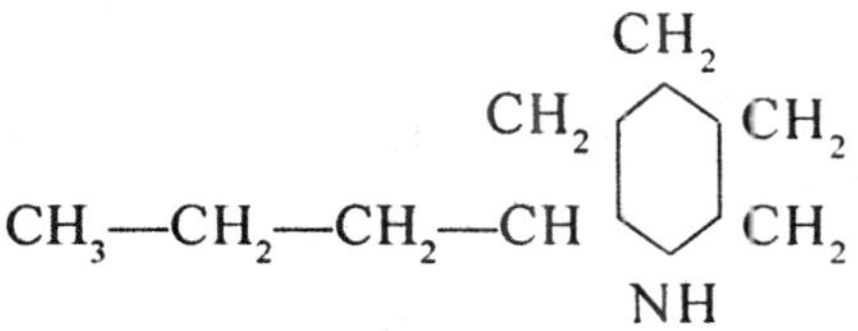

```
                            CH2
                      CH2 /    \ CH2
CH3—CH2—CH2—CH            \    / CH2
                            NH
```

is in the ratio of: 2 : 4 : 8.

In the case of almost every alkaloid which is a tertiary base, the original action disappears and is replaced by a curare -like action. The quaternary methyl and ethyl derivatives of atropine are partially exceptional; they resemble atropine itself in their action on the sympathetic nervous system, but differ from it in their action on the central nervous system and in having a curare-like action. Thus, atropinemethyl-nitrate retains the mydriatic action of atropine, while the action of atropine on the brain is lacking in this substance and also in the alkyl bromides of atropine and the allied alkaloids, hyoscyamine, homatropine, and scopolamine. The effect of the presence of hydroxyl groups, carboxyl groups, unsaturated linkages, etc., has already been dealt with in a previous chapter, and the special importance of the esterification of hydroxyl groups by acid radicles in alkaloids will be discussed in connection with cocaine, heroin, etc.

Of the alkaloids which are used in medicine, caffeine, theobromine, and the other purine derivatives are often considered separately, on account of their close relationship to various non-alkaloidal substances, such as uric acid, xanthine, etc. The morphine group, the isoquinoline derivatives, and the derivatives of tropine will be considered in the following sections, and there remain quinine, pilocarpine, and strychnine as being important from the medicinal point of view. Nicotine and many other alkaloids are also of great interest, but they are not much used in medicine, and comparatively little work has been done on them during recent years.

It has been shown by the work of Jowett J and of Pinner that pilocarpine probably has the constitution represented by the formula and that isopilocarpine is probably a stereo-isomeride. This alkaloid has not yet been synthesized, nor have synthetic substances of similar constitution and physiological action been introduced into medicine, and therefore it need not be considered at length. Pilocarpine illustrates the difficulty often encountered in correlating chemical constitution and physiological action. Pharmacologically, it is very similar in many respects to muscarine, to which it bears no particular chemical relationship. On the other hand, it may be pointed out that pilocarpine resembles nicotine in some of its physiological effects, and that both substances contain a five-membered ring containing nitrogen. It is true that nicotine also contains a six-membered nitrogen ring, but its physiological action is probably due more to the five-membered pyrrolidine ring.

With regard to quinine, our knowledge is more satisfactory, and although its synthesis has not yet been accomplished, a great deal of synthetic work has been carried out in connection with it, and therefore a more extended account of this alkaloid will not be out of place.

CH_2 CH
CH—CH
CH—CH_2
N C CH CH_2
CO N—CH CH_2
CH_2 CH
OH

Strychnine

The therapeutic value of quinine arises chiefly from the fact that it appears to have a specific action in malaria, apparently being far more poisonous to the protozoal parasites than to the cells of the host.

Becently it has been shown that the specific action of quinine is closely paralleled by various related compounds, namely, the alkyl ethers of dihydrocupreine, amongst which different homologues were found to have a maximum specific effect on different pathogenic organisms.For example, ethyldihydrocupreine or Optoquin was found to have a specific action on pneumococci, but it is too poisonous for clinical use. Though pneumococci were found to be most sensitive to this ethyl compound, cholera vibrios were found to be most sensitive to the iso-amyl compound known as Eucupin, whilst most other pathogenic organisms, such as diphtheria bacilli, streptococci, and staphylococci, were most easily destroyed by the iso-octyl derivative known as Vuzin.

Quinine has also been widely used as a febrifuge, but this is referred to in the chapter on antipyretics, and at present it is proposed to consider its properties from the same point of view as those of the other alkaloids.

Quinine was isolated, together with cinchonine, by Pelletier and Caventou in 1820, and as a result of the labours of many chemists, of .whom Skraup, Konigs, von Miller, and Ehode may be specially mentioned, the following formula has been generally accepted to represent its constitution:

H
|
HO—CH——C
CH_3O N H_2C
N — CH_2 — CH_2 — C — H
CH_2 CH — CH = CH_2

that of cinchonine only differing from it by the absence of the methoxy group, (—OCH_3), and that of cupreine only in having this group replaced by a hydroxyl group, (OH). The earlier attempts to prepare substances which should re semble quinine in their properties were based on the assump tion that the quinoline nucleus was the active portion of the molecule, but Fraenkel brings forward many facts in support of the view that it is the piperidine ring, the so-called " loiponic acid portion " of the molecule, which is the true " pharmaco phore ". Thus, it is pointed out that in nicotine

CH_2——CH_2
— CH CH_2
N N
|
CH_2

the contraction of the blood-vessels is almost certainly due to the pyrrolidine and not to the pyridine ring, seeing that this action is not shown by pyridine at all, but is produced by piperidine

$$\begin{array}{ccc} & CH_2 & \\ CH_2 & & CH_2 \\ | & & | \\ CH_2 & & CH_2 \\ & NH & \end{array}$$

the last named closely resembling nicotine in this respect.In the same way it is considered that the action of quinine is due to the reduced piperidine ring, which is rendered still more active by the presence of the unsaturated group, $CH{=}CH_2$. The weak and uncertain action of cinchonine, compared with that of quinine, is probably owing to the absence of the paramethoxy group leaving the molecule without an " anchoring " group. Owing to this, the hydrogen atom of the quinoline nucleus probably has to become oxidized to OH with the formation of cupreine before it can exert the characteristic quinine action.

Many attempts have been made to prepare derivatives of quinine which should be free from some of its unpleasant byeffects, especially its bitter taste, and also to prepare derivatives which should be more soluble than the generally used sulphate, and so be suitable for hypodermic injection. Of the latter class, mention should be made of the double chloride and sulphate of quinine, which is very soluble in water, and of the double chloride of quinine and caffeine, which is readily soluble and suitable for hypodermic injection. Cinchonine is less bitter than quinine, but its action is weak and untrustworthy, and so it is not a suitable substitute. If quinine is converted into insoluble derivatives its taste is diminished, and a favourite derivative of this class is the tannate, which is practically tasteless, but suffers from the drawback that it is only slowly split up in the intestine into its components, and is therefore lacking in promptitude and certainty.

Other derivatives have been prepared by esterification of the hydroxyl group, the most important of these being the esters of carbonic acid. The diquinine ester of carbonic acid

$$C_{20}H_{23}N_2O{-}O{-}CO{-}O{-}C_{20}H_{23}N_2O$$

is known as Aristoquinine, and is comparatively tasteless and fairly insoluble.Euqidnine is the ethyl carbonate of quinine

$$CO\begin{cases} O{-}C_2H_5 \\ O{-}C_{20}H_{23}N_2O \end{cases}$$

and is prepared by the action of ethyl chloro-formate, $Cl.COOC_2H_5$, on quinine. It is practically tasteless, and does not have a bitter after-taste. The hydrochloride, on the other hand, is not as good as the free base, and has no advantage over quinine itself. Phenetidine quinine carbonic ester —

$$CO \begin{cases} NH—C_6H_4—OC_2H_6 \\ O—C_{20}H_{23}N_2O \end{cases}$$

is known as Quinaphenin. Saloquinine is the salicylic acid ester of quinine

$$C_6H_4 \begin{cases} OH \\ CO—OC_{20}H_{23}N_2O \end{cases}$$

and is tasteless.

Quinaphthol is quinine β-naphthol-sulphonate, and Quinaform is quinine formate.

In all these compounds, absence of taste is conditioned only by insolubility, their more soluble derivatives having the characteristic bitter taste of quinine. The more soluble hydrochloride of euquinine is an instance of this.

Morphine Alkaloids

Morphine is the most important of the opium alkaloids, and is the one to which most of the physiological effects of opium are due. The principal alkaloids present in opium may be divided into two well-defined groups:

(1) The Morphine Group, consisting of morphine, codeine, and thebaine, all of them very poisonous substances, and containing a phenanthrene nucleus, as well as a nitrogen ring.

(2) The Papaverine Group, consisting of narcotine, papavarine, narceine, laudanosine, oxynarcotine, etc'. These substances, which are derivatives of isoquinoline, are far less physiologically active than those of the first group. These alkaloids and some of their derivatives will be considered in the next section, and for the present attention will be directed to the members of the first group.

All the opium alkaloids produce nervous depression, beginning in the psychic centres of the brain, and extending downwards through the various cerebral centres in the reverse order of the development, and they also have a strychninelike action on the cord, giving rise to convulsions. In some of these alkaloids, such as morphine, the depressant (i.e. narcotic and analgesic) action predominates, twitchings or convulsions being extremely rare, while in others, such as thebaine, the convulsant action is far stronger, and practically masks the weak narcotic action. Codeine stands between the two, having marked narcotic properties, but in a weaker degree than morphine, while on the other hand, it is more liable than morphine to give rise to increased reflexes and spasmodic twitchings.

The following table indicates which of these effects pre dominates :

— Morphine (most narcotic),

— Papaverine,

— Codeine,

— Narcotine,

— Thebaine,

— Laudanine (most convulsant).

Morphine has never been synthesized, and its structure is not even known with certainty, so that this section will deal mainly with the efforts that have been made to synthesize derivatives of morphine, which should differ from it in certain respects with regard to their physiological action, and also with the attempts that have been made to obtain synthetic products, which should possess a similar action to morphine owing to the presence of similar groupings in the molecule. These attempts are hampered by the fact that we do not know the exact constitutional formula of morphine, nor which portion of the molecule plays the chief part in determining the physiological action.

Nevertheless, it is necessary to indicate the basis on which our knowledge of the structure of morphine rests. The formula of morphine is $C_{17}H_{19}O_3N$; it is a tertiary base, and contains two hydroxyl groups, one of which is phenolic, and the other alcoholic in character. The alkaloid, codeine, $C_{18}H_{21}O_3N$, differs from morphine by CH_2, and as it contains one hydroxyl group, it therefore appeared probable that it was morphine in which the hydrogen of one of the hydroxyl groups had been replaced by CH_3. This assumption was shown to be highly probable by the work of Matthiessen and Wright, and confirmed in 1881 by Grimaux, who converted morphine into codeine by direct methylation. Therefore the constitution of codeine and that of morphine can be considered together, the former being $C_{17}H_{17}ON(OH)(OCH_3)$, and the latter $C_{17}H_{17}ON(OH)_2$; the complex $C_{17}H_{17}ON$ being the same in both, it is clear that a knowledge of this complex will reveal the structure of both alkaloids.

The study of this question has occupied the attention of many chemists, and thanks mainly to the labours of Vongerichten, Schrotter, Knorr, and Pschorr, we have gained a tolerably clear insight of the nature of this grouping. Vongerichten and Schrotter by distillation of morphine with zinc dust obtained phenanthrene, together with pyrrol, pyridine, trimethylamine, $N(CH_3)_3$ and ammonia.

An account the recent work on this subjet is given by fatis and the reader desiritnng further information is reffered to this and to Picet's Vegetable alkaloids. As a result of his own and other investigations, Knorr has suggested the following formula.

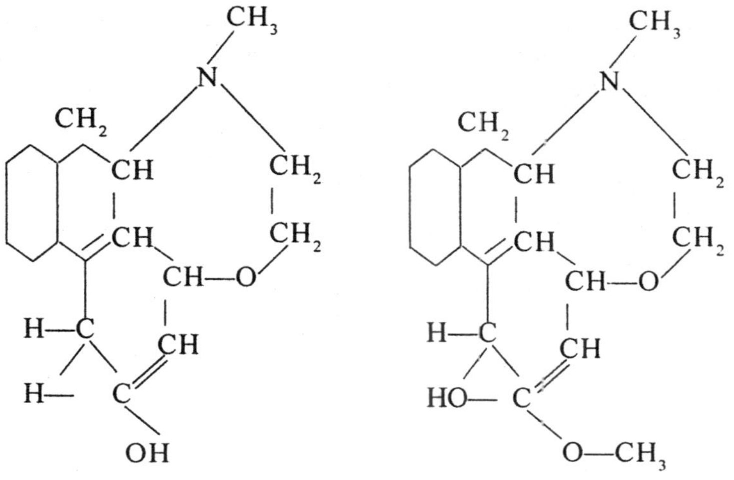

Subsequently Knorr somewhat modified this formula, as shown below, and Pschorr, Jaeckel, and Fecht have suggested a formula for morphine which depicts it as a derivative of N-methylpiperidine. This formula is based on a study of apomorphine, $C_{17}H_{17}O_2N$, a dehydration product of morphine

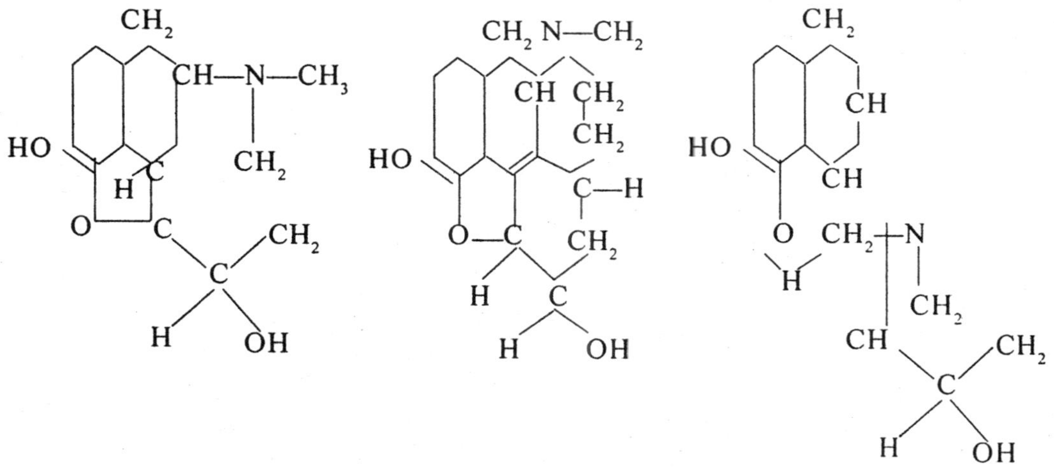

Knorr's later formula *Pschorr's formaula* *Bucherer's formula, modified by Knorr.*

An important difference between these two formulae and Knorr's earlier formula is that they show the O and N in separate ring systems. Bucherer's formula, as modified by Knorr, is a link between the two types.

More recently other workers have cast doubts on the presence of an ethylenic linkage in morphine, and formulae have been suggested in which an additional "bridged ring" of carbon atoms takes the place of the " double bond ". Faltis suggests a somewhat similar formula in which, however, the " double bond " is retained. In all of these formulae codeine is the same but for the phenolic OH being replaced by OCH_3. The alkaloid thebaine has been shown by the work of many investigators to be closely related to morphine and codeine

$$(C_{16}H_{14}ONCH_3)(OH)_2$$

Morphine

$$(C_{16}H_{14}ONCH_3)(OCH_3)$$

Codiene

$$(C_{16}H_{12}ONCH_3)(OCH_3)_2$$

Thebaine

It differs from morphine and codeine in having both the hydroxyl groups replaced by methoxy (OCH_3) groups, and in having two atoms of hydrogen less in the rest of the molecule. Accordingly, if we accept Knorr's formula for codeine, we obtain the following formula for thebaine:

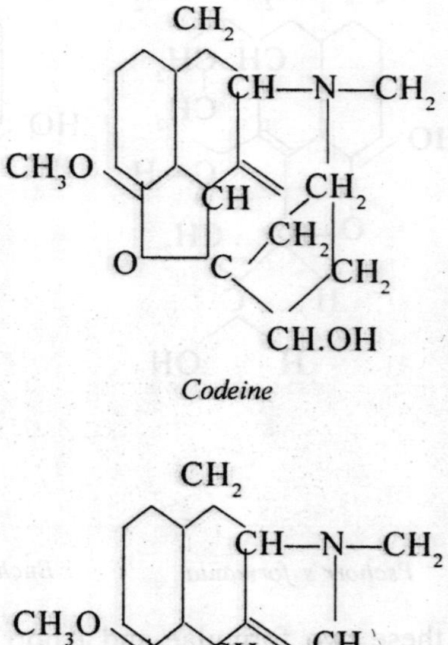

CH_2

CH—N—CH_2

CH_3O CH CH_2

O C CH_2 CH_2

CH.OH

Codeine

CH_2

CH—N—CH_2

CH_3O CH CH_2

O C CH_2

$COCH_2$

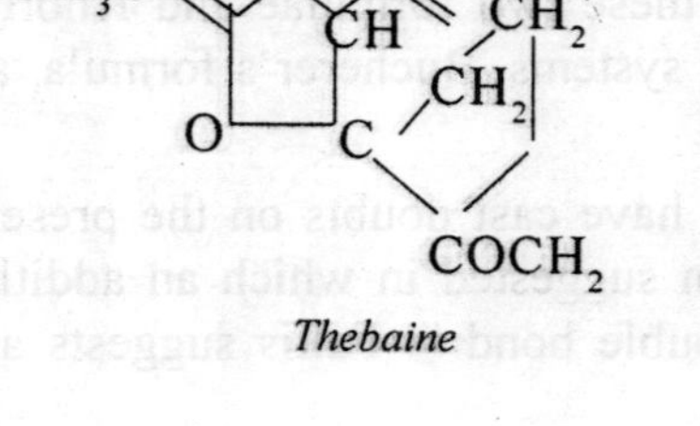

Thebaine

and on the basis of Pschorr's formula for codeine, the corre sponding formula for thebaine is as shown below

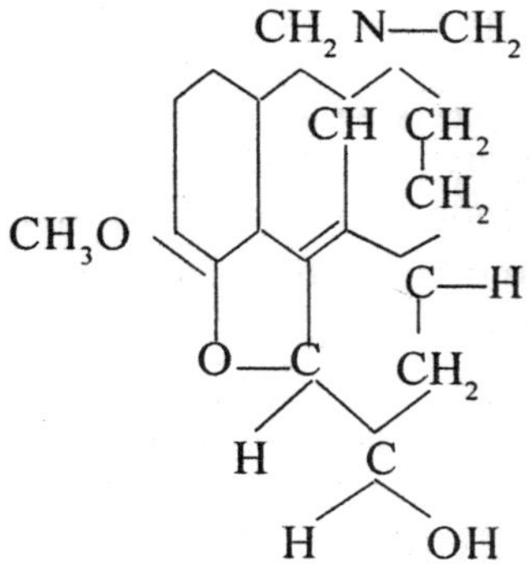

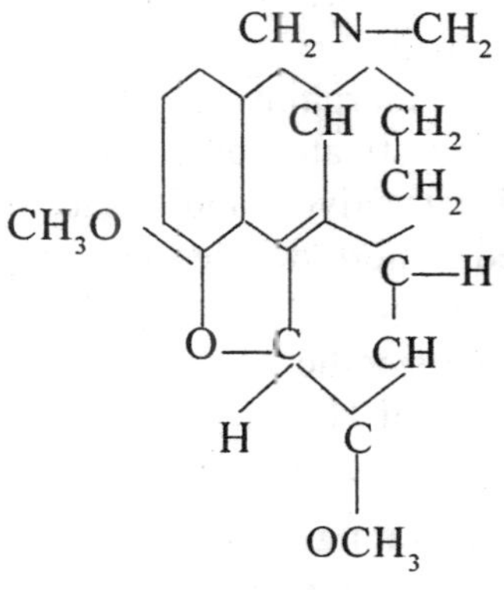

Of the various derivatives prepared from morphine which are used in medicine, the naturally occurring alkaloid codeine is the one which most closely resembles morphine. As already stated, it was first prepared directly from morphine by Grimaux, who also prepared ethylmorphine (codethyline), but since then many other manufacturing methods have been de vised in order to obtain better yields. It was first prepared on a large scale by Knoll, and Pechmann's method of methyla tion by means of diazo-methane has also been applied to the preparation of codeine from morphine, but by this method also the yields do not appear to be very satisfactory. Methyl sul phate is at the present time a favourite methylating agent, and Merck has devised a means of preparing codeine by the action of methyl sulphate on morphine in the presence of alcohol and sodium. The neutral alkyl esters of phosphoric and nitric acids can also be used in the same way as the sulphate.

The pharmacology of morphine, codeine, monacetyldiacetyl-, and benzoyl-morphine has been investigated by Stockman and Dott, and that of the homologues of codeine, together with a large number of morphine derivatives by Mering. The action of the higher homologues of codeine is similar to but somewhat weaker than that of codeine itself. Codeine methyl bromide, $C_{18}H_{21}O_3NCH_3Br$, or Eucodeine is said to be less toxic than codeine. Ethyl morphine is somewhat exceptional in having a stronger and more prolonged action than codeine, and is recommended against irritant coughing. Its hydrochloride has been introduced into therapeutics under the name of Dionine.

The benzyl (C_6H_5—CH_2) derivative of morphine also re sembles codeine in its action, and has been introduced by Mering in the form of its hydrochloride, under the name of Peronine. It is obtained by the action of sodium ethoxide and benzyl chloride on morphine in alcoholic solution.

The carbolic acid esters of morphine are very unstable, but the acyl derivatives are quite stable, and some of them have attained great practical importance. The acyl derivatives of morphine, in which only the phenolic hydrogen is replaced, monoacetyl-, propionyland benzoyl-morphines closely resemble morphine itself in their physiological action, being intermediate between it and the diacetyl compounds to be described later. (It should be pointed out that in the case of codeine, dionine, and peronine, it is also the hydrogen of the phenolic hydroxyl which is replaced by CH_3, C_2H_5 and $C6H_5$ .CH_2 respectively. These are phenolic ethers containing the group OR, but monoacetyl-, propionyl-, and benzoylmorphine are phenolic esters, containing the group O—CO—R, and are more readily hydrolyzed than the ethers.) The derivatives of morphine, in which both hydroxyl groups are esterified by acid radicles, have also been investigated by Mering. He investigated the diacetyl, di-propionyl, di-isobutyryl, and divaleryl derivatives, and found, in confirmation of the work of Stockman and Dott, and these possessed a more decided narcotic action on dogs than codeine, and a stronger tetanic action than morphine. Clinically, they have proved valuable in lowering reflex irritability and calming spasmodic coughing, but in checking pain they are less active than morphine. Under the names of Heroin and Acetomorphine, the diacetyl derivative has attained considerable practical importance in checking coughing arising from irritation, etc.

By the action of dehydrating agents, such as concentrated HC1 at 140°C., on morphine, a substance is formed which differs from morphine by the elements of water, and is called apomorphine. It differs very greatly from morphine in its physiological action, being a most powerful emetic, and in large doses stimulating the respiratory centre, with consequent quickening of the rate of breathing. Its action is therefore quite different from that of morphine. The emetic action of this substance is not local, but is of central-nervous origin through stimulation of the medulla. In therapeutics the hydrochloride is usually used, and is given hypodermically, the dose being soon followed by vomiting without harmful by-effects. The fact that apomorphine can be given in this manner constitutes one of its chief advantages over the peripheral emetics, which act locally on the alimentary canal.

With regard to the chemical nature of apomorphine, it was first thought that one of the oxygen atoms was present in a hydroxyl group, and the other in an ether group, as only a monoacetyl derivative had been prepared, but Pschorr, Jaeckel, and Fecht have shown that both oxygen atoms are present as hydroxyl groups

HO
CH_2
HO
N—CH_2
CH_2
CH_2 CH_2

They also prepared a monomethyl ether of apomorphine, which has been shown 3 to be identical with the so-called " pseudo "apocodeine, obtained by heating codeine with anhydrous oxalic acid at 150°C.

Apocodeine, codeine, apomorphine, and morphine all produce purgation in dogs and cats, but this action is greatest in apocodeine, and diminishes in the order given down to morphine. In the case of apocodeine, the purgative action is greater than the vomiting, and it is stated, when given hypodermically in suitable doses, to produce purgation without vomiting, and its use as a hypodermic purgative has been suggested by Dixon. Such a substance is greatly needed, and should apocodeine come into use for this purpose, it would be one of the most important of artificial drugs.

The endeavours that have been made to synthesize substances analogous to morphine do not appear to have been very successful. This is not surprising considering that, apart from any doubt as to the actual constitution of morphine, there is still no means of judging which part of the molecule is most intimately connected with the action of the substance.

The early work of Knorr lead him to the conclusion that morphine might be represented by the formula

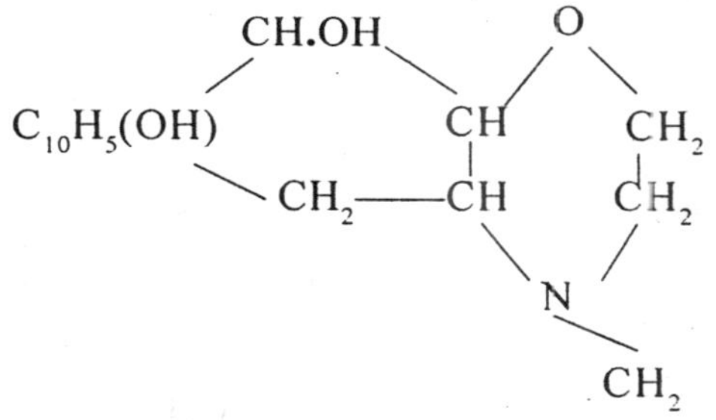

CH.OH, O

$C_{10}H_5(OH)$ CH CH_2

CH_2—CH CH_2

N

CH_2

a conclusion which was subsequently disproved by his own work. Knorr gave the name morpholine to the grouping

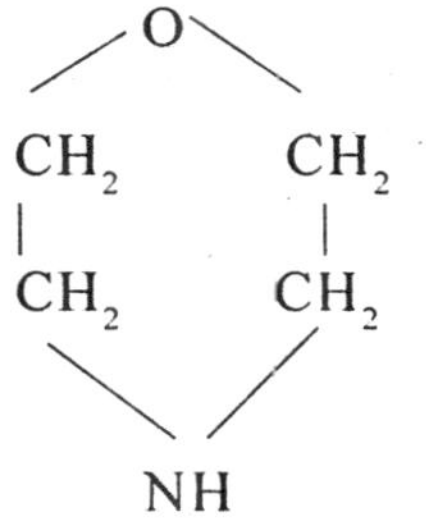

O

CH_2 CH_2

| |

CH_2 CH_2

NH

as he regarded it as the parent substance of morphine. A general method of preparing morpholine and its derivatives which was devised by Knorr, consists in splitting off water from dihydroxyethylamines by means of condensing agents, such as 70 per cent, sulphuric acid, acetic anhydride, etc. The dihydroxyethylamines are prepared by the condensation of ethylene-oxide with amines

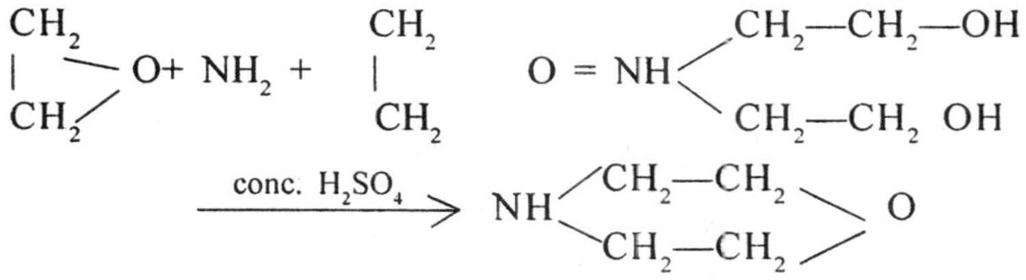

CH_2 | CH_2 >O + NH_2 + CH_2 | CH_2 O = NH<(CH_2—CH_2—OH)(CH_2—CH_2 OH)

$\xrightarrow{\text{conc. } H_2SO_4}$ NH<(CH_2—CH_2)(CH_2—CH_2)>O

The parent substance, morpholine itself, was prepared in this way, using HC1 at 160 as the condensing agent. By using various primary amines of the type KNH instead of ammonia, substituted morpholines of the type

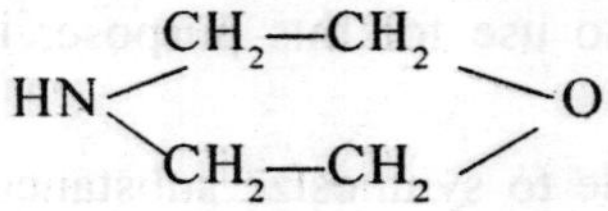

$$HN\begin{matrix} CH_2—CH_2 \\ CH_2—CH_2 \end{matrix}O$$

may be obtained, and by using substituted derivatives of ethylene-oxide derivatives of morpholine can be obtained, some of which are not unlike morphine in their properties. As an example, tetrahydro-naphthalene morpholine may be mentioned, as it resembles morphine most closely. It was obtained by condensing an alicyclic derivative of ethylene-oxide, namely, tetrahydro-naphthylene-oxide with hydroxyethylamine

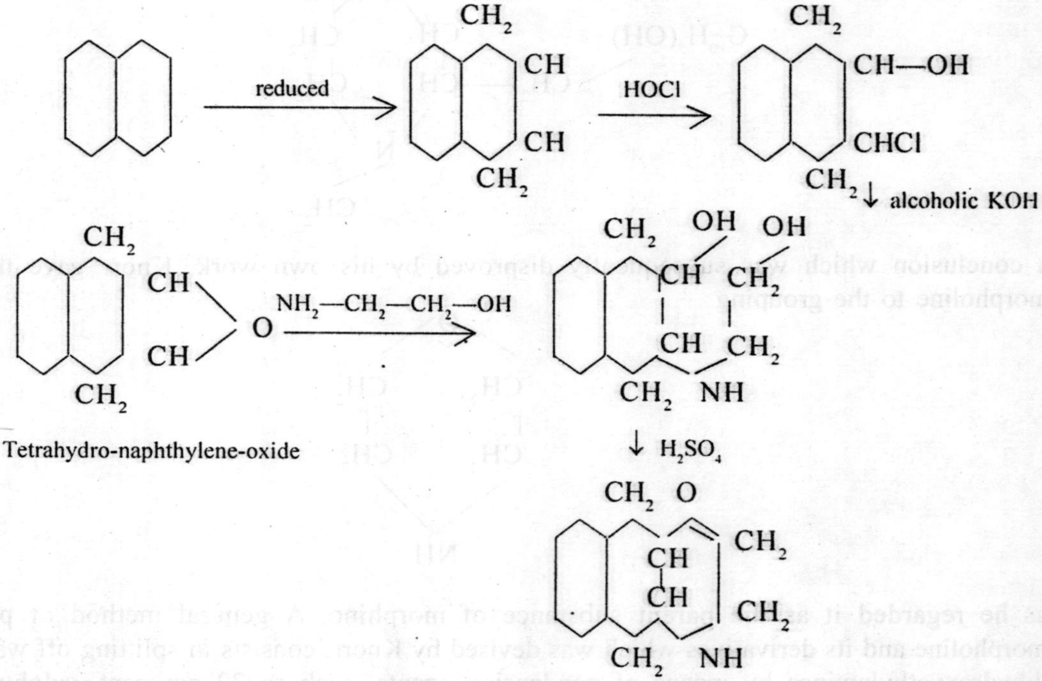

reduced; HOCl; alcoholic KOH

$NH_2—CH_2—CH_2—OH$

Tetrahydro-naphthylene-oxide

H_2SO_4

Another method of preparing derivatives of tetrahydronaphthalene-morpholene which has been devised by Knorr is by means of the action of hot dilute, sulphuric acid on the hydramines of the naphthalene series

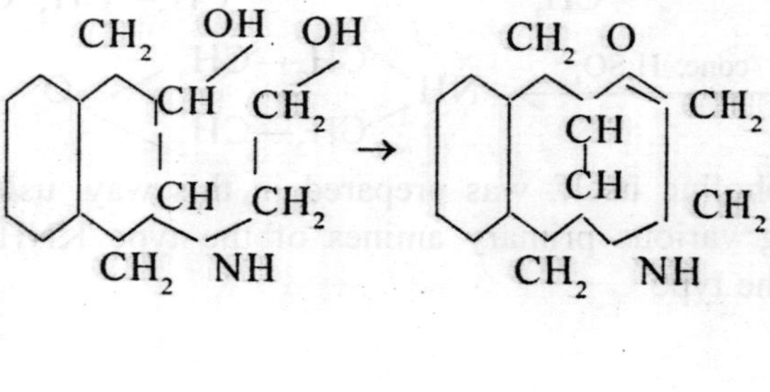

These hydramines can be obtained by the action of the ethanolamines on the chlorhydrin, or on the oxide of dihydronaphthalene

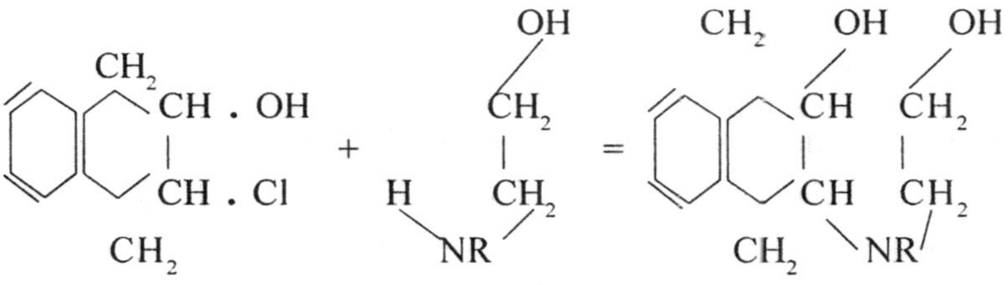

CH_2 CH . OH | CH . Cl CH_2 + OH CH_2 | H CH_2 NR = CH_2 OH OH CH CH_2 | | CH CH_2 CH_2 NR

A somewhat different method of obtaining a morpholine derivative has been devised by Stormer, and consists in reducing a boiling alcoholic solution of ortho-nitro-phenacetol with tin and hydrochloric acid, whereby 2-methyl-pheno morpholine is obtained

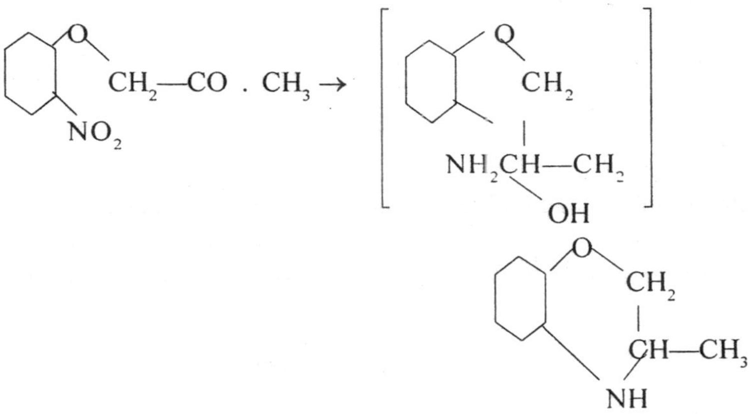

O CH_2—CO . CH_3 → NO_2 [O CH_2 | NH_2CH—CH_2 OH]

O CH_2 | CH—CH_3 NH

Nitro-phenacetol is obtained by the action of a mono-halogen ketone on the sodium salt of o.-nitrophenol

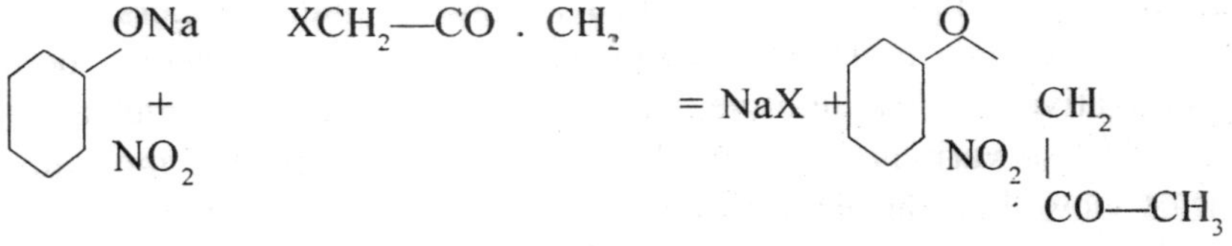

ONa + NO_2 XCH_2—CO . CH_2 = NaX + O CH_2 NO_2 | CO—CH_3

In a similar way a naphtho-morpholine may be obtained from

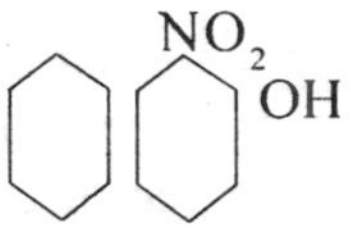

NO_2 OH

On the other hand, Vahlen differs from Knorr, and considers that it is the phenanthrene portion of the molecule which is of greatest importance in determining the physiological action of morphine. (Overton has shown that phenanthrene itself has a narcotic action on tadpoles.) Vahlen considers that the pharmacophore of morphine is the grouping

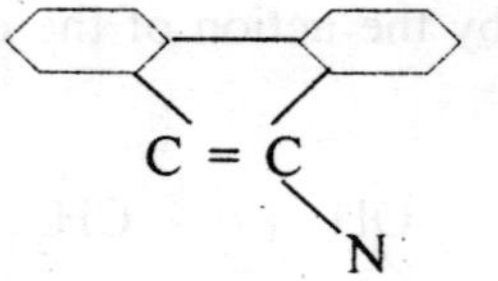

C = C

N

and he prepared the hydrochloride of 9-amino 10-hydroxy phenanthrene

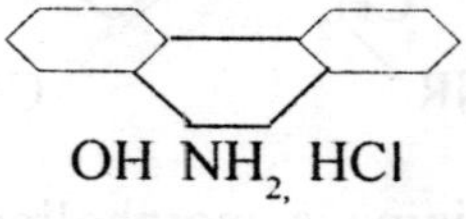

OH NH_2, HCl

which he called "morphigenine," and from this he obtained eposine—

N N—CH_3

CH

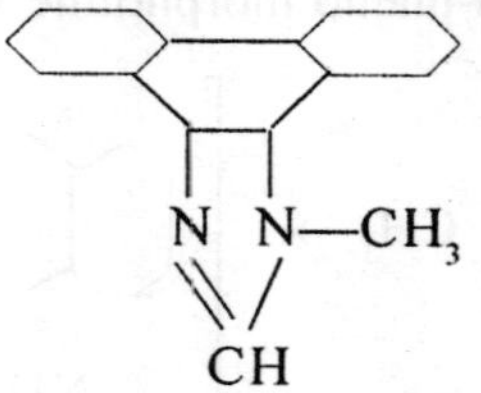

by heating it with sodium acetate, alcohol, and methylamine under pressure. Epiosine, which to a certain extent resembles morphine and codeine in some of its physiological effects, is identical with ethyl-diphenylene-iminazole, which has also been prepared by other investigators.

Isoquinoline Alkaloids

The most important of the opium alkaloids have already been dealt with, but there are several others of some importance, most of which are derivatives of isoquinoline. These include narcotine, papaverine, narceine, laudanosine, and laudanine. The physiological action of these alkaloids is generally not so marked as that of the members of the morphine group. The important alkaloids of Hydrastis canadensis, namely, hydrastine, berberine, and canadine, are also derivatives of isoquinoline, as also is corydaline, the chief alkaloid present in Corydalis cava. Of these various alkaloids, hydrastine is probably the most important from the medical point of view, but various artificial alkaloids have been prepared from many members of this series by simple processes such as oxidation, many of which are of considerable therapeutic importance (e.g. cotarnine).

In considering these alkaloids, it will be found that our knowledge of their constitution is far more satisfactory than that of the alkaloids of the morphine group. In the majority of cases, not only is the constitution known with a high degree of certainty, but in recent years it has, in many cases, been confirmed by the synthesis of the alkaloid. Of the alkaloids of this group found in opium, narcotine and papaverine are present in the largest quantities. The consti tution of the latter has been determined chiefly by the work of Goldschmidt, which leaves no doubt that the structure of this alkaloid is represented by the formula

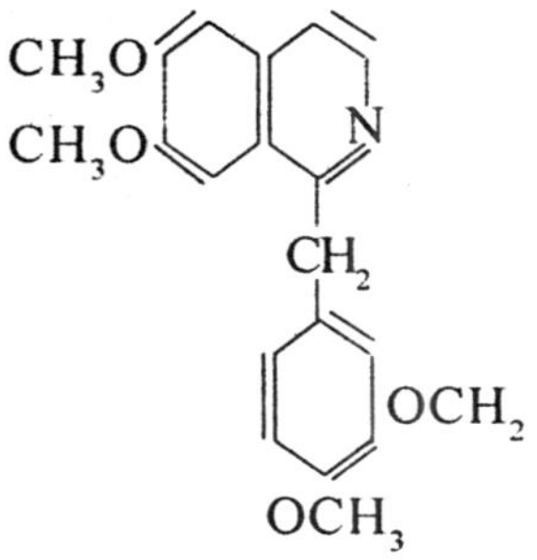

This has been confirmed by the brilliant synthesis of papaverine by Pictet of Geneva.

Pictet had already shown a that substances of the type

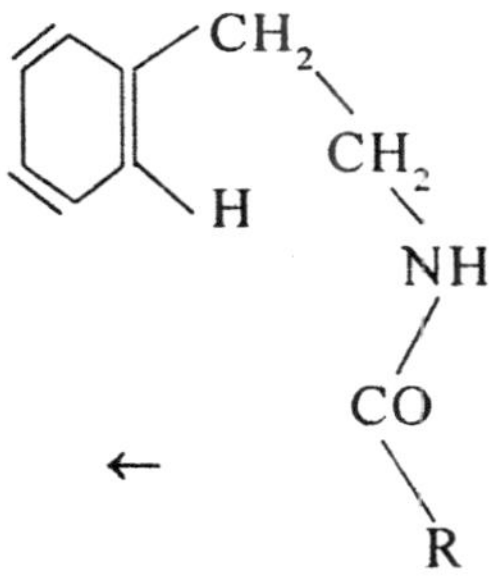

on treatment with dehydrating agents, lose a molecule of water, giving dihydro isoquincline derivatives of the type

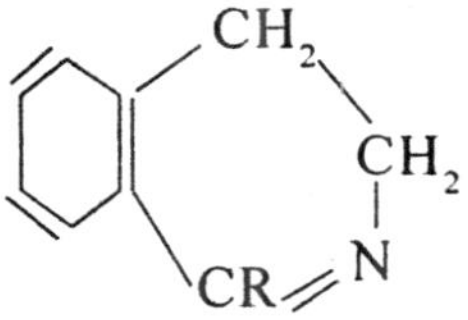

In this way he succeeded in obtaining dihydro-papaverine

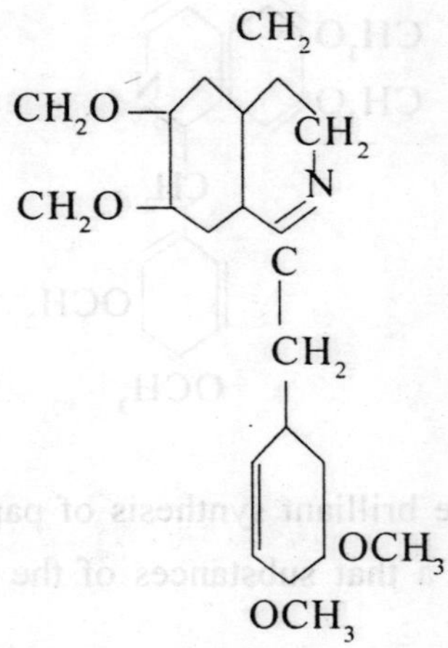

but this cannot be converted into papaverine by oxidation, as any attempt to do this leads to the disruption of the whole molecule. Papaverine could not be obtained directly by using $(CH_3O)_2C_6H_8$. CH=CH . NH_2 instead of $(CH_3O)_2C_6HS$ .CH_2 . CH_2 . NH_2 at the beginning of the synthesis, as the former substance is too unstable for this purpose. The difficulty was finally overcome by preparing the substance A

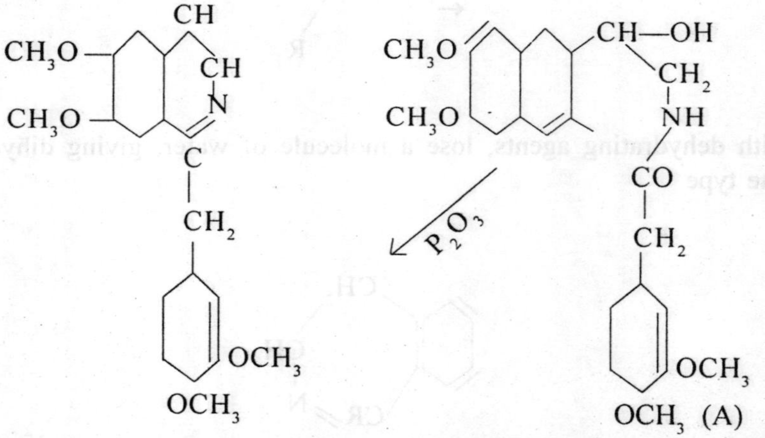

P_2O_5

(A)

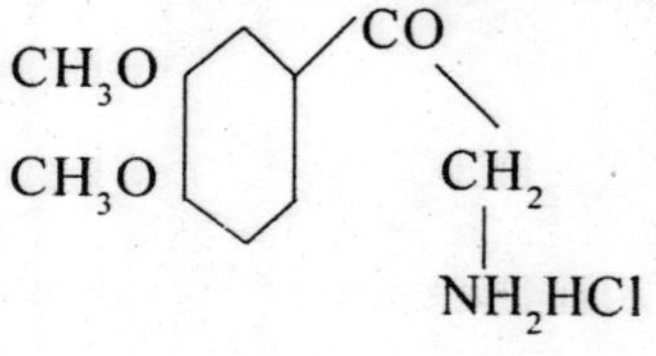

which on treatment with phosphorus pentoxide in xylene solu tion loses two molecules of water with the formation of papaver ine. The product thus obtained was in every way identical with papaverine obtained from opium. The first stages of the synthesis are concerned with the preparation of

and

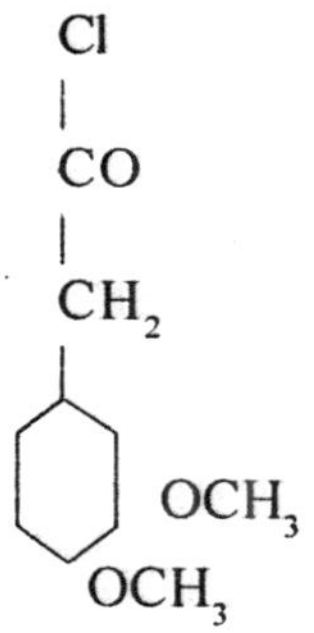

Cl
|
CO
|
CH_2
|
OCH_3
OCH_3

The first mentioned is prepared according to the following scheme :

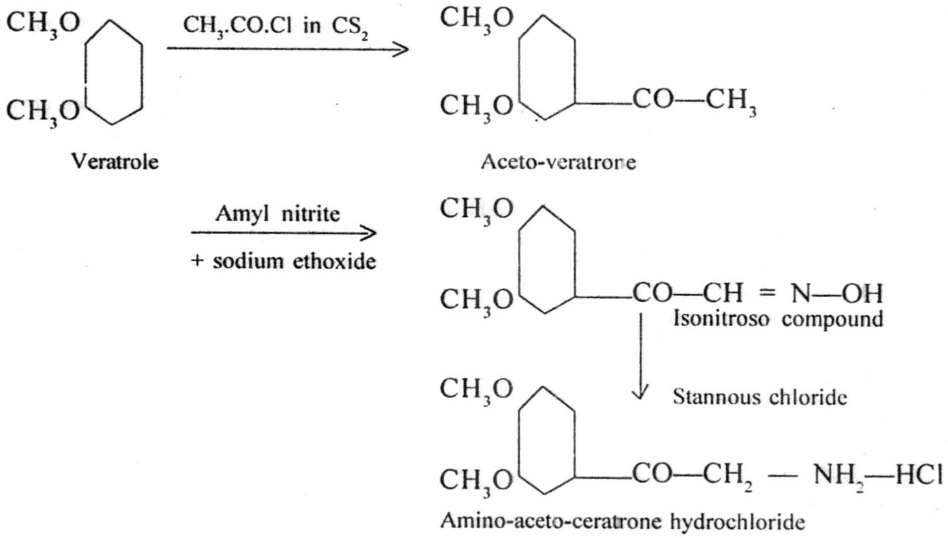

CH_3O / CH_3O — $CH_3.CO.Cl$ in CS_2 → CH_3O / CH_3O —$CO—CH_3$

Veratrole — Aceto-veratrone

Amyl nitrite + sodium ethoxide → CH_3O / CH_3O —$CO—CH = N—OH$

Isonitroso compound

↓ Stannous chloride

CH_3O / CH_3O —$CO—CH_2 — NH_2—HCl$

Amino-aceto-ceratrone hydrochloride

The starting-point of the other half of the synthesis is vanillin

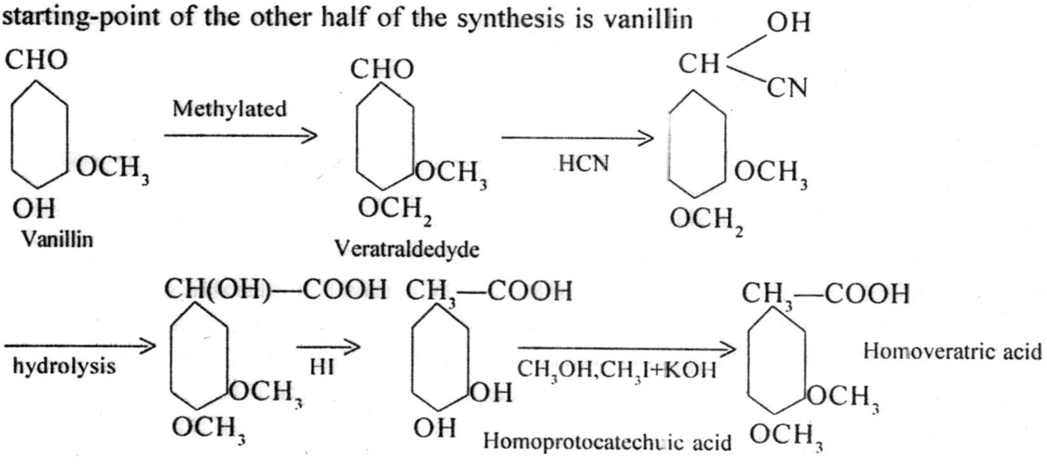

CHO / OCH_3 / OH (Vanillin) — Methylated → CHO / OCH_3 / OCH_2 (Veratraldedyde) — HCN → CH(OH)(CN) / OCH_3 / OCH_2

hydrolysis → $CH(OH)—COOH$ / OCH_3 / OCH_3 — HI → $CH_3—COOH$ / OH / OH (Homoprotocatechuic acid) — CH_3OH, CH_3I+KOH → $CH_3—COOH$ / OCH_3 / OCH_3 (Homoveratric acid)

This homoveratric acid, treated with PCl_5, yields the chloride

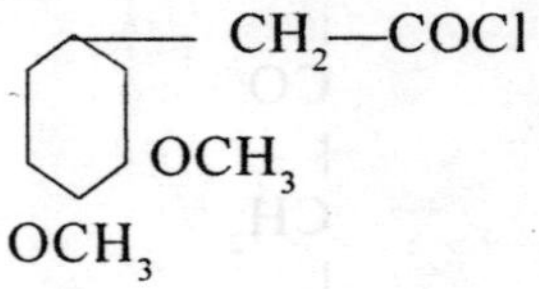

CH_2—COCl
OCH_3
OCH_3

This on shaking with

CH_3O CO—CH_2—NH_2 . HCl
CH_3O

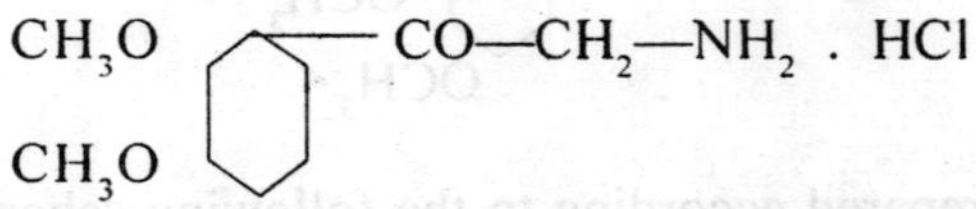

In cold potash solution gives, by loss of HCl

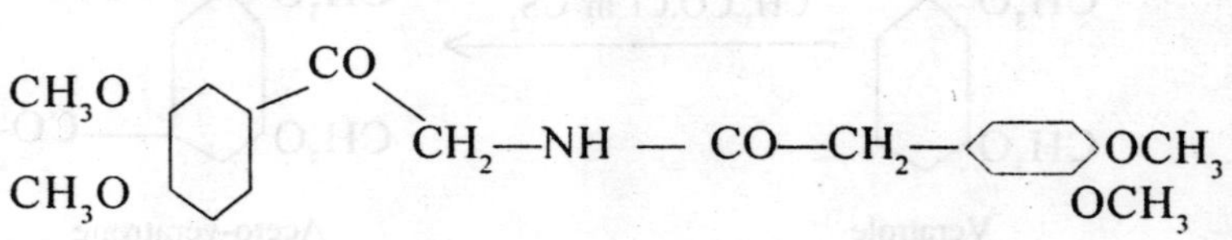

CO
CH_3O CH_2—NH — CO—CH_2— OCH_3
CH_3O OCH_3

which by reduction with sodium amalgam in alcohol yields

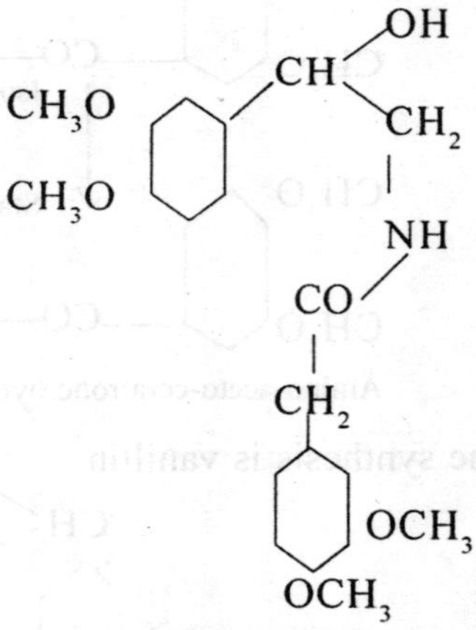

OH
CH
CH_3O CH_2
CH_3O NH
CO
CH_2
OCH_3
OCH_3

which on dehydration gives papaverine, as previously described. Papaverine has only a slight narcotic action, being intermediate in this respect between morphine and codeine. According to Bernard, the chief opium alkaloids stand in the following order in regard to their power of causing convulsions : thebaine, papaverine, narcotine, codeine, and morphine. Of these, thebaine is the only one producing strong tetanic convulsions. Of the alkaloids not mentioned in this list, laudanosine and laudanine have this property strongly marked, the former being slightly less and the latter slightly more active than thebaine itself in this respect.

Laudanosine closely resembles papaverine in its structure being the methyl derivative of tetrahydro-papaverine:

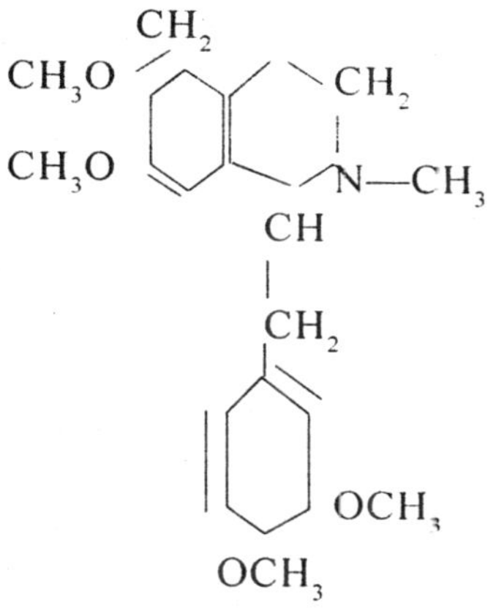

CH_2
CH_3O CH_2
CH_3O $N—CH_3$
CH
CH_2
OCH_3
OCH_3

This was shown by Pictet and Athanescu, who obtained it by the reduction of the metho-chloride of papaverine with tin and hydrochloric acid, and resolution of the racemic compound thus obtained with quinic acid. The dextro-compound was found to be identical with laudanosine. It has also been obtained by the reduction of papaverine to tetrahydro-papaverine, and methylation of the latter, but its complete synthesis, which has been recently accomplished by Pictet and Finkelstein, is of greater interest. Dihydro-papaverine was synthesized by Pictet's general method for the preparation of this type of isoquinoline base (cf. previous pages), and the metho-chloride of this compound gave by reduction racemic laudanosine, which was then resolved as described above. This synthesis was carried out a short while before that of papaverine, and was therefore the first complete synthesis of an opium alkaloid.

Narcotine and its decomposition products have provided material for an enormous number of researches. These are too numerous to be described here, but they point to the probability of the structure shown in the accompanying formula

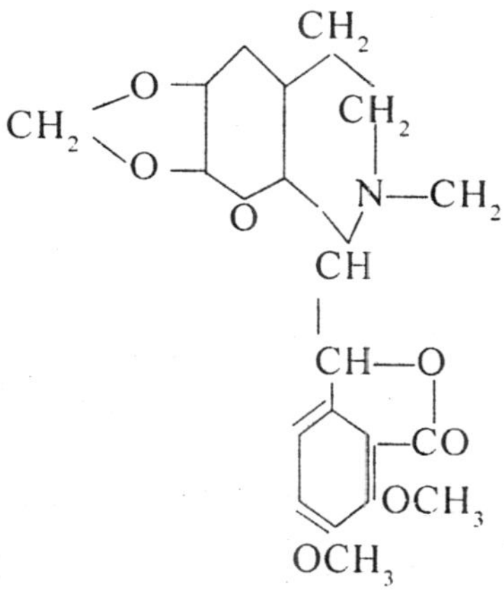

CH_2
O CH_2
CH_2
O $N—CH_2$
O
CH
$CH—O$
CO
OCH_3
OCH_3

Recently narcotine has been synthesized by Perkin and Eobinson, who combined cotarnine and meconine thereby obtaining racemic narcotine (gnoscopine), which was then resolved into its optically active constituents. Meconine was synthesized by Fritsch in 1898, and the synthesis of cotarnine, has more recently been accomplished by Salway, so that the synthesis of narcotine is now complete.

The physiological action of narcotine is similar to that of morphine, but less intense. Chemically it is closely related to hydrastine, the alkaloid to which the physiological action of Hydrastis canadensis is chiefly due. This latter alkaloid, which is an astringent and styptic, is used in uterine haemorrhage, etc.

It was first isolated in a pure condition by Perrins in 1862, and the first observations bearing on its structure were made by Power in 1884. Since then the numerous investigations of Freund and of Schmidt have fully established the structure of this alkaloid. They have shown that it is represented by the formula

which only differs from that of narcotine by the absence of one methoxy group.

The alkaloid present in largest quantities in hydrastis is berberine, but its physiological action is not very marked, and hence it does not play so important a part as hydrastine. It is also found in a large number of other plants, and was first discovered in 1826 in the bark of the prickly ash. It has the formula $C_{20}H_{17}NO_4 + H_2O$.

This is related to that of hydrastine and narcotine, and is still more closely related to that of corydaline, the chief alkaloid in Corydalis cava. Corydaline was discovered in 1826 by Wackenroder, and the formula shown herewith was sug gested for it by Dobbie and Lauder

An alkaloid which may be regarded as an isoquinoline de rivative, although it does not actually contain the isoquinoline ring, is narceine. It is found in opium, but does not have a well-marked physiological action. It can be obtained from narcotine by heating the metho-chloride of the latter with alkali

$$C_{22}H_{23}O_7N,CH_3Cl + NaOH = NaCl + C_{23}H_{27}O_8N.$$

This fact led Freund and Frankforter to suggest the formula

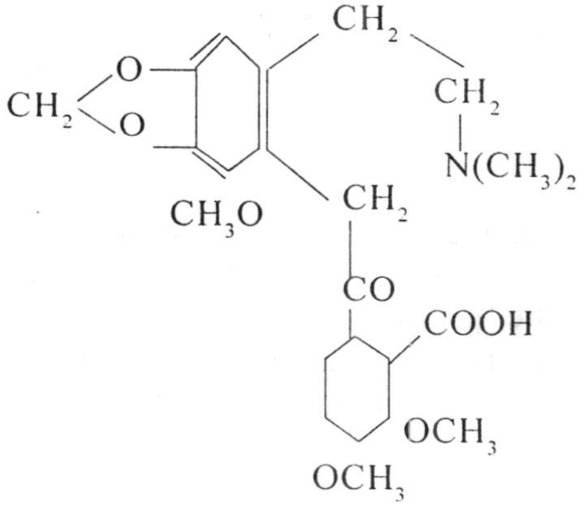

CH₂ / CH₂ / N(CH₃)₂ / CH₂O₂ / CH₃O / CH₂ / CO / COOH / OCH₃ / OCH₃

for narceine, which indicates that it is formed from narcotine by the rupture of both the lactone and the pyridine rings.

One of the most important " artificial alkaloids " that are obtained from the true alkaloids of the isoquinoline series is cotarnine. It was first obtained, together with opianic acid, by Wohler l in 1844, by the oxidation of narcotine with manganese dioxide and sulphuric acid. This substance resembles narceine in the fact that in the free base the pyridine ring is opened out, but in the salts which are formed with elimination of water, the ring closes and derivatives of dihydro-isoquinoline are formed

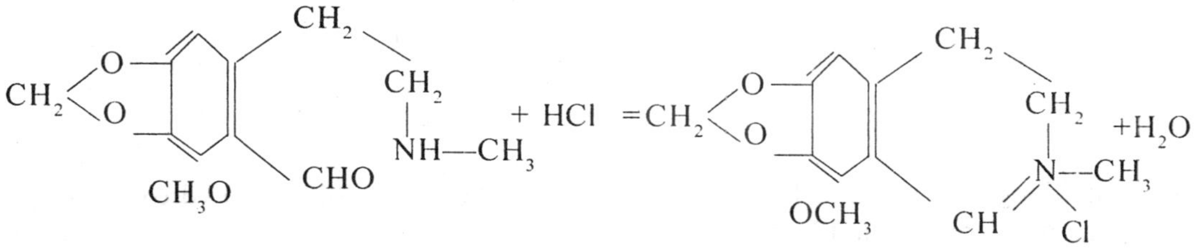

CH₂O₂ / CH₂ / CH₂ / NH—CH₃ / CH₃O / CHO + HCl = CH₂O₂ / CH₂ / CH₂ / N--CH₃ / Cl / OCH₃ / CH + H₂O

There is some evidence that the free base also exists as an isoquinoline derivative of this quaternary ammonium form when dissolved in alcohol. This alkaloid is of importance in medicine as a styptic and as a uterine sedative, and has been brought into the market under the name of " Stypticine," and its phthalic acid salt under the name of "Styptol".

Hydrastine, on oxidation with MnO_2 and H_2SO_4, behaves in a very similar way to narcotine, yielding opianic acid and a basic substance, hydrastinine, which corresponds to cotarnine

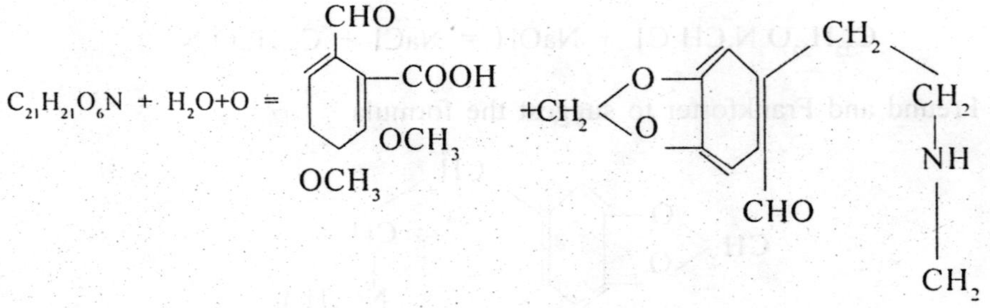

$C_{21}H_{21}O_6N$ + H_2O+O =

CHO
COOH
OCH$_3$
OCH$_3$
+CH_2 <O O>
CH$_2$
CH$_2$
NH
CHO
CH$_2$

Physiologically, hydrastinine resembles cotarnine very closely. It has been shown by Pyman, that laudanosine, on oxidation with MnO_2 and H_2SO_4, behaves in a precisely similar way to narcotine and hydrastine. The products are a basic substance, 4-5 dimethoxy-2 / ^methylaminoethyl-benzaldehyde, and vera traldehyde, the former being analogous to cotarnine and hydrastinine, and the latter corresponding to opianic acid

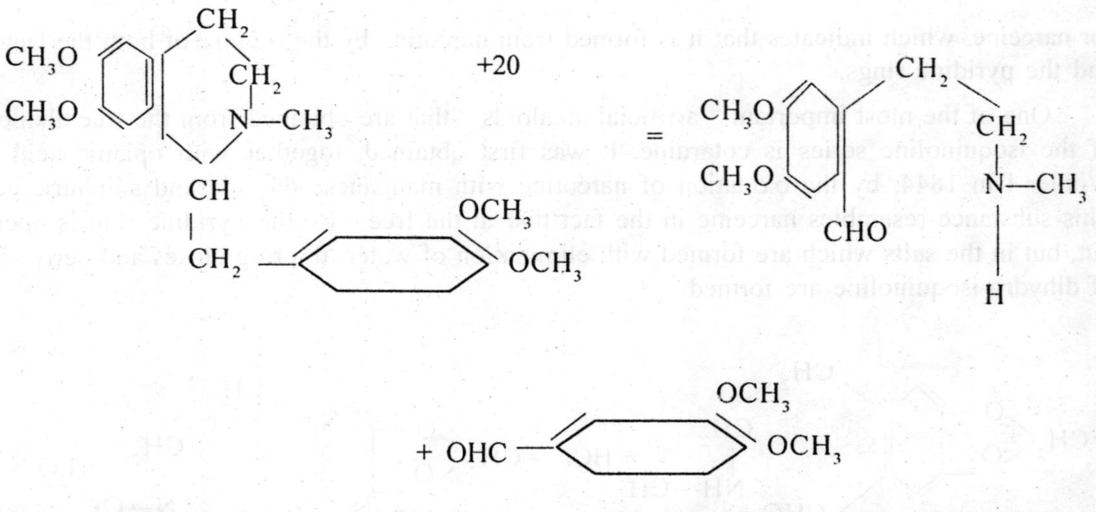

CH$_2$
CH$_3$O
CH$_2$
CH$_3$O
N—CH$_3$
CH
OCH$_3$
CH$_2$ — OCH$_3$

+20 =

CH$_2$
CH$_3$O
CH$_2$
CH$_3$O
N—CH$_3$
CHO
H

OCH$_3$
+ OHC — OCH$_3$

The former combines with acids to form salts of the isomeric isoquinolinium hydroxide, in just the same way as do cotarnine and hydrastinine. Of these, the chloride, which is 6-7 dimethoxy-2 methyl 3-4-dihydro-isoquinolinium chloride, has been introduced into practice under the name "Lodal". It causes a rise of blood-pressure, and renders the. heart-beat slower and stronger.

It is of interest to note that the N-methyl derivatives of tetrahydro-isoquinoline (narcotine, laudanosine, and.hydrastine) are convulsant poisons, while those derived from dihydro-isoquinoline (cotarnine, " Lodal," and hydrastinine) are not. This relation does not hold for those derivatives, such as papaverine, which contain no methyl group attached to the nitrogen.

More recently it has been shown that emetine, $C_{29}H_{40}O_4N_2$, and cephaeline, $C_{28}H_{33}O_4N_2$, the chief alkaloids of ipecacuanha, are derivatives of isoquinoline, as the former on oxidation gives 6-7-dimethoxyisoquinoline-l carboxylic acid,

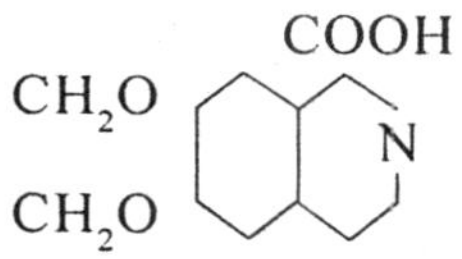

COOH
CH_2O
N
CH_2O

which is also an oxidation product of papaverine. It was also shown that emetine is the monomethyl ether of cephaeline, the former being $C_{25}H_{28}N_2(OCH_3)_4$ and the latter, $C_{25}H_{28}N_2(OCH_3)_3$. These alkaloids have attracted considerable attention recently owing to the valuable effect of emetine hydrochloride in the treatment of amoebic dysentery. It appears to have a specific effect on the protozoal parasites just as quinine has on those of malaria.

According to Pyman the protozoacidal effects on this parasite (Entamceba histolytica) in vitro of emetine, cephaeline, Nmethyl-emetine, and N-methyl-cephaeline are practically equal. Various other interesting facts concerning the relation between chemical constitution and physiological action in this field are given in the same paper. In some cases the use of the double iodide of emetine and bismuth has been found to be more advantageous than that of emetine hydrochloride itself.

Synthesis of Ergot Alkaloids

Alkaloids of therapeutic value are usually derived from plants,but the ergot alkaloids are produced by fungi.The term ergot is used to refer both to the fungus and to the alkaloids which are produced by the fungus. These alkaloids are classified among the indole alkaloids and are derived from the tetracycline ergoline ring system. EA constitute the largest known group of nitrogenous fungal metabolites. Most EA can be separated into two basic groups:

i. *clavines*: The derivatives include clavines ergoline ring system. Their derivatives include clavines (Chanoclavine I, Festuclavinc" Agroclavine, Elymoclavine, Penniclavine. and setoclavine

ii. *ergoline acids*: [(+)-lysergic acid, (+)-isolysergic acid, and paspalic acid].

Ergot is sclerotium of the Pyrenomycetes, Claviceps purpurea, which infects rye and, less commonly, other grains. In infected plants, normal development of the grain is suppressed, and a hardened, purple-black sclerotium develops in place of the grain.

Ergot alkaloids (EA) seem to be a never-ending story. Known since the middle ages as drugs, they continue to play an important role in medical therapy. Because the supply of naturally grown ergots is not sufficient, the biotechnological production of EA has gained in importance. EA continue to be of great interest as drugs themselves, or as prototype compounds from which new drugs can be derived. The biosynthesis of EA lies at level of the genetic information apparatus and continues at the level of physiological realization. Genetic manipulation and knowledge of physiological control are the dominant tools of modern EA production. Scientific research into EA has moved closely to industrial application and become more sophisticated. The production of EA is today largely carried out through submerged Ctaviceps cultures.

EA is produced from genus Claviceps of the class Ascomycetes.C. Purpurea and C. fusiformis paspali produce these alkaloids. The Claviceps adopts a parasitic way of life. The mycelium grows rapidly with the formation of sclerotium (purple). The sclerotia contain 30% fat and 0.01-1% alkaloids. Other species producing alkaloids include Balansia, Penicillium and Aspergillus. Lysergic acid alkaloids also occur in a higher plants i.e. Rivea cormybosa. However, EA have also been found in the Aspergillus and Penicillium genera, and in the Convolvulaceae (morning-glory) family.

Genetic Considerations

The DNA template of Claviceps contains sequence elements specifying the start of transcription; other neighboring sequences are involved in binding of the RNA polymerase complex. In addition, groups of transcription factors are necessary. The RNA Polymerase are suggested to be present in the nuclei in two forms, namely an active and a less active form. The less active form has associated a small peptide inhibitor and has low affinity toward the transcription initiation complex. Addition of stimulatory peptides, a" happens when the cells are inoculated on fresh broth medium, competes out the inhibitory peptide. As a result, we observe the high increase in transcription rate in newly diluted cells.

Ergot Alkaloid Enzymes

Enzymes that are repressed or inhibited under conditions of rapid growth generally produce The EA The Claviceps fungus possesses a great variety of enzymic activity and is versatile in its ability to utilize different compounds as growth substrates. It has also evolved. Efficient regulatory systems to ensure that the enzymes of specific metabolic pathway are synthesized only_ when required and in proper amounts for efficient metabolism. The hypothesis appears to be that common regulatory moments may be participating in the: synthesis of both alkaloids and lipids. synthesizing enzymes are present in mycelia grown in high levels of phosphate, are activated by decreasing the intracellular phosphate level, and do not undergo rapid turnover, in contrast to both alkaline and acid intracellular phosphates.

Commercial Production in Bioreactors

Selection and screening of potent strains are required for -highly yield of suitable alkaloids. The screening may be done in. in shake flasks and -after mutagenic treatment. Claviceps strains of high producers are unstable. The selected strain should be in heterokaryotic state because nomokaryotic parents have rapid growth or abundant sporulation. The stock culture maintenance procedure is important for high yielding strains of Claviceps purpurea, because, these high yielding strains proved to be genetically unstable.

The media may be of complex natural compounds or synthetic media. The production medium includes a mixture of glucose, 5% malt extract, 2%; a mixture of glycerol, 10% gelatin, 1.25% and NaCl, 0.7%; and milk sucrose are typical adaptive protective media for storage of vegetative hyphae or conidia. Conidia from agar slants or liquid cultures or hyphae developed in a preculture medium are suitable for storage. Agar slant cultures can be stored at temperature below 25°C. They are stored under liquid nitrogen or lyophilized. The sealed freeze-dried tubes can be stored at 4°C.

A close relationship exists between the production organism and the environment in every fermentation. The adjustment of medium composition and physico-chemical properties helps in the maintenance of a maximum rate and a desirable direction of the required process. Cultivation conditions suitable for the metabolism of primary substances are not suitable for EA metabolism. The EA fermentations are done in the temperature range 20-30°C (favorable 23-26°), in a pH range of 4.6-6.3 (at best 5.2-5.8) in a medium containing slowly metabolized saccharides (as carbon source).

Asparagines is a superior nitrogen source for saprophytic clavine-alkaloid production in C. purpurea. It enters the cell without degradation and its transport into the cell exceeds the cell's biosynthetic need for asparagines. Thus asparagine disturbs the relative pool sizes of various amino acids. This pattern results in a change in the genetically determined ratio at which amino acids are utilized for protein synthesis. The concentration of spores used to seed the inoculums is critical factor. The importance of the inoculums as a factor determining the efficiency of EA production is well recognized industrially.

Preparation of Inoculum

Conidia are the propagation units and sporulation occurs on agar media (enriched with grains of cereals and in shaken liquid media). Conidiogenesis (formation of conidia) is enhanced by limited aeration.

Media are inoculated with spores or with vegetative cells in the growing phase, incubated at temperatures ranging from 21-27°C for a period of 5-8 days (for liquid cultures) and 10-16 days (for surface cultures).

For inoculum development, spores from heavily sporulated working stock (special agar sporulation media with enriched grains of cereals and in shaken liquid media) are used. The starting materials were sclerotia of ergot from rye and different grasses. Sclerotia containing desired alkaloids in relatively high amounts were selected.

The media contained in a flask on rotary shaker is inoculated with conidia (2-105 conidia m/l) of Claviceps purpurea strain at 24°C.

The germination (up to 5lh day) results into production of numerous conidiophores at the end of growth phase before the exhaustion of carbon and nitrogen source and reaches maximum after 7th day (i.e., 12 to 16 days).

Production of Alkaloid

The hyphae in proliferation stage are transferred to a production in a fermenter containing growth medium. The fermenters are with multiple impellers, suitable for propagation and for the production of Claviceps sp.

The fermenter should be under strict control of factors such as:

— temperature (24°C),

— backpressure, pH (5.5),

— agitator speed, air supply and foam formation.

— osmotic pressure should be between 10-20 bars. It is accomplished by high sugar concentration (200-300 g/L of disaccharides. The high pressure inhibits conidiogenesis and induces differentiation of sclerotial cells.

The production medium has to be inoculated with 10-20% v/ pre-culture at the end of growth phase, that allows optimal alkaloid production.

The aeration rate of the production culture requires progressive increase from O.5c1.5 v/v/m speed of the impellers from 75-200 rpm. The cultivation lasts up to three weeks for most strains 12-16 days.

Peptide Alkaloids

The active principles in ergot that are responsible for ergotism are the alkaloids ergometrine, ergornetrinine, ergotamine, and ergotaminine, which occur in the sclerotium.

Ergotamine Production

For production of ergotamine following medium is used.

Table 1: Amici's production medium for Shake-flask culture

Chemical	Amount
Sucrose	300 g
Citric acid	15 g
$Ca(NO_3)_2.4H_2O$	1g
KH_2PO_4	0.5g
$MgSO_4.7H_2O$	0.25g
Yeast extract	0.1g
KCL	0.2g
$FeSO_4.7H_2O$	7mg
$ZnSO_4.7H_2O$	6mg
Aqueous ammonia PH	5.2
Tap water add	1L

Total yield After 12 days was 1.1-h5 g/L containing 80% of ergotamine is obtained. During the production phase, that lasts 6th/7th days to 12th/,14th days, the alkaloids are actively synthesized. The required energy is obtained from oligosaccharides and after 16 days, the autolysis commences.

Ergotoxine Production

Amici reported the production of ergotoxine ill shake-flask culture. The media employed was similar to that employed for ergotamine production. Kobel and Sanglier introduced new medium for submerged cultivation. They isolated the strain that sporulated well on agar media into a sclerotium-like morphological form of the mycelium.

Table 2. Kobel and Sanglier medium for submerged cultivation

Chemical	Amount
Sucrose	240g
Ammonium oxalate	9.6g
Urea	1.7g
KH2PO4	0.125g
MgSO4.7H2O	0.625g
KCI	0.31g
FeSO4.7H2O	25mg
ZnSO.7H2O	10mg
Aqua dest. Add pH adjusted to	1000mi and 6.2

Production of Simple Lysergic Acid Derivatives

Lysergic acid is produced by the strain Claviceps paspali. It. synthesizes (+)-lysergic acid-a-hydroxy-ethylamide in submerged culture. This substance can be converted into (+)-lysergic acid amide, and on alkaline hydrolysis results into (+)-lysergic acid. Lysergic acid is a key substance, in the production of ergot alkaloids and can be employed in, the chemical synthesis of their derivatives. Paspalic acid functions as the precursor of (+)-lysergic acid.

Table 14.4: The production medium

Chemical	Amount
Sorbitol	50 g
Succinic acid	36g
KH2PO4	2g
MgSO4.7H2O	0.3g
NaOH to PH	5.4
FeSO4.7H2O	10mg
ZnSO4.7H2O	1mg
Water to	1ltr.

The total alkaloid yield in shake flask was 600 mg/L, of which paspalic acid consisted of 86.5%. A mutant strain reported that yields 3g/L of paspalic acid in shake flask. In stainless steel fermenter the production is reduced to 2.3 g/L.

Now lets consider the Isolation Separation And Purification

Peptide Alkaloids

Peptide alkaloid is found in the mycelial cells.

↓

excreted from the medium under alkaline condition traced with ether and chloroform.

↓

extracted with a water solution of tartaric acid or citric acid.

↓

after rendering alkaline they are taken back, into organic solvent,

↓

which is then evaporated to crystallized form.

Paspalic Acid

The paspalic acid is a water-soluble compound. It is obtained from the culture filtrate and is extracted by adsorbing on to cation exchange, resin.

It is subsequently eluted from the resin with dilute ammonia. The elute is evaporated. pH adjusted to 5.5 and finally the substance crystallized.

Clinical Significance

Sometimes they are consumed in large quantities, might happen if the sclerotia are included in milled flour; a disease known as, ergotism (a condition of intoxication which follows the ingestion of excessive amounts of ergot alkaloids; symptoms may include vomiting, diarrhea, thirst and convulsion§, and gangrenous lesions may subsequently develop at the extremities). Cattle in the field may also be poisoned by the sclerotia if they eat infected pasture grasses. Ergotism in humans is marked by vomiting, feelings of intense heat or cold, pain in the muscles of the calf, a yellow color in the face, lesions on the hands and feet, diarrhea, and an impairment of the mental functions.

Therapeutic Applications

Ergot alkaloids have pharmacological properties like-

1. stimulation of the sympathetic nervous system,an effect on the smooth muscles and anatogonism to adrenalin and serotonin. Some lysergic acid alkaloids are used therapeutically in obsterics,including ergotamine,ergobasine,and the semisynthetic methylergobasine, all of which act to stimulate labor and cause the uterus to contract.
2. Are also employed during the period after birth to contract the postpartum uterus.
3. dihydroergotamine is thus used to treat migraine headache
4. mixture of dihydro compounds from ergocristine,ergocryptine and ergocornine is used for treatment of disturbances of the peripheral and central circulation system. i.e., uterine atonia, migraine and other vascular headaches, orthostatic circulatory disturbances, cenile cerebral insufficiency. Hypertension, hyperprolactimenia, acromegaly and Parkinson's.
5. a few naturally occurring alkaloids,particularly the semisynthetic compound lysergic diethylamide (LSD) are very effective Hallucinogens.

Chapter 8

Antipyretics and Analgesics

Antipyretics and analgesics are some of the most important of the synthetic drugs. These substances were introduced on account of their power of reducing the body temperature in fever, but their present importance is due, to an -even greater extent, to their action on the nervous system in soothing pain. The idea of scientists was to produce a compound with properties similar to those of quinine, and this they sought to accomplish by preparing substances the composition of which was closely related to that attributed to that compound. The formula in vogue for quinine was however incorrect, but in spite of this several compounds were produced which had a marked antipyretic effect, but which lacked one very important attribute of quinine, namely, its specific effect against malaria. Quinine was shown to be a derivative of quinoline, and further to differ from the less active cinchonine in containing a methoxy group in the epi (1-6) position to the quinoline nitrogen. Quinoline itself has an antipyretic action, but cannot be used as a drug, owing to its4 unpleasant by-effects.

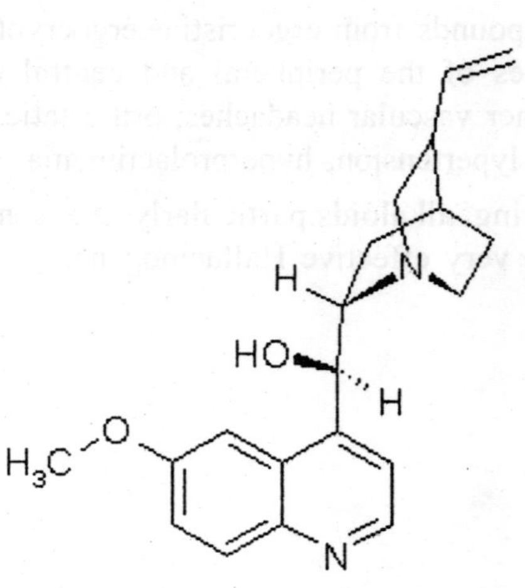

Quinine

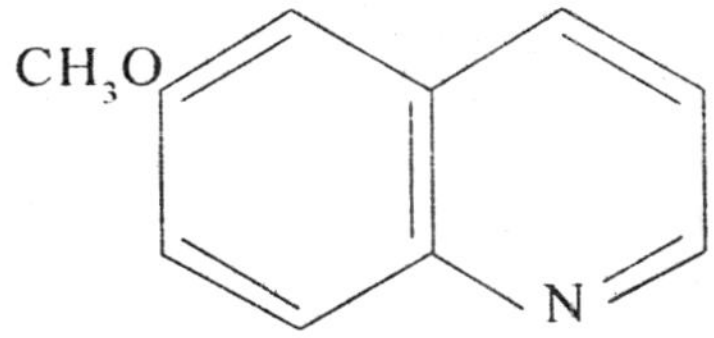

CH_3O … N

Quinoline

Paramethoxyquinoline has a weaker antipyretic action than quinoline, but in accordance with the previously mentioned fact that tetrahydroquinoline has a stronger physiological action than quinoline, it was supposed that quinine was a derivative of epi methoxy-tetrahydroquinoline. This supposition is now known to be incorrect, but nevertheless when 6-methoxy quinoline, is reduced to its tetra-hydro compound, we obtain a substance with a strong antipyretic action, though without any specific effect in malaria. It is not used as a drug, as it has harmful effects on the blood and kidneys. Some earlier experiments had been carried out on N.-alkyl quinoline derivatives, and it was found that the introduction of a hydroxyl group caused the effect to appear more rapidly, but also to cease more quickly and suddenly. For example, kairine, hydroxy-N-ethyl-tetrahydroquinoline, more rapid than Kairoline A or Kairoline B, which are respectively N-ethyl-tetrahydroand N.-methyl-tetrahydroquino lines

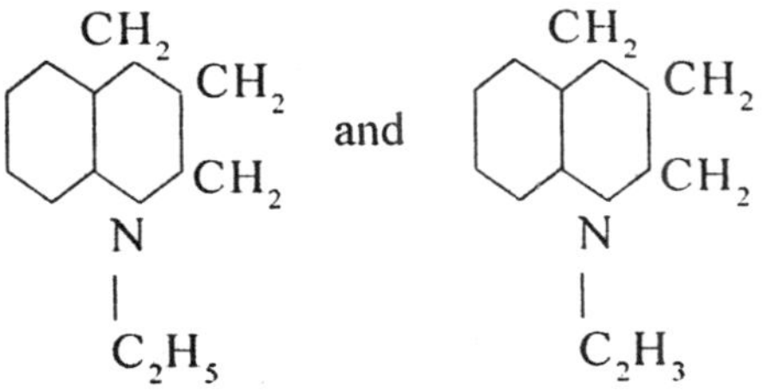

CH_2 CH_2 CH_2 N | C_2H_5 and CH_2 CH_2 CH_2 N | C_2H_3

These substances are all useless, owing to their toxic action on the red blood-corpuscles. Although the artificial quinoline derivatives have proved use less in medicine, a compound produced by Knorr,with the intention of obtaining a substance resembling quinine, met with a great and surprising success. The compound in question was first considered by Knorr to possess a structure similar to that which was then attributed to quinine. Later Knorr, however, showed that it was more closely related to pyrazole

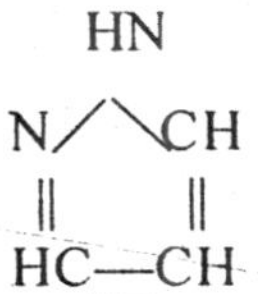

HN

N/\CH

|| ||

HC—CH

in structure, being in fact phenyl-dimethyl-pyrazolon

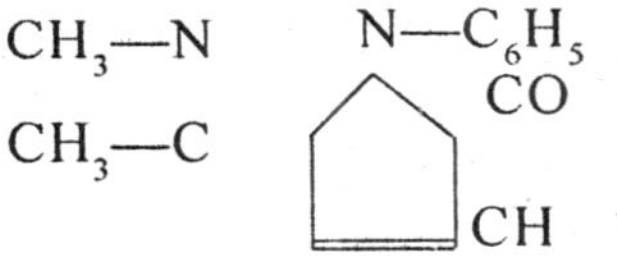

CH_3—N N—C_6H_5

CO

CH_3—C

CH

This compound has attained a very extended use in medicine under the name of antipyrine. It is obtained by the action of aceto-acetic ester on phenyl hydrazine, whereby a hydrazone is formed, which on heating loses the elements of alcohol, with the formation of phenyl methyl-pyrazolone. This is then heated with methyl iodide and methyl alcohol at 100-150°C., which transforms it into antipyrine

Phenylhydrazine: $HN(H)-NH-C_6H_5$ + Aceto-acetic ester: $CH_3-C(HO)=CH-CO-O-C_2H_5$ = → Hydrazone: $CH_3-C(HN-NH-C_6H_5)=CH-CO-O-C_2H_5$ → (Heating) Phenyl-methyl-pyrazolone: $N-C_6H_5$, HN, CO, $CH_3-C=CH$; Phenyl-dimethyl-pyrazolone: $N-C_6H_5$, CH_3-N, CO, $CH_3-C=CH$

This view of its constitution is confirmed by its direct synthesis from $C_6H_5-NH-NH-CH_3$ and aceto-acetic ester.

Its salt with acetylsalicylic acid is known as Acetopyrine.

Antipyrine has a very strong antipyretic action, which is far greater than that of quinine, and it is free from injurious effect on the haemoglobin, but it has no specific action against malaria. On the other hand, it has the very useful property of diminishing neuralgic pains (analgesic action), which has secured for it a great popularity. A large number of deriva tives of antipyrine have been placed on the market, but most of these have no advantage over it, and many of them are very loose compounds of antipyrine and other substances, which act practically like mixtures. One derivative of antipyrine has, however, proved of great value, namely, 4-dimethyl-amino antipyrine. This substance, which is called Pyramidon or Amidopyrine, is about three times as powerful in its action as antipyrine, and also has the advantage of being free from in jurious effects on the heart (Kobert). It is prepared by treating an acid solution of antipyrine with sodium nitrite, whereby nitroso-antipyrine is formed. This on reduction gives amino antipyrine

$N-C_6H_5$, CH_3-N, CO, $CH_3-C=C-NO$ → $N-C_6H_5$, CH_3-N, CO, $CH_3-C=C-NH_2$

which is isolated in the form of its benzylidene derivative (i.e. condensation product with benzaldehyde)

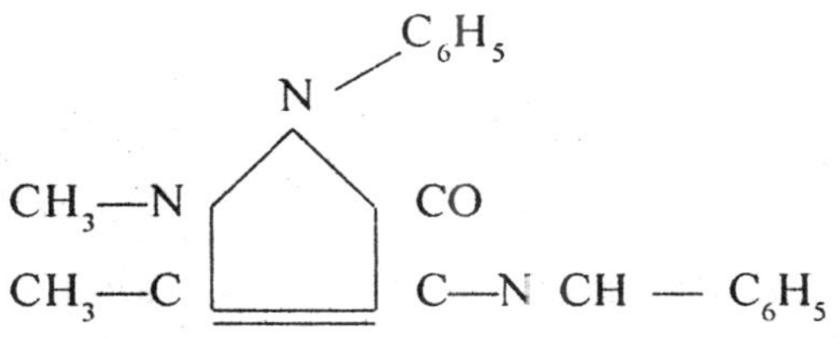

$$\begin{array}{c} \quad\quad N-C_6H_5 \\ CH_3-N \quad\quad CO \\ CH_3-C = C-N\ CH - C_6H_5 \end{array}$$

This is then decomposed with hydrochloric acid and on methylation yields pyramidon

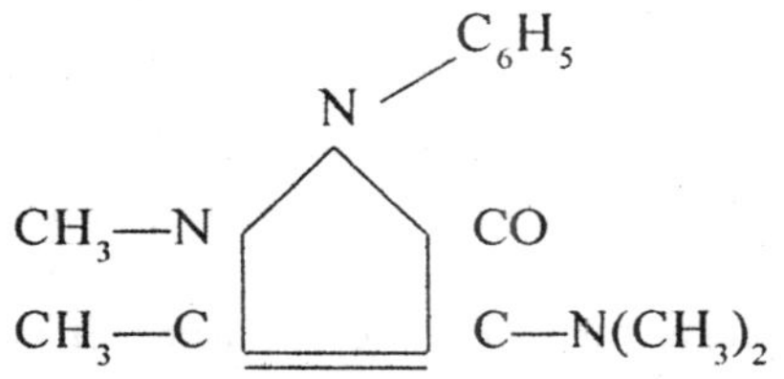

$$\begin{array}{c} \quad\quad N-C_6H_5 \\ CH_3-N \quad\quad CO \\ CH_3-C = C-N(CH_3)_2 \end{array}$$

The success of antipyrine, and the mistaken idea that its action might belong to phenylhydrazine derivatives in general, led to the production of a large number of these, most of which are, however, without therapeutic value. The intense toxic action of phenylhydrazine is weakened by the entrance of acetyl groups,; but the monoacetyl derivative (hydracetin), C_6H_6—NH—NH . CO . CH_3, and the diacetyl derivative, C_6H_5NH—N(CO . CH_3) are both too toxic to be of any use. Other phenylhydrazine derivatives have been prepared, in which the carboxyl group enters the molecule, e.g. the hydra zone of Isevulinic acid

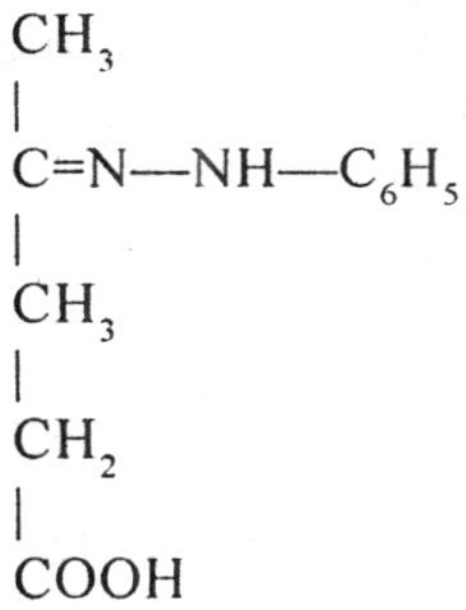

$$\begin{array}{l} CH_3 \\ | \\ C{=}N-NH-C_6H_5 \\ | \\ CH_3 \\ | \\ CH_2 \\ | \\ COOH \end{array}$$

called Antithermin, but this, like Orthin, and other derivatives of this type, is too toxic for use. More success has attended a compound in which acid amide groups are introduced into the molecule, meta-benzaminsemi carbazide,

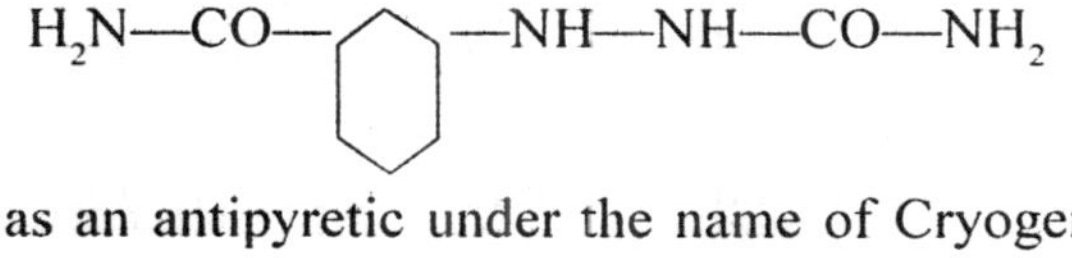

$$H_2N-CO-C_6H_4-NH-NH-CO-NH_2$$

having been extensively used as an antipyretic under the name of Cryogenin.

Maretin is meta-tolylsemicarbazide

$$CH_2—C_6H_4—NH—NH—CO—NH_2$$

The antipyretics already described owe their origin to the endeavour to prepare substances similar to quinine. Those to be dealt with in the present section are based on the discovery of Cahn and Hepp,that aniline and acetanilide have powerful antipyretic and anti-neuralgic properties, and the low price of aniline gave a still further stimulus to the endeavour to find a suitable derivative of it that could rival the more expensive quinine and antipyrine. Aniline and its salts have a strong antipyretic action, but they are readily absorbed, and, owing to their action on the haemoglobin, give rise to toxic symptoms.

By the entrance of an acetyl group, aniline is rendered more resistant and less toxic, so that acetanilide, C_6H_6—NH—NH.CO.CH_3, has been used in medicine under the name of " antifebrin ". It has marked antipyretic properties, and also acts as an analgesic. Though far less toxic than aniline, yet its physiological effect is due to the slow liberation of aniline, and after a time symptoms of aniline poisoning may become apparent. It is only used now on account of its cheapness it is by far the cheapest of all antipyretics and it forms the active ingredient of many of the secret remedies that are advertised for the cure of headaches, etc., and is also used to adulterate other more expensive drugs, such as phenacetin. Small quantities of acetanilide are oxidized in the body to para-aminophenol,and the observation of this fact has led to the introduction of various derivatives of para-aminophenol as antipyretics. Before passing on to these, attention may be given to some other compounds which are more closely akin to acetanilide. In most of these cases the compounds that have been introduced could have no marked advantage over acetanilide, as they likewise depend on the liberation of free aniline for their physiological effect. For example, benzanilide and salicyl-anilide resemble acetanilide, but have to be given in bigger doses, as they are less readily split up in the intestine ; while on the other hand, formanilide, owing to the ease with which it undergoes hydrolysis, is far more toxic than acetanilide. Toluidine derivatives also have not the slightest advantage over acetanilide, as all the three toluidines have, when once in the system, practically the same toxic action as aniline.

Attempts have been made to obtain more soluble derivatives of acetanilide by the introduction of acidic groups. As the sparing solubility of acetanilide does not have any injurious effect on its therapeutic action, these attempts must be regarded as quite unscientific, especially as the resulting compounds are, as was to be expected with acidic substances, quite inert and without the characteristic action of acetanilide. An example of this class is Cosaprin,

Substances containing a sulphonic acid group in the CD position of the acetanilide molecule are said to be an exception to the rule that acidic substances are inert. These derivatives are said to be easily soluble and to have an antipyretic action, but they do not appear to have come into use, and it may therefore be that the original statements as to their antipyretic action were mistaken. C_6H_6—NH—NH . CO . CH_2. SO_3Na is an example, and is prepared as follows : The aniline salt of chloracetic acid is treated with phosphorus pentoxide

The product is then heated with an aqueous solution of sodium sulphite

$$C_6H_5 \cdot NH \cdot CO \cdot CH_2Cl + Na_2SO_2 = NaCl + C_6H_5 \cdot NH \cdot CO \cdot CH_2 \cdot SO_3Na,$$

the sodium salt of the acetanilide sulphonic acid separating on cooling. Methyl-acetanilide, $C_6H_6.N(CH_2)$. CO . CH_3,, exalgin, has toxic properties and is not much used in medicine. An aniline derivative of a different type is phenyl-urethane, Euphorin. It is prepared by the action of aniline on chloro-formic ester

$$C_6H_5 \cdot NH_3 + Cl + COOC_2H = C_6H_5 \cdot NH \cdot COOC_2H_5 + HCl.$$

It is far less toxic than aniline, and has marked analgesic properties, but is irregular in its antipyretic effect.

Para-aminophenol Derivatives

Para-aminophenol is far less toxic than aniline, but it still has an action on the haemoglobin sufficient to be harmful in moderate doses. The replacement of the hydrogen of the amino group by an acetyl group, with the formation of HO⬡NH . CO . CH_3, yields a compound with a lower toxicity than para-aminophenol, and with anti pyretic and anti-neuralgic properties, but still not quite satis factory as a drug. The next step was the replacement of the hydroxylic hydrogen by radicles, and diacetyl-aminophenol, was prepared, but it was found to be inferior in its action to para-ethoxy-acetanilide and para-methoxy-acetanilide, CH_3O⬡NH.CO.CH_3 and CH_3O⬡NH . CO . CH_3, which were named phenacetin and methacetin respectively.

The last-named compound was found to possess antipyretic and anti-neuralgic properties in the highest degree of all, but phenacetin, in which these properties are only slightly less well marked, has the advantage of having less toxic action on the blood. Phenacetin, which was the first drug of this class to be placed on the market, has retained the lead ever since, and is by far the most important compound of this class, and indeed, at the present time, is probably more used than any of the other antipyretic and analgesic substances. The starting-point for the synthesis of phenacetin is paranitrophenol. The sodium salt of this is treated with ethyl bromide, and the product reduced with tin and hydrochloric acid. The para-phenetidine thus formed is then acetylated by boiling with glacial acetic acid

NO_2 (ring) ONa → NO_2 (ring) OC_2H_5 → NH_2 (ring) OC_2H_5 → NH.CO.CH_2 (ring) O.C_2H_5

Para-phenetidine — Acetyl para-phenetidine (phenacetin)

On the commercial scale, an ingenious variation of this process has been devised, whereby one molecule of para-nitrophenol is made to furnish a large number of molecules of phenacetin. This process is adopted on account of the fact that para-nitro phenol is rather difficult to obtain in the pure condition. In this process, the para-nitrophenol is converted into para-phene tidine by the method already described, and the phenetidine is then diazotized and coupled with phenol in the presence of sodium carbonate. The product is then ethylated, and the substance thus obtained is then reduced, whereby two molecules of phenetidine are formed. These can then be converted into phenacetin by acetylation, or can be made to yield a double quantity of phenetidine by repeating the above-described process l of diazotization and coupling with phenol.

OC_2H_5 (benzene ring, N_2Cl) + OH (benzene ring) = OC_2H_5 (benzene ring) — $N{=}N$ — (benzene ring) OH + HCl

↓

C_2H_5O (benzene ring) — $N{=}N$ — (benzene ring) OC_2H_5 → [C_2H_5O (benzene ring) NH_2]

Phenacetin is also obtained by the ethylation of *p*-acetylamino phenol. Phenacetin is only slightly toxic, but when given in very large doses, some methaemoglobin is found in the blood. Its antipyretic action is due to increased heat-loss from the surface of the body, but this drug is used chiefly for its analgesic action, in the treatment of headache, neuralgia, etc. Several other derivatives of the phenacetin group have been prepared, but none of them have any novel therapeutic action, as in all of them this depends on the liberation of para-amino phenol or para-phenetidine in the body. Indeed, it has been shown by Treupel and Hinsberg, that the antipyretic action of aniline and para-aminophenol derivatives is, within certain limits, proportional to the amount of aniline, para-aminophenol or phenetidine, formed in the organism. This statement is based on the fact that, in those cases where an antipyretic action is found, the urine shows the indophenol reaction, and that the intensity of the reaction is roughly proportional to the strength of the antipyretic effect. This reaction is carried out as follows : The urine is acidified with two drops of hydro chloric acid, and two drops of a one-per-cent. solution of sodium nitrite added, whereby the phenetidine is diazotized. On adding an ajkaline solution of β-naphthol, a red coloration is produced, which becomes violet on acidifying with hydrochloric acid. This reaction is given by which resemble phenacetin in their physiological action, but not by

OCH_2 (benzene ring) $NH \cdot CO \cdot CH_3$ OC_2H_5 (benzene ring) $NH \cdot CO \cdot CH_2$ and OC_3H_7 (benzene ring) $NH \cdot CO \cdot CH_3$

which resemble phenacetin in their physiological action. but not by

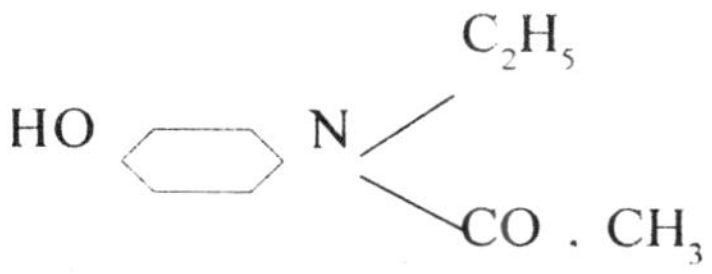

$$HO\langle\rangle N\begin{matrix}\diagup C_2H_5 \\ \diagdown CO\ .\ CH_3\end{matrix}$$

which has no antipyretic action, and no action on the blood. If, however, the hydrogen of the amino group is substituted in phenacetin itself, physiologically active substances are obtained. For example, the methyl derivative, $C_2H_5O\langle\rangle\underset{\displaystyle CH_3}{\overset{}{N}}\ .\ CO\ .\ CH_3$, has greater narcotic and anti-neuralgic properties, but less antipyretic action than phenacetin, while in the ethyl derivative, $C_2H_5O.,\ C_6H_4—\underset{\displaystyle C_2H_3}{N}—CO\ .\ CH_3$

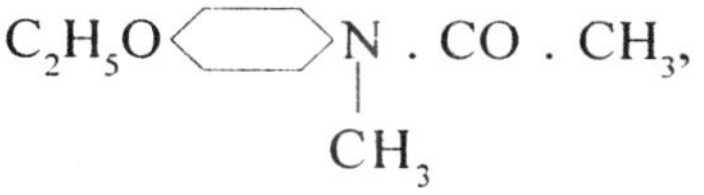

there is an increase in the narcotic power, accompanied by a decrease in the toxic properties, the antipyretic action being retained with slightly diminished intensity. In the higher members of the series, such as the propyl and butyl derivatives, the narcotic and anti-neuralgic properties rapidly diminish with the increase in the molecular weight. In this series the maximum antipyretic and anti-neuralgic effect is shown by the methyl and ethyl derivatives, and the minimum toxicity by the ethyl compound.

Owing to the slight solubility of phenacetin, many attempts have been made to prepare more soluble derivatives, which should nevertheless be sufficiently stable to resist the action of the dilute hydrochloric acid of the gastric contents, and so prevent the liberation of poisonous phenetidine salts. As the slight solubility of phenacetin does not appear to interfere with its physiological effect, these efforts are not of great practical value. The introduction of sulphonic acid or carboxyl groups in order to obtain increased solubility was not likely to meet with much success, as generally the presence of acid groups tend to destroy the physiological activity of a com pound. For example, both the sulphonic and the carboxylic acid of phenacetin are almost inert, but the sodium salt of the former has been introduced under the name of Phesin and is said to have slight temporary antipyretic properties.

Similar compounds to phenacetin have been prepared in which the hydrogen of the amino group is replaced by acid radicles other than acetyl. Para-propionyl-phenetidine (Tripheniri) is similar to phenacetin, but less soluble and hence less physiologically active.

Lactyl-phenetidine (Lactopheniri)

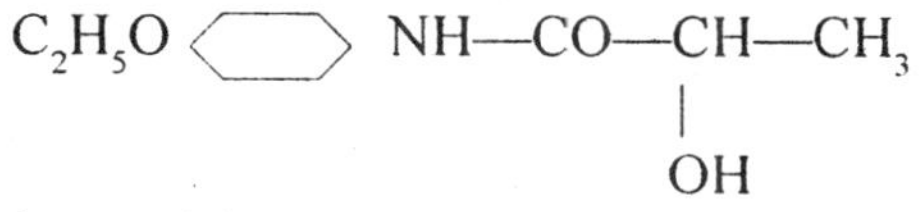

$$C_2H_5O\langle\rangle NH—CO—\underset{\displaystyle OH}{CH}—CH_3$$

is more soluble than phenacetin, and has less antipyretic action, but has well-marked anti-neuralgic and narcotic properties. It is more liable to liberate toxic phenetidine hydrochloride. Para-ethoxy-phenyl-succinimide (Pyrantin)

$$C_2H_5O\langle\quad\rangle N\langle{CO—CH_2 \atop CO—CH_2}$$

and diacetyl-phenetidine, have been prepared, but have no value.

Salicyl-phenetidine is stable, insoluble, and practically inert, as would be expected, and amygdophenine

$$C_2H_5O\langle\quad\rangle —NH—CO—\overset{H}{\underset{OH}{C}}—C_6H_5$$

is also rather insoluble and inert. It is the mandelic acid derivative of phenetidine.

It has been noticed that phenyl urethane, Euphorin, is partially oxidized by the organism into para-hydroxy-phenyl-urethane, and some of its derivatives have been prepared for pharmaceutical purposes.

The acetyl derivative of the last-named above has been introduced under the name "Thermodin". It has a gradual antipyretic action, is not toxic, and has no depressant action on the heart or respiration. It is insoluble except in acid media. Various derivatives of this type have been prepared by Merck, by passing carbonyl chloride into a solution of para-oxyphenyl urethane, or acid derivatives of para-phenetidine in presence of alkalies

$$2C_6H_4\langle{OH \quad (1) \atop NH\,.\,COH\ (4)} + COCl_2 = CO\langle{O—C_6H_5—NH\,.\,COR \atop O—C_6H_4—NH\,.\,COR} + 2HCl$$

where $R = CH_3,\ C_2H_5,\ C_6H_5,\ OC_2H_5,\ \text{or}\ OC_3H_7.$

If the reaction is carried out in alcoholic solution in presence of sodium ethoxide, mixed carbonates are formed

$$C_6H_4\langle{O(H \atop NH\,.\,CO\,.\,R} + Cl)\,.\,CO\,.\,(Cl + H)\,OC_2H_5$$

$$= CO\langle{O—C_2H_5 \atop O—C_6H_4—NH\,.\,COR} + 2HCl$$

By varying the alcohol, methyl or propyl groups can replace the ethyl group.

Among other derivatives of phenetidine which have been prepared are certain condensation products with aldehydes. Mention may be made of Malakin

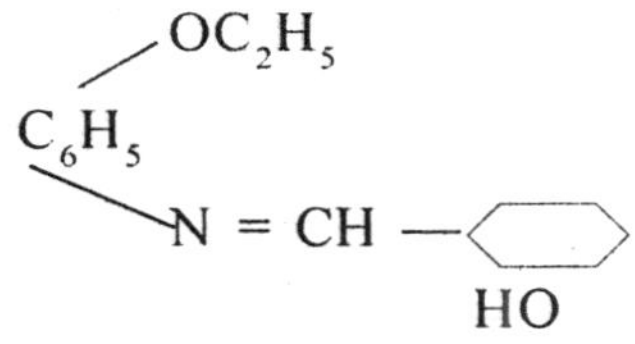

$$C_6H_5\begin{cases} OC_2H_5 \\ N = CH — \bigcirc \\ \qquad\qquad HO \end{cases}$$

prepared by condensing phenetidine with salicylic aldehyde, Malarin, the citrate of methyl benzylidene phenetidine, vanillin-phenetidine

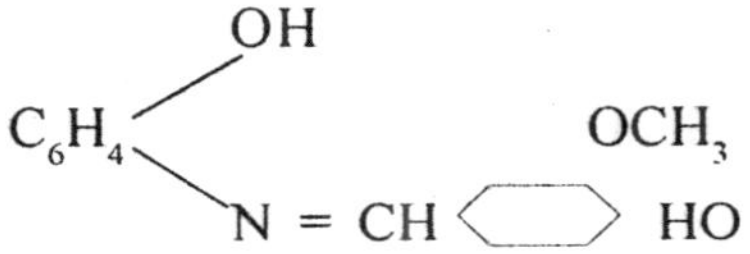

$$C_6H_4\begin{cases} OH \\ N = CH \bigcirc \begin{matrix} OCH_3 \\ HO \end{matrix} \end{cases}$$

(said to be a good styptic) and the para-phenetidine derivative of vanillin ethyl carbonate

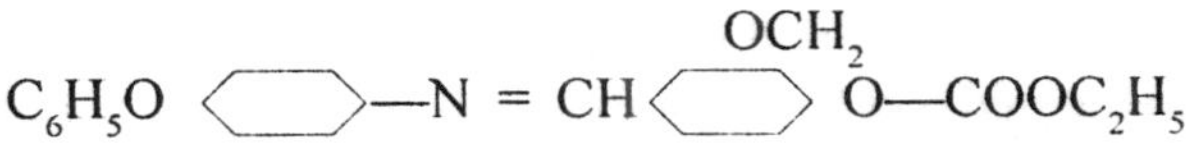

$$C_6H_5O \bigcirc —N = CH \bigcirc \begin{matrix} OCH_2 \\ O—COOC_2H_5 \end{matrix}$$

which has been named Eupyrin. None of these are, however, of any particular value.

A soluble derivative of phenacetin has been prepared by Schmidt and Majert, by the action of ammonia on bromacetylphenetidine (bromo-phenacetin)

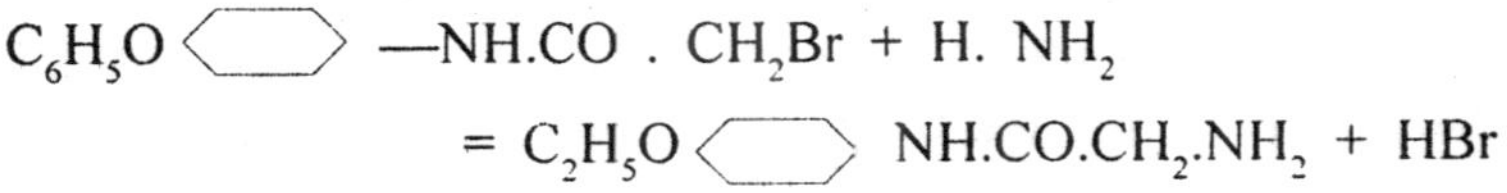

$$C_6H_5O \bigcirc —NH.CO \,.\, CH_2Br + H.\ NH_2$$
$$= C_2H_5O \bigcirc \ NH.CO.CH_2.NH_2 + HBr$$

The amino-phenacetin thus formed is named Phenocoll. Its hydrochloride is readily soluble in water, and has a similar action to phenacetin, but the action appears and disappears more rapidly. It is said to have a greater analgesic action, and also to have antiseptic properties, and to be a good substitute for salicylic acid as an antipyretic in rheumatic fever. Its only insoluble salt is the salicylate, Salocoll.

So far, mention has only be made of phenetidine derivatives in which the hydrogen of the hydroxyl group has been replaced by simple alkyl groups, but several derivatives have been prepared where this hydrogen has been replaced by more complex groups.

Chapter 9

Halogen Antiseptics

The halogens or halogen elements are a series of nonmetal elements from Group 17 of the periodic table, comprising fluorine, F; chlorine, Cl; bromine, Br; iodine, I; and astatine, At. The undiscovered element 117, temporarily named ununseptium, may also be considered a halogen.The group of halogens is the only group which contains elements in all three familiar states of matter at standard temperature and pressure.

Owing to their high reactivity, the halogens are found in the environment only in compounds or as ions. Halide ions and oxoanions such as IO_3- can be found in many minerals and in seawater. Halogenated organic compounds can also be found as natural products in living organisms. In their elemental forms, the halogens exist as diatomic molecules, but these only have a fleeting existence in nature and are much more common in the laboratory and in industry. At room temperature and pressure, fluorine and chlorine are gases, bromine is a liquid and iodine and astatine are solids; Group 7 is therefore the only periodic table group exhibiting all three states of matter.

The halogens show a number of trends when moving down the group - for instance, decreasing electronegativity and reactivity, increasing melting and boiling point.

Halogen	Standard Atomic Weight (u)	Melting Point (K)	Boiling Point (K)	Electronegativity (Pauling)
Fluorine	18.998	53.53	85.03	3.98
Chlorine	35.453	171.6	239.11	3.16
Bromine	79.904	265.8	332.0	2.96
Iodine	126.904	386.85	457.4	2.66
Astatine	210	575	610	2.2

Halogens are highly reactive, and as such can be harmful or lethal to biological organisms in sufficient quantities. This high reactivity is due to their atoms being one electron short of a full outer shell of electrons. They can gain this electron by reacting with atoms of other elements. Fluorine is the most reactive element in existence, attacking such inert materials as glass, and forming compounds with the heavier noble gases. It is a corrosive and highly toxic gas. The reactivity of fluorine is such that, if used or stored in laboratory glassware, it can react with glass in the presence of small amounts of water ɔ form SiF_4. Thus fluorine must be handled with substances such as Teflon, extremely dry glass, or metals such as copper or steel which form a protective layer of fluoride on their surface.

Both chlorine and bromine are used as disinfectants for drinking water, swimming pools, fresh wounds, dishes, and surfaces. They kill bacteria and other potentially harmful microorganisms through a process known as sterilization. Their reactivity is also put to use in bleaching. Chlorine is the active ingredient of most fabric bleaches and is used in the production of most paper products.

The halogens all form binary compounds with hydrogen, the hydrogen halides, HX (HF, HCl, HBr, HI), a series of particularly strong acids. When in aqueous solution, the hydrogen halides are known as hydrohalic acids. HAt, or "hydrastatic acid", should also qualify, but it is not typically included in discussions of hydrohalic acid due to astatine's extreme instability toward alpha decay.

The halogens react with each other to form interhalogen compounds. Diatomic interhalogen compounds (e.g. BrF, ICl, ClF) bear resemblance to the pure halogens in some respects. The properties and behaviour of a diatomic interhalogen compound tend to be intermediate between those of its parent halogens. Some properties, however, are found in neither parent halogen - Cl_2 and I_2 are soluble in CCl_4 but ICl is not, since it is a polar molecule due to the relatively large electronegativity difference between I and Cl.

Many synthetic organic compounds such as plastic polymers, and a few natural ones, contain halogen atoms; these are known as halogenated compounds or organic halides. Chlorine is by far the most abundant of the halogens, and the only one needed in relatively large amounts (as chloride ions) by humans. For example, chloride ions play a key role in brain function by mediating the action of the inhibitory transmitter GABA and are also used by the body to produce stomach acid. Iodine is needed in trace amounts for the production of thyroid hormones such as thyroxine. On the other hand, neither fluorine nor bromine are believed to be essential for humans, although small amounts of fluoride can make tooth enamel resistant to decay.

In drug discovery, the incorporation of halogen atoms into a lead drug candidate results in analogues that are more lipophilic and less water soluble. Consequently, halogen atoms are used to improve penetration through lipid membranes. However, there is an undesirable tendency for halogenated drugs to accumulate in lipid tissue.

The chemical reactivity of halogen atoms depends on both their point of attachment to the lead and the nature of the halogen. Aromatic halogen groups are far less reactive than aliphatic halogen groups, which can exhibit considerable chemical reactivity. For aliphatic carbon-halogen bonds the C-F bond is the strongest and usually less chemically reactive than aliphatic C-H bonds. The other aliphatic-halogen bonds are weaker, their reactivity increasing down the periodic table. They are usually more chemically reactive than aliphatic C-H bonds. Consequently, the most popular halogen substitutions are the less reactive aromatic fluorine and chlorine groups.

Clorine Compounds

The strong germicidal properties of chlorine and of hypochlorous acid have long been known. Bleaching powder, which is essentially a compound of calcium chloride and hypochlorite, has been used extensively for many years as a disinfectant, for which its cheapness renders it very suitable. Sodium hypochlorite, NaOCl, is readily soluble in water, but its caustic and irritant properties prevented its use as an antiseptic for the dressing of wounds. By the addition of boric acid to bleaching powder, a mixture may be obtained which contains free hypochlorous acid and calcium biborate, and is very well adapted as a dressing for wounds and the treatment of septic conditions. This powder has been given the name of Eupad, and the corresponding solution the name of Eusol, Another form of hypochlorite solution known as Dakin's solution is prepared by the action of a solution of sodium carbonate and sodium bicarbonate on bleaching powder, and care must be taken that it contains from 0.45 per cent, to 0.5 per cent, of sodium hypochlorite. It has been very largely used with great success for the irrigation of wounds.

Subsequently, organic chloramines were introduced by Dakin as substitutes for sodium hypochlorite. These substances were first discovered and investigated by Chattaway, and are prepared by the action of hypochlorite solutions upon organic compounds containing the imino (—NH) or amino (—NH_2) groups, whereby chloramines containing the (NCl) group, and dichloramines containing the (NCl_2) group, are produced. Physiologically, their action resembles that of hypochlorite, but therapeutically they possess many advantages over it. For example they are far less irritant, and are stable solids which can be dissolved in water to give solutions of a definite strength, whereas sodium hypochlorite cannot be kept in the solid state, and its solutions are always of somewhat uncertain strength.

The best-known and most used of these chloramines is the substance introduced under the name of Chloramine T. It is sodium *p*-toluenesulphonchloramide, and is prepared from *p*-toluenesulphonyl chloride, a by-product in the manufacture of saccharin. This by treatment with ammonia is converted into *p*-toluenesulphonamide, which on warming with sodium hypochlorite solution is converted into Chloramine T.

CH_3

SO_2—NClNa, $3H_2O$,

It is also prepared by dissolving Dichloramine T. in hot caustic soda solution. This substance is also known by the trade name of Tolamine.

It is a stable crystalline compound which is soluble in water, and has been of great value in a variety of septic conditions, especially in military surgery for the treatment of infected wounds. Its freedom from irritant properties has rendered it particularly useful in the treatment of injuries to the mouth and jaw. It is also used as a disinfectant lotion in cases of infectious disease, such as scarlet fever, measles, etc.

The corresponding dichloramine, *p*-toluenesulphondichlor amide, is also used as an antiseptic under the name of Dichloramine T. It is readily prepared by dissolving *p*-toluenesulphonamide in bleaching powder solution, or by the action of hypochlorous acid upon Chloramine T. It is used for the treatment of wounds in the form of a solution in solvents such as chlorcosane (chlorinated paraffin wax) and chlorinated eucalyptol. Its oil solution has also been used as a naso-pharyngeal sterilizer in cases of meningitis, and complete sterilisation of the carrier has been claimed by this means.

Another dichloramine derived from *p*-toluenesulphonamide is *p*-sulphondichloraminobenzoic acid.

COOH

SO_2—NCl_2

This substance is known as Halazone, and is specially recommended for the sterilization of drinking water. It is prepared by heating *p*-toluenesulphonamide with bichromate and sulphuric acid, whereby the methyl group is oxidized to the carboxyl group with formation of *p*-sulphonamino-benzoic acid. This acid is then dissolved in caustic soda solution, and treated with chlorine gas in the cold to form the desired dichloramino acid.

$$CH_3 . C_6H_4 . SO_2 . NH_2 \rightarrow COOH . C_6H_4 . SO_2 . NH_2 \rightarrow COOH . C_6H_4 . SO_2 . NCl_2.$$

Iodine Compounds

Iodoform

Iodoform has been very extensively used in surgery as a dressing for wounds, as it seems to promote healing as well as exerting an antiseptic action. In laboratory experiments in vitro, iodoform has so slight a bactericidal power as to indicate that it would be useless for such a purpose, and its antiseptic action appears to be due to the free iodine liberated when it comes into contact with fat, putrefactive material, etc.

Iodoform is prepared according to a well-known method by the action of free iodine on a warm mixture of acetone or alcohol and sodium hydroxide or carbonate solution. In this process, part of the iodine is used up in forming alkaline iodide from which free iodine has to be again regenerated before it can be used for the manufacture of a further quantity of iodo form. For this reason, various electrolytic methods for pre paring iodoform have been devised. One of these consists in electrolyzing in a current of carbon dioxide a warm solution of potassium iodide to which alcohol has been added.

The same method can be applied to the preparation of chloro form and bromoform, but in these cases the stream of carbon dioxide is unnecessary. Iodoform is also manufactured by passing ozone through a solution of potassium iodide and sodium carbonate in 30 per cent, alcohol, at a temperature of 50°C. The ozone is passed into the liquid until all the potassium iodide has been decomposed.

Iodoform, in spite of its many valuable properties, suffers from some serious drawbacks, the chief of these being its powerful odour and the fact that it often has an irritant action on the skin, giving rise to a kind of eczema. It also has toxic properties, which sometimes give rise to definite symptoms of poisoning. In order to mask the odour of iodoform, many preparations have been introduced in which it is mixed with strong and pleasantly scented substances, but these need not be mentioned here. On the other hand, by combining iodoform with inodorous substances, its volatility can be lowered and its odour destroyed.

For example, a compound of iodoform with hexa-methylenetetramine is inodorous, and is known as iodoformin and a similar compound with hexamethylene-tetramine-ethyl iodide was introduced under the name iodoformal but both of these compounds are decomposed by water into their components, and hence in practice they possess no advantage over a mere mixture.

Of far greater importance are the efforts that have been made to produce other insoluble compounds which should possess the valuable properties of iodoform in promoting the healing of wounds, and have the further advantage of being less irritant and toxic than iodoform, and free from any powerful odour.

Many substances have been prepared in attempting to meet these requirements, and these comprise derivatives of iodine as well as those of other halogens and of various metals such as bismuth. Of the various iodine derivatives, pure and simple, that have come into use, only one of them, iodol (tetraiodopyrrol), resembles iodoform in the fact that its action is due to the actual liberation of iodine.

In aromatic compounds it is true that the antiseptic power is generally increased by the replacement of hydrogen by iodine, but nevertheless the latter is too firmly united to the nucleus to be liberated by the action of the tissues, and therefore these compounds cannot be compared in their action with iodoform. In the case of the pyrrol ring, however, the iodine is in a more labile state, and tetraiodo-pyrrol

```
I—C        C—I
   ||      ||
I—C        C—I
    \     /
      NH
```

resembles iodoform in its action, and has the advantage of being odourless and non-irritant as well as insoluble. It is used as an iodoform substitute under the name of iodol, and is prepared by the action of iodine on an alkaline solution of pyrrol, the latter being obtained from "bone-oil."

There are, however, many other organic derivatives of iodine which possess antiseptic properties, and are used as iodoform substitutes, although their mode of action is quite different. For example, by the addition of a solution of iodine in potassium iodide to an alkaline solution of a phenol, compounds are pro duced containing two atoms of iodine, one of which replaces a nuclear hydrogen atom, and the other the hydrogen of the hydroxyl group. By the action of sodium sulphite, the iodine atom can be removed from the OI group and the monoiodo substituted phenol formed. A different type of iodophenol is produced by the action of iodine in potassium iodide on phenyl-carboxylic acids in presence of an exact amount of alkali, whereby the COOH group is removed by loss of CO_2, and iodine enters the nucleus but not the hydroxyl group.

The iodoxyl compounds, containing the group —OI, possess marked antiseptic and anti-syphilitic properties, but the iodophenols containing iodine in the nucleus do not differ very much from the phenols in their action, but they are sometimes used, as they are convenient crystalline insoluble antiseptics.

The best known of the iodoxyl compounds is Aristol, dithymol-diodide

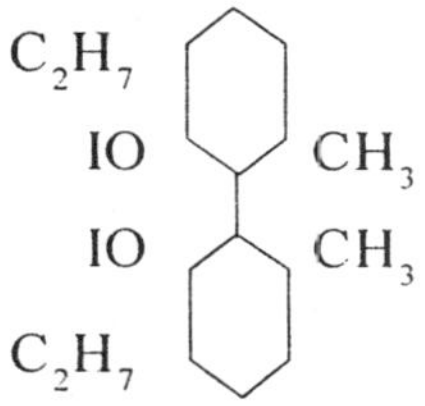

$$\begin{array}{ccc} C_2H_7 & \bigcirc & \\ IO & | & CH_3 \\ IO & | & CH_3 \\ C_2H_7 & \bigcirc & \end{array}$$

which is a favourite antiseptic, but suffers from the drawback of being rather unstable and expensive.

Europhen—

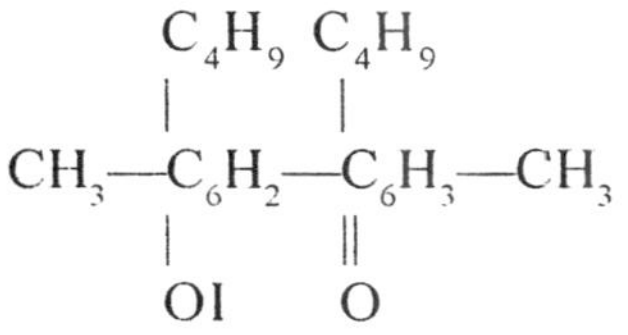

$$\begin{array}{c} \quad\quad C_4H_9 \quad C_4H_9 \\ \quad\quad | \quad\quad\quad | \\ CH_3—C_6H_2—C_6H_3—CH_3 \\ \quad\quad | \quad\quad\quad \| \\ \quad\quad OI \quad\quad O \end{array}$$

is another substance of this type which serves as an odourless and non-irritant iodoform substitute.

Tri-iodo-cresol, known by the fancy-name of Losophan, is an example of a phenolic compound in which the hydrogen atoms of the nucleus are replaced by iodine. These substances possess the antiseptic properties of phenols in an enhanced degree, but they cannot be regarded as iodoform substitutes, as they do not possess the characteristic iodine action, and are also too corrosive in their action on the skin.

Tetra-iodo-phenolphthalein has been prepared by many different methods. Phenolphthalein itself

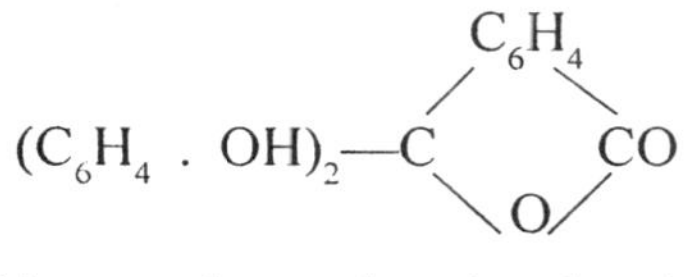

$$(C_6H_4 \cdot OH)_2—C\begin{array}{c} \diagup C_6H_4 \diagdown \\ \diagdown O \diagup \end{array}CO$$

when taken internally has a mild purgative action, but is otherwise physiologically inert ; but the tetra-iodo-derivative

$$(C_6H_2I_2OH)_2\text{—}C\overset{C_6H_4}{\underset{O}{\diagdown\!\!\diagup}}CO$$

called Nosophen, has powerful antiseptic properties. As this substance contains two hydroxyl groups, salts can be prepared with the heavy metals such as zinc, iron, mercury, and bismuth, but these have not come into use, although they combine the antiseptic action of nosophen with the useful properties of the metal.

Isoform is the name that has been given to para-iodoxy-anisol, CH_3, a colourless insoluble powder that has been recommended as a dry antiseptic, and which is said to be especially valuable in the treatment of mercurial stomatitis. It is obtained by the oxidation of para-iodo-toluene, $CH_3\langle\quad\rangle I$, whereby $CH_3\langle\quad\rangle ICl_2$ or $CH_3\langle\quad\rangle IO$ is formed, either being easily oxidized to $CH_3\langle\quad\rangle IO_2$ Commercially, it is not met with in the pure state, but owing to its explosive nature is mixed with calcium phosphate or glycerol.

Sozoiodol, $C_6H_2I_2$ (OH)(SO_3H), probably owes its antiseptic action partly to its strongly acid character. Its zinc and mercury salts are useful preparations for the administration of these metals.

Loretin is iodo-hydroxy-quinoline sulphonic acid

[Structure: quinoline ring with SO_3H, I, N, OH]

prepared by Glaus by the sulphonation of hydroxy-quinoline with fuming sulphuric acid, and treating the reaction-product with iodine. This substance is said to be an iodoform substitute free from unpleasant effects. Its sodium salt has been used in tuberculosis under the name of Griserin. Other similarly constituted compounds with the same kind of action have been prepared, and iodo-chlor-hydroxy-quinoline has been introduced as a non-poisonous iodoform substitute under the name of Vioform.

Substitutes for Alkaline Iodides

Another important class of iodine compounds includes those substances which have been prepared with the object of producing substitutes for the alkaline iodides. The latter are largely used in medicine, but are liable to cause unpleasant symptoms, to which the term " iodism " is given, and for this reason a very large number of drugs have been introduced with the object of furnishing substances which should possess the therapeutic value of the alkaline iodides without giving rise to the symptoms of iodism.

Iodoform has been used internally, being for the most part converted by the organism into alkaline iodides, but it has no real value in this respect. Compounds of iodine (and of bromine) with fats which gradually split off their iodine appear to be very good substitutes for the alkaline iodides, compared with which they are said to possess many advantages.

lodipin is one of the best -known substances of this type, and is a combination of iodine and sesame-oil.

Iothion is di-iodo-hydroxypropane, $C_3H_5I_2OH$, and is used in the form of ointments as a substitute for tincture of iodine.

Iodival is mon-iodo-isovaleryl-urea

$$\begin{matrix} CH_3 \\ CH_3 \end{matrix} \!\!> CH—CHI—CO—NH—CO—NH_2,$$

and is administered internally as a substitute for iodine and the iodides. It is said to be specially useful in nervous diseases, and in syphilis.

Para-iodo-guaiacol is decomposed into its components in the intestine, and has been recommended as a substitute for the iodides.

The thyroid gland contains iodine in organic combination, and various attempts have been made to prepare a similar substance to the so-called iodo-thyrine by combining iodine with proteins. The products thus obtained do not resemble iodo-thyrine in their physiological effects, but they form useful substitutes for the alkaline iodides. Compounds of iodine with proteins are readily obtained by the addition of iodine in potassium iodide solution to aqueous protein solutions, or by adding finely powdered iodine to warm aqueous solutions, and coagulating the product with acetic acid. One of the best-known compounds of this type is lodoglidine, a compound of iodine with gliadin, the vegetable protein obtained from wheat. It behaves in a similar manner to potassium iodide, but is said also to increase the metabolism, thus resembling iodo-thyrine in this respect. Another compound of iodine and protein is lodalbin, which is also used as a substitute for the alkaline iodides.

Bromine Derivatives

As in the case of iodine, the entrance of chlorine or bromine into aromatic compounds increases their antiseptic power, but the chloroand bromo-phenols have so strong an irritant action that derivatives of that type have found no therapeutic application.

On the other hand, the alkaline bromides are of very great medicinal value, being especially important in the treatment of certain nervous disorders. Potassium bromide has some drawbacks precisely similar to those attending the use of potassium iodide, and so efforts have been made to find substitutes which will possess the useful sedative properties of potassium bromide without its drawbacks.

Bromipin is analogous to iodipin, and is said to be a valuable substitute for potassium bromide. Valerobromine, is said to be free from the drawbacks of potassium bromide, while retaining its sedative properties. It is also said to derive a sedative action from the presence of the valerianic acid grouping, but this seems doubtful. It is obtained by the action of bromine on valerianic acid, whereby bromo-valerianic bromide is obtained, which is then hydrolyzed with water and neutralized, giving Valerobromine

$$\begin{matrix} CH_3 \\ CH_3 \end{matrix}\!\!>CH—CH_2—CO\,.\,OH \rightarrow \begin{matrix} CH_3 \\ CH_3 \end{matrix}\!\!>CH—CNBr—CO—BR$$

$$\begin{matrix} CH_3 \\ CH_3 \end{matrix}\!\!>CH—CNBr—CO—ONa$$

Other derivatives of a-bromoisovalerianic acid which are used as nervous sedatives are its ester with borneol $C_4H_8BrCOOC_{10}H_{17}$, and its urea derivative

$$C_4H_8BrCOOC_{10}H_{17},$$

$$\begin{matrix} CH_3 \\ CH_3 \end{matrix}\!\!>CH—CHBr—CO—NH—CO—NH_2$$

The former is known as Brovalol and Eubornyl. The latter, which is exactly analogous to lodival in constitution, is known by the names of Bromurul and Dormigene.

Another bromine derivative of urea which is used as a safe hypnotic is Adalin, bromdiethylacetylurea

$$\begin{matrix} C_2H_5 \\ C_2H_5 \end{matrix}\!\!>CBr\,.\,CO\,.\,NH\,.\,CO\,.\,NH_2.$$

In this compound, the mild hypnotic action of the —CBr. CO. NH. CO. NH_2 group is probably enhanced by the presence of the two ethyl groups united to it.

Chapter 10

Ethylamine Derivatives

Ethylamine is a chemical compound with the formula $CH_3CH_2NH_2$. It has a strong ammonia-like odor. It is miscible with virtually all solvents and is considered to be a weak base, as is typical for amines. Ethylamine is widely used in chemical industry and organic synthesis. Ethylamine, like some other amines, has the unusual property of dissolving lithium metal to give the ion and the solvated electron. Evaporation of these solutions, gives back lithium metal. Such solutions are used for the reduction of unsaturated organic compounds, such as naphthalenes and alkynes.

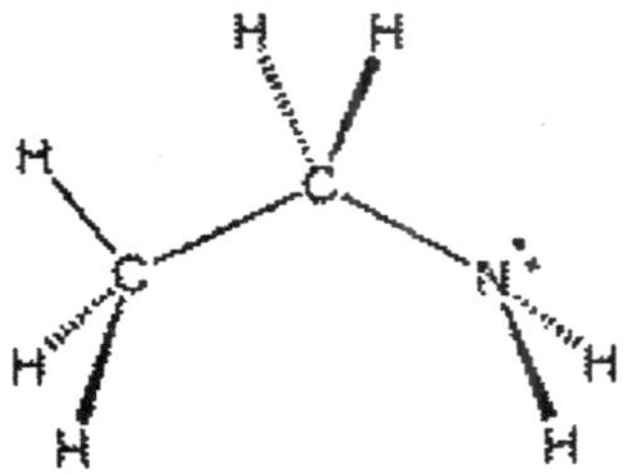

Ethylamine

A large number of ethylamine derivatives, possessing powerful physiological action, have been isolated from various plant and animal sources. Many of these are derivatives of para-hydroxyphenylethylamine, HO⟨ ⟩CH_2—CH_2—NH_2 and have the property of producing effects very similar to those produced by stimulation of the sympathetic nervous system, one of the most notable of these being rise of blood-pressure.

Para-hydroxyphenylethylamine is present in aqueous extracts of ergot, and other closely related substances are also found in various plants, but the most important compound of this class is adrenaline, HO⟨ ⟩CH(OH)—CH_2—NH—CH_3 an active HO principle which has
HO
been isolated from the suprarenal glands. It was first obtained in the impure condition by Abel and Crawford, in 1897, and in a more pure condition as the benzoyl derivative by Abel, in 1899. It was called epinephrine by these investigators, and it was also isolated by von Fiirth,

who gave it the name suprarenine. The name adrenaline was first given to it by Takamine, who was also the first to obtain it in a crystalline condition. He proposed the formula $C_{10}H_{13}O_3N$ from the results of his analyses, and also made the first observa tions throwing some light on its constitution, as he obtained substances from it supposed to be catechol and protocatechuic acid, $C_6H_4(OH)(OH)$ and $(OH)(OH)C_6H_3COOH$.

Shortly afterwards it was isolated by Aldrich by means of a somewhat different method, and he gave to it the formula $C_9H_{13}O_3N$, which is now universally accepted. Abel still pre ferred the formula $C_{10}H_{13}O_3N, 1/2H_2O$, which corresponds to almost identically the same composition as $C_9H_{13}O_3N$, but he brought forward no evidence to show that it contained water of crystallization. Pauly confirmed the formula $C_9H_{13}O_3N$, and suggested that it contained the groupings—

$$-\underset{OH}{\overset{H}{C}}-CH_2-NH-CH_2 \quad \text{or} \quad -\underset{CH_2-OH}{\overset{H}{C}}-N\begin{matrix} H \\ CH_3 \end{matrix}$$

Von Fiirth had already suggested the formula $(HO)_2C_6H_3$- $[C_2H_3(OH)NH—CH_3]$, as he had found that it did not con tain a methoxy group, and that it yielded methylamine salts on treatment with concentrated acids. Jowett confirmed the formula $C_9H_{13}O_3N$, and from the products of a potash fusion isolated a small quantity of a substance believed to be protocatechuic acid, but obtained more positive evidence by methylation and subsequent oxidation, by which means veratric acid, $C_6H_3(OCH_3)(OCH_3)COOH$, was obtained, and he therefore suggested the COOH alternative formulas—

I. $(OH)(OH)C_6H_3-\underset{|}{C}H(OH)-CH_2-NH-CH_3$ (H—C—OH, CH₂—NH—CH₃)

II. $(OH)(OH)C_6H_3-CH_2-CH(OH)-NH-CH_3$ (CH₂, HO—CH—NH—CH₃)

III. $(OH)(OH)C_6H_3-CH(NH-CH_3)-CH_2-OH$ (H—C—NH—CH₃, CH₂—OH)

with a preference in favour of I., which is the formula at present accepted as representing the constitution of adrenaline.

In 1904 Friedmann showed that the benzenesulphonyl derivative of adrenaline on oxidation yielded the corresponding derivative of a ketone adrenalone, which he also obtained by the action of methylamine on chloracetyl-catechol—

$$\text{HO}\langle\text{C}_6\text{H}_3\rangle\text{—CO—CH}_2\text{—Cl, (HO)}$$

and which, therefore, has the constitution—

$$\text{HO}\langle\text{C}_6\text{H}_3\rangle\text{—CO—CH}_2\text{—NH—CH}_2\text{, (HO)}$$

and this confirms Jowett's formula fora adrenaline.

From this time onwards, the chemical investigations of adrenaline have had as their chief object the synthesis of this important and valuable substance. The. adrenalone of Friedmann can be obtained by the action of an excess of methylamine on chloracetyl-catechol, the resultant base being precipitated by ammonia.

OH, OH (ring) — Catechol

$\xrightarrow[\text{Chloracetyl-chloride}]{\text{Cl . CO. CH}_2\text{Cl}}$

OH, OH (ring), CO—CH_2—Cl — *Chloracetyl-catechol*

$\xrightarrow[\text{Methylamine}]{\text{H—NH—CH}_3}$

OH, OH (ring), CO—CH_2—NH—CH_3

Methylamino-aceto-catechol
(adrenalone)

The reduction of this ketone to the corresponding secondary alcohol (adrenaline) presented great difficulty, but it was suc cessfully accomplished by electrolytic reduction or by the action of aluminium amalgam on the sparingly soluble sulphate of the ketonic base—

OH, OH (ring), CO—CH_2—NH—CH_3 $\longrightarrow$ OH, OH (ring), H—C(OH)—CH_2—NH—CH_3

Another method for the synthesis of adrenaline starts from protocatechuic aldehyde. This on treatment with hydrocyanic acid yields the cyanhydrin, which is then reduced to an amine which can be converted to adrenaline by methylation. This method is not used commercially—

The product thus obtained is the racemic form of adrenaline, which is less active than the naturally occurring Isevo-rotatory variety. The resolution of racemic adrenaline into its two optically active components has been accomplished by means of the fractional precipitation of its salts with tartaric acid, and also by means of Penicillium glaucum. The synthesis of natural adrenaline is therefore complete. The methylene and dimethyl ethers of adrenaline have been synthesized by means of a different method, but unfortunately these could not be transformed into adrenaline itself—

Piperonal (CHO; O, CH_2, O) —CH_2—Mg—I→ (CH . OH | CH_3) —heating→ (CH || CH_2) —Bromine→ (CHBr | CH_2Br) —aqueous acetone→

The methylene ether of adrenaline thus formed has a physiological action similar to that of adrenaline itself. Similarly

(O, CH_2, O; CH(OH)—CH_2Br) —methylamine→ (O, CH_2, O; CH(OH)—CH_2—NH—CH_2)

veratradehyyde, (OCH_3, OCH_3, CHO) was transformed into the dimethyl-ether of adrenaline, (OCH_3, OCH_3, CH(OH)—CH_2—NH—CH_3)

although this was only obtained as a syrup. Although by the action of phosphorus pentachloride on the methylene ether, Barger and Jowett were unable to obtain a chlorinated compound which should yield the dihydroxy-compound with water—

$$C_6H_3(O{-}CH_2{-}O)\text{—}CHBr\text{—}CH_2Br \rightarrow C_6H_3(O{-}CCl_2{-}O)\text{—}CHBr\text{—}CH_2Br \rightarrow C_6H_3(OH)_2\text{—}CH(OH)\text{—}CH_2Br$$

Bottcher states that an excess of phosphorus pentachloride gives, $C_6H_3(OH)_2\text{—}CH(OH)\text{—}CH_2Br$

which with aqueous methylamine yields a physiologically active base, supposed to be adrenaline. Pauly, however, considers that Bottcher has not proved the formation of adrenaline by this method.

Considerable light has been thrown on this reaction by Mannich, who showed that a mere direct replacement of the halogen by the methylamino group does not take place in chloror bromhydrins of the types—

$$CH_2\langle O_2 \rangle C_6H_3\text{—}CH(OH\text{—}CH_2Br$$

and $(CH_3O)_2\ C_6\ H_3$—CH(OH)—CH_2 Br. Under the influence of methylamine, hydrobromic or hydrochloric acid is lost, result ing in the formation of an unstable oxide of the type—

$$(CH_3O)_2C_6H_2\text{—}CH\text{—}CH_2 \text{ (with O bridging CH and } CH_2),$$

which then reacts with more methylamine to form bases of the adrenaline series. e.g,

$$(CH_3O)_2C_6H_2\text{—}CH\langle {OH \atop CH_2\text{—}NH\text{—}CH_2}$$

or of the isoadrenaline series. e.g,

$$(CH_3O)_2C_6H_3—CH\begin{matrix} ⁄ NH—CH_3 \\ \backslash CH_2—OH \end{matrix}$$

Mannich and Jacobsohn succeeded in preparing the methyl ene and dimethyl ethers of adrenaline in a pure state, and they also obtained the methyl ether of β-methyl-adrenaline by the action of bromine on methyl-isoeugenole—

$$\underset{CH_3O}{CH_3O\langle\;\;\rangle}— CH = CH—CH_3 \rightarrow \underset{CH_3O}{CH_3O\langle\;\;\rangle} CHBr—CHBr—CH_3$$

$$\rightarrow \underset{CH_3O}{CH_3O\langle\;\;\rangle}—CH(OH)—CHBr—CH_3 \rightarrow CH_3O\langle\;\;\rangle—CH(OH)—\overset{CH_3}{\overset{|}{CH}}—NH—CH_3$$

This then gives β-methyl-adrenaline

$$\underset{HO}{HO\langle\;\;\rangle}\; CH(OH)—\underset{CH_3}{\underset{|}{CH}}—NH—CH_3$$

on treatment with hydriodic acid. According to Robert this substance has not the physiological action of adrenaline.

Meanwhile an extended physiological investigation of adrenaline was carried out by Elliott, who confirmed and extended the earlier work which Schafer and Oliver, and Langley had carried out with extracts of the gland. In general, the effect of adrenaline on any structure is similar to that which follows excitation of the sympathetic nerves supplying the tissue. If administered subcutaneously or locally to mucous surfaces, it causes very marked constriction of the blood-vessels, and so arrests bleeding. In moderate doses, subcutaneous injections produce no general systemic effect, very much larger doses being needed to raise the blood-pressure by this means.

It is of interest to note that there is a very marked difference in the physiological activity of the naturally occurring laevo-adrenaline and the dextro-isomeride, this being one of the very best examples of difference in physiological behaviour between stereo-isomerides. It was shown by Cushny that natural laevo-adrenaline acts approximately twice as strongly in increasing the bloodpressure as synthetic racemic-adrenaline, and presumably also on the other organs affected by adrenaline. From this it was inferred that *d*-adrenaline was inactive on these tissues, and this view was apparently confirmed by an examination of partly isolated *d*-adrenaline. The general character of these results was confirmed by other workers, but subsequently Cushny showed that *d*-adrenaline is not quite inert in this respect, but has an activity about one-twelfth of that of *l*-adrenaline, so that the action of *r*-adrenaline is almost entirely due to the *l*-adrenaline contained in it.

The synthetic racemic r ˙xture can be completely converted into the desired active component. Either the dextro or the laevo compound can be racemized by treatment with acids ; the inactive mixture is then resolved into its components, and the isomeride not required can then be again racemized and subsequently resolved, this process being repeated as often as required. The authors of this method state that the dextro compound also possesses valuable therapeutic properties.

The therapeutic uses of adrenaline are very numerous, and a mere list of the references to the literature of this subject would fill pages ; an account of many of these publications is given in Merck's Eeports during the last fifteen yearsAdrenaline is largely used in conjunction with cocaine and novocaine, as it produces a localized anaemia, and so checks bleeding, and it also appears to neutralize the toxic effect of cocaine. The action of adrenaline in producing ischaemia finds application in a variety of complaints, hay-fever being an example.

Adrenaline is also met with under the names of hemisine, adrenine, epinephrine, suprarenine, etc.

Recently various substances chemically related to adrena line, and to a large extent resembling it in their physiological action, have been isolated from various plant and animal sources. Of these, para-hydroxyphenylethylamine, which may be regarded as the mother substance of the series, is the most important. It was first prepared in small quantities

by heating tyrosine, $HO\langle\rangle CH_2—CH(NH_2)—COOH$ and it has since been obtained in small quantities from various animal sources. Putrid meat has for some time been known to produce a rise of blood-pressure (pressor action), and in 1909 it was found that this action was due to a number of amines, of which para-hydroxyphenylethylamine had the most powerful action. The amines which showed this action in a weaker degree were iso-amylamine, $(CH_3)_2$ CH—CH_2—CH_2 NH_2, and phenylethylamine. It is almost certain that these bases are produced in the process of putrefaction by loss of carbon dioxide from the corresponding amino acids :

Para-hydroxyphenylethylamine, $HO\langle\rangle CH_2—CH—NH_2$ (with NH_2 below), From tyrosine, HO CH_2—CH<(COOH, NH_2)

Phenylethylamine, $C_6H_5—CH_2—CH_2—NH_2$, from phenyl-alanine, $C_6H_5—CH_2—CH$<(COOH, NH_2)

And isoamylamine, $(CH_3)CH—CH_2—CH_2—NH_2$, from leucine, $(CH_3)_2—CH—CH_2—CH$<(COOH, NH_2)

Putrid placental extracts had also been shown to produce a pressor action, and para-hydroxyphenylethylamine has been isolated from such extracts.

The drug ergot has long been used on account of its thera peutic properties, but it is very variable in its activity, and requires to be physiologically standardized owing to the un satisfactory state of our knowledge of the alkaloids present in ergot. Recently, however, a great deal of light has been thrown on the chemistry of ergot by the work of Barger, Dale, and others. The amorphous alkaloid ergotoxine is physiologically active, but it does not possess all the character istics of the action of ergot, and the small amount of this alkaloid present in most pharmacopoeial preparations of ergot led to the postulation of an active principle soluble in water.

The physiological properties of para-hydroxyphenylethylamine suggested that it might be the expected active principle, and this expectation was realized when it was shown to be present in aqueous extracts of ergot, and to be the chief cause of their physiological action. Not only has this substance been isolated from ergot, but synthetic methods for this preparation have been devised which have rendered practicable its introduction into therapeutics. These syntheses will be discussed together with those of other compounds of this series, but mention should be made at this point of another base isolated from ergot. This is β-iminazolyl-ethylamine

```
NH—CH
|     ≫C—CH2—CH2—NH2,
CH=N
```

and it is formed from the ammo-acid, histidine

```
                      NH2
NH—CH ≫               |
|         C—CH2—CH—COOH
CH=N
```

by loss of carbon dioxide in just the same way as p.-hydroxyphenylethylamine is formed from tyrosine.

The action of ergot in producing gangrene of the cock's comb is regarded as being due to the alkaloid ergotoxine, and the rise in blood pressure is attributed to p.-hydroxyphenylethylamine, while the powerful action of ergot in stimulating the isolated uterus to tonic contraction is caused by β-iminazolyl-ethylamine. This substance, although it has a very powerful physiological action, differs from all the other active derivatives of ethylamine described in this chapter in causing a lowering instead of a rise of blood-pressure.

It has been introduced into medicine under the names of Histamine and Ergamine.

The original method for the synthesis of p.-hydroxy phenylethylamine was by the reduction of p.-hydroxyphenyl acetonitrile, HO⟨ ⟩CH_2—CNand subsequently two other methods of synthesis were described. One of these is by the nitration of benzoyl-phenylethylamine, reduction of the result ing para-nitro compound to the amine, which yields the ben zoyl derivative of the desired product when diazotized in boiling solution. This benzoyl derivative is then hydrolyzed—

$$\langle\rangle CH_2—CH_2—NH—COC_6H_5 \rightarrow NO_2 \langle\rangle CH_2—CH_2—NH—COC_6H_5$$

$$\rightarrow NH_2 \langle\rangle CH_2—CH_2—NH—COC_6H_5$$

$$\downarrow HNO_2$$

$$HO \langle\rangle CH_2—CH_3—NH—COC_6H_5$$

$$HO \langle\rangle CH_2—CH_2—NH_2 \leftarrow$$

The other method starts from anisic aldehyde, $CH_3O\langle\rangle CHO$, which, by the method of Perkin and Robinson, yields the acid, $CH_3O\langle\rangle CH_2—CH_2—COOH$. This is then converted into the chloride, and thence into the amide, $CH_3\langle\rangle CH_2—CH_2—CO—NH_2$, which by the Hofmann reaction is made to yield the amine $CH_3O\langle\rangle CH_2—CH_2—COOH$. By means of strong hydrobromic acid, the methoxy group is converted into hydroxy, giving para-hydroxyphenylethylamine.

p-Hydroxyphenylethylamine has also been prepared from anisaldehyde by Rosenmund, who condensed the latter substance with nitromethane to prepare β-nitro-p.-methoxystyrene. This is then reduced to *p*-methoxyphenylethylamine, which is demethylated with hydriodic acid—

$$CH_3O \langle\rangle CH(O+H_2)\ CH—NO_2 = CH_3O \langle\rangle CH — CH—NO_2$$

$$\downarrow$$

$$CH_3O\langle\rangle CH_2—CH_2—NH_2$$

$$\downarrow$$

$$HO \langle\rangle CH_2—CH_2\ —NH_2$$

This substance has been introduced into practice under the name of " *Tyr amine*" as the chief active pressor principle of aqueous extracts of ergot, and being a pure chemical compound it has the advantage in being certain and uniform in its action.

Tyramine and adrenaline are representatives of two im portant subdivisions of these pressor amines, the former being the simplest member of those derivatives of phenylethylamine containing one phenolic hydroxyl group in the para position to the ethylamine group, and the latter being a representative of the compounds containing two phenolic hydroxyl groups in the 3-4 position. Many other compounds of this type have been obtained by synthetic methods, and their physiological action determined. The syntheses of some of the most important will now be considered, and the physiological action dealt with subsequently.

The simplest member of the adrenaline series is 3-4 dihydroxyphenylethylamine,

$$HO \langle\rangle CH_2—CH_2—NH_2.$$
$$HO$$

This compound has been synthesized from eugenol methyl ether, which by oxidation with ozone in benzene solution in presence of water gives the aldehyde of homoveratric acid. This is converted into the oxime, which on reduction with sodium amalgam and glacial acetic acid yields the dimethyl ether of the desired compound. This ether is then converted into the dihydroxy compound with hydriodic acid in the usual manner.

OCH_2, OCH_2 (ring) — CH_2—CH=CH_2 $\xrightarrow{\text{Ozone}}$ OCH_2, OCH_2 (ring) — CH_2—CHO $\xrightarrow{NH_2OH}$ OCH_2, OCH_2 (ring) — CH_2—CH=NOH

$\xrightarrow{\text{Sodium amalgam}}$ OCH_2, OCH_2 (ring) — CH_2—CH_2—CH_2—NH_2 $\xrightarrow{HI}$ OH, OH (ring) — CH_2—CH_2—NH_2

By the reduction of the oximes of other aldehydes and ketones, other members of these two series may be obtained. For example, homoanisic aldehyde, CH_30⟨ring⟩CH_2CHO, gives p.-hydroxyphenylethylamine ; and para-m'ethoxybenzyl methyl-ketone, CH_3O⟨ring⟩CH_2—CO—CH_3, gives, when treated in the same way, p.-hydroxyphenyl-isopropylamine—

HO⟨ring⟩—CH_2—CH(CH_3)—NH_2

3-4 dihydroxyphenylethylamine (A) differs from adrenaline by the absence of a methyl group attached to the nitrogen and of an aliphatic hydroxyl group on the side chain. An interesting intermediate compound between these two is 3-4 dihydroxyphenylethylmethylamine (B)—

(A): OH, OH (ring) — CH_2—CH_2—NH_2

(B): OH, OH (ring) — CH_2—CH_2—NH—CH_3

Adrenaline: OH, OH (ring) — CH—OH—CH_2—NH—CH_3

This, it will be observed, differs from adrenaline only by the absence of the aliphatic hydroxyl group, and it is of interest in being a connecting link between that substance and the pressor derivatives obtained from isoquinoline alkaloids. It differs very little from adrenaline in the qualitative nature of its physiological action, but the rise of blood-pressure, although not so intense, is more prolonged. This substance has been introduced into therapeutics under the name of "*Epinine*". It was obtained from l-keto-6-7 dimethoxy-2 methyl-tetrahydroisoquinoline by heating it with hydrochloric acid at 170-175°. The reaction probably takes place in the following stages :

CH_2, CH_3O, CH_2, CH_3O, N—CH_3, CO ⟶ CH_2, HO, CH_2, HO, N—CH_3, CO

⟶ HO, CH_2, HO, NH, CO, OH, CH_3 ⟶ CH_2, CH_2, HO, HO, NH, H, CH_3

The corresponding propyl and ethyl derivatives were prepared in a precisely similar fashion. The other derivatives of 3-4 dihydroxyphenylethylamine to be considered, are mostly prepared by the same method as that used for preparing the ketone adrenalone and adrenaline itself. In that method, which has been already described, the methylamine used in the last stage of the synthesis of adrena lone may be replaced by other amines or by ammonia. In this way Dakin and Stolz obtained the ketone

HO
HO$\langle\quad\rangle$—CO—CH_2—NH_2

and its reduction product—

HO
HO$\langle\quad\rangle$—CH—CH_2—NH_2
|
CH

and also the substituted ketones of the type $(HO)_2C_6H_3$. CO . CH_2 . NR_1R_2. Before considering the physiological action of these substances with two phenolic hydroxyl groups, attention must be given to those synthetic products containing one phenolic hydroxyl group in the para position (i.e. those related to tyramine). Most of the compounds of this type resemble tyramine in being derivatives of phenylethylamine, C_6H_5—CH_2—CH_2—NH_2, rather than of phenylethanolamine, C_6H_5—CH(OH)—CH_2—NH_2, but one member of the latter class, and also the corresponding ketone, have been prepared and physiologically examined. These are p.-hydroxyphenylethanolamine, HO⟨ ⟩CH(OH)—CH_2—N_2, and p.-hydroxy-w-amino-acetophenone, HO⟨ ⟩—CO—CH_2—NH_2, which could not be prepared by the method employed in the adrenaline series by using phenol instead of catechol, but which have been prepared instead by the following synthesis:

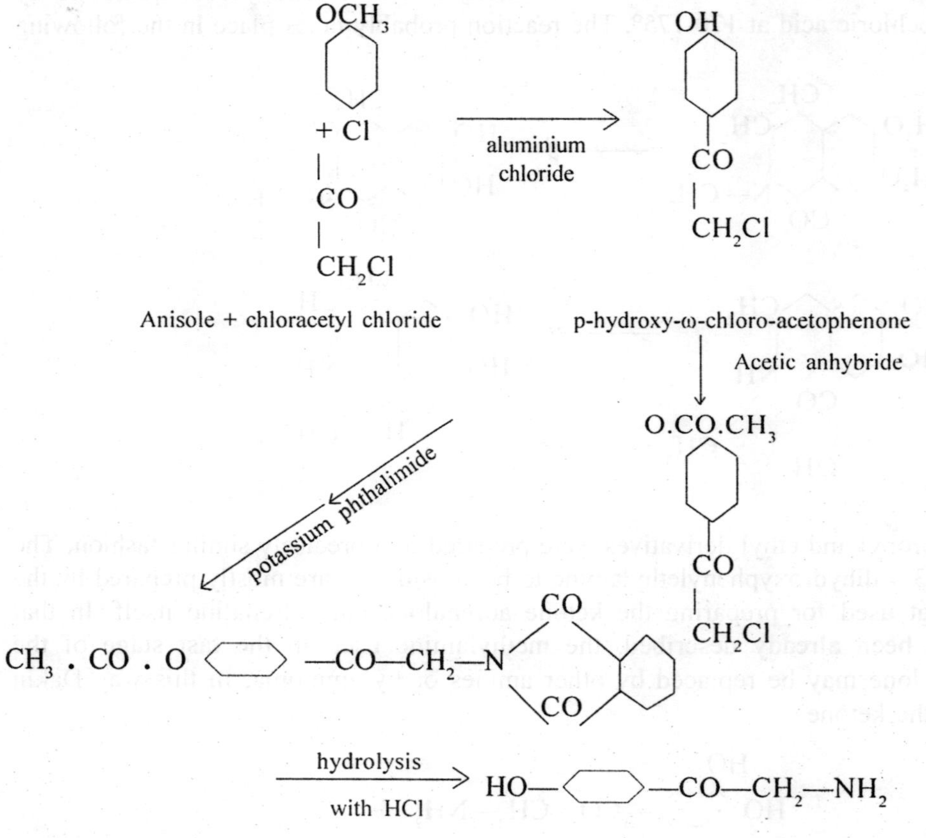

OCH_3-C₆H₄ + Cl—CO—CH_2Cl —(aluminium chloride)→ OH-C₆H₄-CO—CH_2Cl

Anisole + chloracetyl chloride → p-hydroxy-ω-chloro-acetophenone

↓ Acetic anhybride

$O.CO.CH_3$-C₆H₄-CO—CH_2Cl

↓ potassium phthalimide

CH_3 . CO . O⟨ ⟩—CO—CH_2—N(CO)₂C₆H₄

—(hydrolysis with HCl)→ HO—⟨ ⟩—CO—CH_2—NH_2

p.-Hydroxy-w-chloro-acetophenone could not be condensed with ammonia, but its acetate could be condensed with potassium phthalimide, and the resultant compound, on removal of the phthalic acid by hydrolysis, yields the desired compound.

The ketone then gives the corresponding secondary alcohol, HO⟨ ⟩CH(OH)—CH_2—NH_2, by reduction with sodium and alcohol.

The other compounds of this series which have been physiologically examined differ from tyramine only in having one or more of the hydrogen atoms of the amine group replaced by alkyl groups. Of these, hordenine, HO⟨ ⟩CH_2—CH_2—$N(CH_3)_2$, is an alkaloid, first obtained from barley germs, but it cannot be obtained from tyramine by methylation, as on attempting to do this, no compound but the completely methylated quaternary substance, hordenine-methiodide, $HO.C_6H_4.CH_2CH_2 . N(CH_3)_4$, could be isolated. A fortiori this method of direct methylation is inapplicable to the preparation of p.-hydroxyphenylethylmethylamine, $HO.C_6H_4.CH_2—CH_2—NH.CH_3$. Both these compounds have, however, been prepared by synthetic methods. The synthesis of hordenine was first accomplished according to the following scheme, which is self-explanatory:

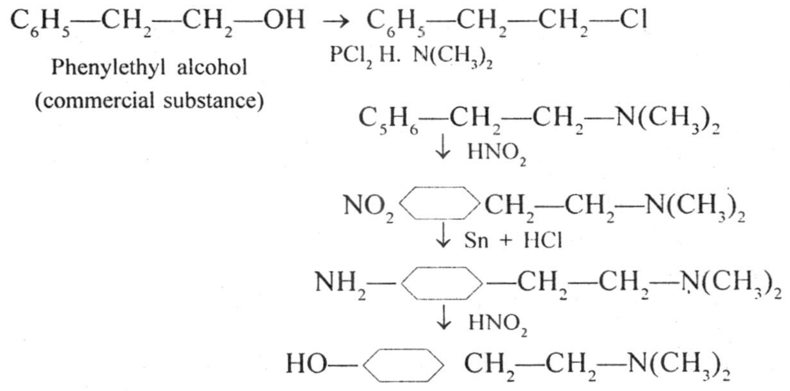

$$C_6H_5—CH_2—CH_2—OH \rightarrow C_6H_5—CH_2—CH_2—Cl$$

Phenylethyl alcohol (commercial substance)

PCl_2 H. $N(CH_3)_2$

$$C_5H_6—CH_2—CH_2—N(CH_3)_2$$

$\downarrow$ HNO_2

$$NO_2\langle\;\rangle CH_2—CH_2—N(CH_3)_2$$

$\downarrow$ Sn + HCl

$$NH_2—\langle\;\rangle—CH_2—CH_2—N(CH_3)_2$$

$\downarrow$ HNO_2

$$HO—\langle\;\rangle\; CH_2—CH_2—N(CH_3)_2$$

Very shortly afterwards, Bosenmund obtained hordenine by the direct methylation of p.-methoxyphenylethylamine with alcoholic potash and methyl iodide, and separation of the primary, secondary, tertiary, and quaternary compounds thus obtained.

The methoxy group is then converted into hydroxyl with hydriodic acid. Hordenine was also obtained by this investigator by the action of hydriodic acid on p.-methoxy phenyl-trimethylammonium iodide—

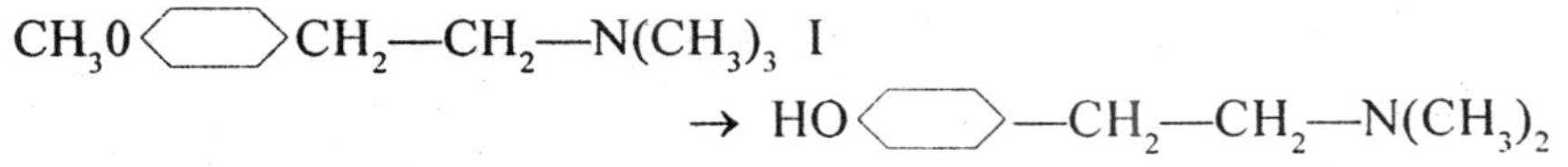

$$CH_30\langle\;\rangle CH_2—CH_2—N(CH_3)_3\ I$$

$$\rightarrow HO\langle\;\rangle—CH_2—CH_2—N(CH_3)_2$$

Hordenine is manufactured by the methylation of p.-hydroxyphenylethylamine with methyl chloride (CH_3Cl).

The p.-hydroxyphenylethylmethylamine— $HO\langle\;\rangle CH_2—CH_2—NH—CH_3$ and p.-hydroxyphenylethylethylamine— $HO\langle\;\rangle CH_2—CH_2—NH—C_2H_5$ were prepared by the methylation and ethylation respectively of the acetyl or benzene-sulphonyl derivatives of p.-methoxyphenylethylamine, the former by both methods, the latter only by means of the benzene-sulphonyl derivatives. p.-Methoxyphenylethylamine is prepared as indicated on pages 139-140, and the rest of the synthesis is easily understood from the accompanying scheme:—

$$CH_3O\langle\;\rangle CH_2—CH_2—NH_2 \rightarrow \text{(acetic anhydride)} \rightarrow$$

$$CH_3O\langle\;\rangle CH_2—CH_2—NH—CO.CH_3\ \text{(I)}$$

or ↘ Benzene-sulphonyl chloride

$$CH_3O\langle\;\rangle CH_2—CH_2—NH—SO_2—C_6H_5\ \text{(II)}$$

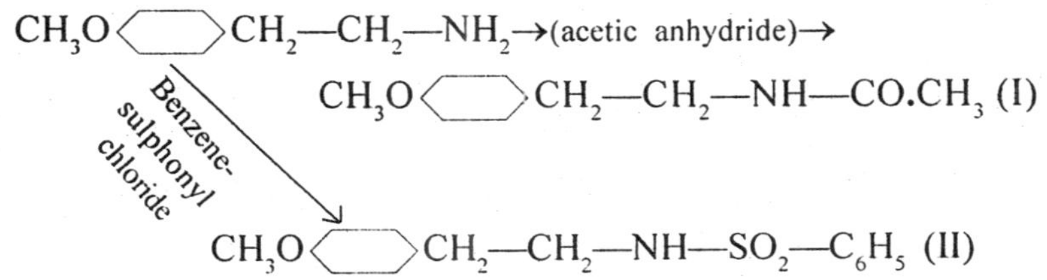

In the case of the acetyl derivative (I.), the substance (III.)— is formed by 'the action of sodium and methyl iodide, and this on treatment Hydrolyzed with HI gives (IV.), which then loses its acetyl group when hydrolyzed with concentrated HC1 in sealed tubes. The benzene-sulphonyl derivative is treated in a similar manner, and the corresponding ethyl-amine is obtained by using ethyl iodide instead of methyl iodide.

The relation between the chemical structure and the physiological (sympathomimetic) action of amines has formed the subject of an extended investigation by Barger and Dale. Dakin and Loewi and Meyer had examined many of the ketones of the general formula $(HO)_2C_6H_3$—CO—CH_2—NR_1R_2, and the cor responding secondary alcohols of the type—$(HO)_2C_6H_3$—CH(OH)—CH_2—NR_1R_2 and they found that in most cases, as for example with adrenaline itself, reduction to the secondary alcohol greatly increased the adrenaline-like action, but where R_1 and R_2 represent relatively complex radicles, Dakin found no such increase of activity on reduction. The physiological effects examined by Barger and Dale were not confined to rise of blood-pressure, but included dilatation of the pupil, action on the cat's uterus, etc., for the details of which the original paper should be consulted. Besides the various compounds that have been described in the preceding sections of this chapter, various other amines were examined.

Of the various aliphatic amines that were investigated, the only ones which w"ere found to produce a marked rise of bloodpressure were the higher open-chain primary amines, such as amylamine , C_5H_{11} . NH_2 , and hexylamine, C_6H_{13} . NH_2 . Of these, the normal compounds with unbranched side chains were found to be more active than the corresponding iso compounds with branched side chains. Trimethylamine, $N(CH_3)_3$, has practically no pressor action, and neither has tetraethylammonium iodide, $N(C_2H_5)_4I$. Cadaverine, NH_2 . $[CH_2]_5$. NH_2, the only diamine examined, was found to have the opposite action (depressor instead of pressor).

A large number of fatty-aromatic amines without a phenolic hydroxyl group were also investigated, and it was found that marked sympathomimetic action was associated only with those containing an amino group attached to the second carbon atom of the side chain. β-phenyl-ethylamine, C_6H_5—CH_2—CH_2—NH_2, for example, produces all the characteristic sympathomimetic effects. In the series containing two phenolic hydroxyl groups in the 3-4 position, the introduction of an aliphatic hydroxyl in the β-position of the side chain, and the methylation of the amino group have an important effect in intensifying the action, but in the present series, this does not hold, methyl-phenyl ethylamine, C_6H_5—CH_2—CH_2—NH . CH_3, phenylethanolamine, C_6H_5—CH(OH)—CH_2—NH_2, and methyl-phenylethanolamine, C_6H_5 CH(OH)—CH_2—NH. CH_3, differing but little in their action from phenylethylamine itself. *Ac.*-tetrahydro-β-naph thylamine—

```
              CH2
     ____    /
    /    \__/  CH—NH2
   |      |    |
    \____/\   CH2
           \ /
           CH2
```

which may be regarded as a derivative of this type, is more active than phenylethylamine in producing a rise of bloodpressure, but is less active than p.-hydroxyphenylethylamin^ (tyramine), the simplest member of the next series, namely:— *Amines with one Phenolic Hydroxyl Group*. The sources and preparation of many of these have already been men tioned. Methylation of the amino group produces very little increase in the activity, $HO—C_6H_4—CH_2—CH_2—NH. CH_3$, being only very slightly more active than the primary amine, while the ethyl derivative— $HO—C_6H_4—CH_2—CH_2—NH—C_2H_5$ is less active than either the methyl derivative or the parent sub stance. The tertiary base, $HO—C_6H_4—CH_2—CH_2—N(CH_3)_2$, which is the alkaloid hordenine, has a relatively very weak action, but the quaternary b#se, hordenine methiodide, $HO—C_6H_4—CH_2—CH_2—N(CH_3)_3I$, although it has no sym pathomimetic action, is of interest as it is one of the few exceptions to the rule of Crum Brown and Fraser that qua ternary bases have a curare-like action. Instead, its action resembles that of nicotine, which is a physiological antagonist of curare.

Destruction of the basic property is accompanied by loss of activity, acetyl p.-hydroxyphenylethylamine— $HO—C_6H_4—CH_2—CH_2—NH . CO . CH_3$, for example, being inactive. Tyrosine ethyl ester

$$HO—C_6H_4—CH_2—CH\begin{cases} COOC_2H_5 \\ NH_2 \end{cases}$$

is also inactive. A phenolic hydroxyl group in the 3 position is about as active as in the 4 position, the meta-hydroxy compound, closely resembling the corresponding para-derivative (tyramine), but in the 2 position it has no effect, ortho hydroxyphenylethylamine, , being no more active than phenylethylamine itself.

Amines with two Phenolic Hydroxyl Compounds

The following compounds in which the two hydroxyl groups are in the 3-4 position were tested:

(a) Derivatives of aceto-catechol (ketones).

(1) Amino-aceto-catechol,

(2) Methylamino-aceto-catechol

(3) Ethylamino-aceto-catechol

(4) Propylamino-aceto-catechol

(5) Trimethylamino-aceto-catechol chloride

(b) Derivatives of ethyl-catechol

(6) Amino-ethyl-catechol,

(7) Methylamino-ethyl-catechol

(8) Ethylamino-ethyl-catechol

(9) Propylami'no-ethyl-catechol

(10) Trimethylamino-ethyl-catechol chloride

(c) Derivatives of ethanol-catechol (secondary alcohols)

(11) Amino-ethanol-catechol

(12) Methylamino-ethanol-catechol (adrenaline)

And also

(13) 2-4 dihydroxy-w-amino-acetophenone

It was found that qatechol itself has no sympathomimetic action, although it produces a rise of blood-pressure. The following table shows the comparative strength of the action of the various amines in causing a rise of blood-pressure :

	Substance (numbered as before).	Ratio of activity
(1)	$(HO)_2C_6H_3—CO—CH_2—NH_2$	1-5
(2)	$(HO)_2C_8H_3—CO—CH_2—NH—CH_3$	----
(3)	$(HO)_2C_6H_3—CO—CH_9—NH—C_2H_5$,	2-25
(4)	$(HO)_2C_6H_3—CO—CH_2—NHC_3H_7$	0-25
(5)	$(HO)_2C_6H_3—CH_2—CH—NH_2$	1-0
(6)	$(HO)_2C_6H_3—CH_2—CH_2—NH—CH_3$	5-0
(7)	$(HO)_2C_6H_3—CH_2—CH_2—NH—C_2H_5$	1-5
(8)	$(HO)_2CfiH_3—CH_2—CH_2—NH—C_3H_7$	0-25
(9)	r-$(HO)_2C_6H_3—CH(OH)—CH_2—NH_2$	50
(10)	r-$(HO)_2C_6H_3—CH(OH)—CH_2—NH—CH_3$ (r-adrenaline)	35

The ratios shown are only approximate, and vary to some extent with the sensitiveness of the animal.

Chapter 11

Platinum Anti-tumour Drugs

Platinum amine coordination compounds are very interesting from a medical biological point of view. By studying the growth of *E.coli* bacteria under the influence of an electric field, Rosenberg *et al.*, initiated the renewed interest in these classical compounds. This field was generated between two "inert" platinum electrodes and aqueous NH_4Cl was used as an electrolyte. The bacteria showed a strong filamenteous growth which initially was not understood. However, subsequent experiments by the same group soon made clear that the filamenteous growth was not caused by the electric field, but by the presence of small amounts of dissolved Pt(II) and Pt(IV) compounds in the—corroding -NH_4Cl solution. Detailed investigations showed that the compounds present in solution were, among others, cis-$PtCl_2(NH)_2$, *trans*-$PtCl_2(NH_3)_2$ and some Pt(IV) compounds. In subsequent microbiological studies the cis-Pt(II) species—now called cisplatin—turned out to be the most active in causing filament formation. After these findings, tests were undertaken to study the growth-reducing capacity of cisplatin on animal tumours, such as Sarcoma 180 in Swiss white mice. The results were so positive—in many cases total regression of the tumours was observed—that clinical trials were soon thereafter scheduled and performed. The first studies, demonstrated a remarkable anti-tumour activity for cisplatin. However, large-scale applications had to wait for some more years, since toxic side effects were severe. To be mentioned are: nausea, vomiting, bone marrow toxicity, neurotoxicity and renal toxicity.

The clinical application of cisplatin has increased enormously, mainly as a result of improved administration procedures and its use in combination therapy, i.e. the simultaneous application of a variety of synergistic anti-tumour drugs. Nowadays, usually a dosage of about 100 mg of cisplatin per m^2 body surface area, dissolved in saline, is given intravenously, e.g. every month, by standard protocols. Final approval in the USA in 1979, has led to an increased number of applications of cisplatin and as a result it has become the leading and most widely used anticancer drug. The drug is also registered widely in many other countries. Complete remissions are obtained for testicular cancers in more than 85% of all treated patients, whereas high effectivity in the treatment of ovarian and bladder cancer and considerable activity in osteogenic sarcoma, head and neck cancer, endometrial and cervical cancer and non-small cell lung cancer have been reported. It should be noted that the corresponding *trans* isomer of cisplatin exhibits no anti-tumour activity, even though it has comparable chemical and binding properties to DNA. The working mechanism of cisplatin has been of great interest for a variety of research groups, involving chemists, biochemists, biologists and medical researchers. The first studies already made it clear that reaction of platinum compounds with nucleic acids plays an important role in the biological effect.

Several reviews have appeared during the last decade, describing a wide variety of aspects of platinum-anti-tumour compounds and of binding of metal compounds to nucleic acids and fragments In his recent study on the mechanism of action of platinum amine compounds, used as antitumour drugs., Jan Reedijk, Department of Chemistry, Gor1aeu Laboratories, State University Leiden, discussed more deeply about the relationship between biological acitivity and structure of the compounds, with special attention to the nature of the ligands coordinated to platinum. In this chapter we will discuss his findings on mechanism of action of platinum anti-tumour drugs in detail.

Derivatives

Soon after the first reports about the biological activity of the classical compound *cis*-PtCl2$(NH_3)_2$, an extensive investigation was started searching for analogs with similar activities and—hopefully—improved properties as a drug. Studies by several groups have resulted in a large variety of analogs with similar biological activities; some of these are used in extensive clinical trials in many countries. Indications are available that some or these analogs have superior properties in the treatment of tumours in patients who do not respond to cisplatin. Some of the most promising compounds, considered today, are $Pt(NH)_2$(CBDCA), (CBDCA = 1, 1-dicarboxylatocyclobutane), now also called paraplatin, and CHIP (cis-,cis-,*trans*-dichlorodihydroxobis(isopropylamine)platinum(IV).) Some examples are presented in Figure 1. Especially the CBDCA derivative is promising because of its low toxicity.

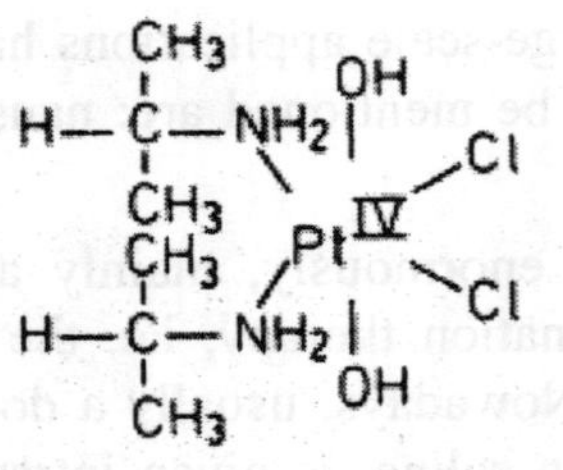

CHIP

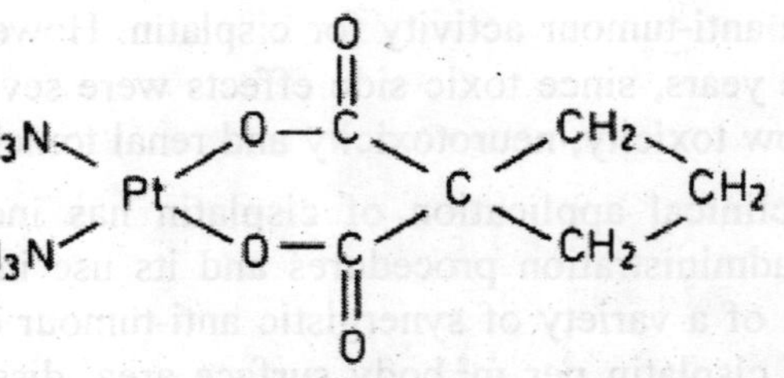

PARAPLATIN

Fig.l. Schematic structure of derivatives of cisplatin, used in clinical trials and tumour treatment.

The studies with a large number of Pt(II) and Pt(IV) compounds, prepared during the last decade, have made clear that Pt compounds with anti-tumour activity have to fulfil all of the following structural requirements:

1. The two amine ligands in the Pt-compound should be in a cis-orientation. In a didentate chelating ligand, this geometric requirement is automatically fulfilled. The general formulae should be cis-Pt(II)$X_2(Am)_2$ and cis-Pt(IV)$Y_2X_2(Am)_2$. However, the number of variations studied in the case of PtCIV) is still limited.
2. The ligands X, usually anions, should consist of groups that have intermediate binding strength to Pt(II), or that are—for other reasons—easily leaving (i.e. by enzymatic action). Examples are: Cl^-, SO_4^{2-}; citrate(3-), oxalate(2-) and other carboxylic acid residues. For the Pt(IV) compounds, the Y group is often OH^- (with the two I ligands in *trans* orientation).
3. The amine ligands, either monodentate or didentate, should have at least one N-H group, i.e. possess a hydrogen-bond donor function. All compounds with both amine ligands lacking such a H-bond donor property, were found to be inactive. The role of this N-H group in the biological activity, however, is far from being understood. It could be either kinetic (i.e. play a role in the approach of the DNA), or thermodynamic (e.g. give an additional (de)stabilisation after binding to the biological target DNA; vide infra). However, also steric effects and/or a role in transport through the cell wall cannot be excluded.

Knowledge of these basic requirements for an active drug, narrows the area in which one could—successfully—search for new anti-tumour active platinum compounds. Up till now, too little is known about the origin of the sometimes severe side effects. These effects cannot yet be predicted on the basis of molecular geometry of the Pt(II) compounds and even the targets (blood, organs, tumour tissue) show a different sensitivity to the drug. Much work remains to be done and is ongoing, e.g. to improve the administration protocol. Research towards new compounds is also focussing on readily soluble derivatives; some of them being Pt(IV) compounds, and have been described in detail.

Action Mechanism

Hydrolysis Reaction

The interactions that may occur in living tissue and during the transport of drugs through the body are of great importance to understand the mechanism of action. Most of the present knowledge is based on the behaviour of the classical compound cisplatin in aqueous solution, which may be hydrolysed depending upon the C1 concentration and pH in the medium. The hydrolysis reaction scheme of Martin has been redrawn in Fig. 2. Under conditions with a high Pt-concentration and high pH values, formation of dinuclear and trinuclear hydroxo-bridged Pt-species does occur. However, these species are unlikely to be formed under biological conditions. In aqueous solution cisplatin will loose Cl, and aqua or/and hydroxo (at high pH) species are formed. In the body, the Cl concentration outside cells is rather high, (about 100 mM), and therefore hydrolysis is largely prevented there, but inside cells the Cl concentration is about I mM, allowing the hydrolysis process. According to calculations by Martin, the aquated species in the body fluids is present only for a few percent, whereas it amounts to about 50% of the total present inside the cell membrane.

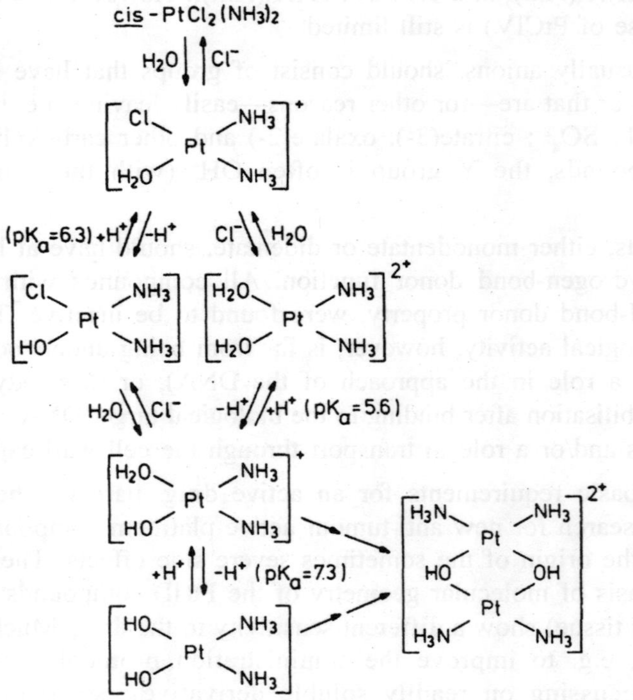

Fig. 2. Hydrolysis reaction scheme of cisplatin.

The kinetics of these processes *in vivo* are not known in detail, but some information is available from in vitro studies. In these investigations cisplatin and analogs were reacted with e.g. nucleotides (such as 5'-GMP). As one would expect, the aqua-species, Lois-$Pt(NH_3)_2(H_2O)_2]^{2+}$, reacts much faster with nucleic acids than the dichloro species. Quantitative kinetic information about these processes is still lacking, however.

Generally, dilute solutions with a concentration of the Pt compound of about 0.1—1 mM in the dichloro form, need at least 24 hours for completion of the reactions with nucleic acid fragments under ambient conditions (pH 7, room temperature), whereas in case of the diaqua form, the reactions may be completed within a few hours. Preliminary investigations with analogs of cisplatin have shown, that in the case of carboxylate ligands as leaving groups, reactions with nucleic acids in vitro proceed much more slowly, although these compounds are good drugs.

Much less is known about the kinetics of the Pt(IV) drugs, such as CHIP, although evidence is accumulating\ that binding to DNA only occurs after reduction of Pt(IV) to Pt(II) species. It is evident that much more experimental knowledge in this analyticalphysical chemistry area is required.

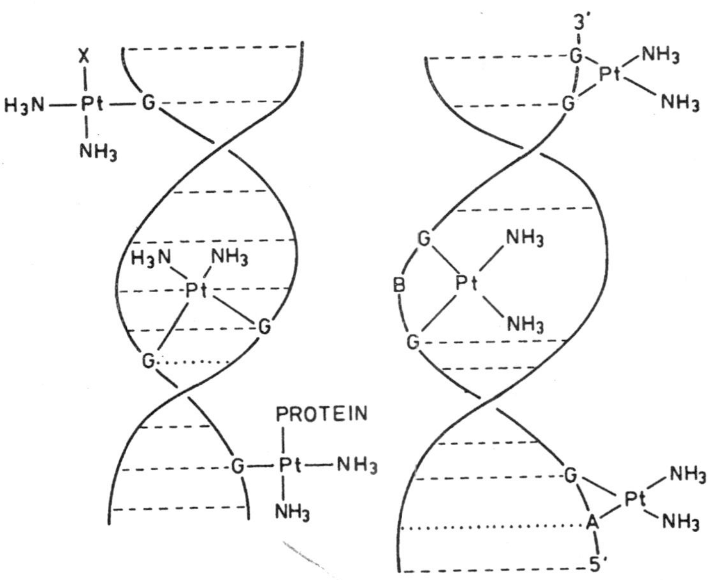

Fig. 3. Schematic binding possibilities for cisplatin to double-stranded DNA.

It is generally accepted that DNA is the most important intracellular target of cisplatin. However, DNA is by far not the only target. Binding to proteins and RNA also occurs, as has been shown by many investigators. In addition, it cannot be excluded that damage to cell membranes is also relevant, although no detailed information is as yet available. The possibilities concerning the cisplatin binding to DNA are schematically depicted in Figure 3. Even though there is good evidence for DNA being the critical target, this does not necessarily prove that this type of binding is the only reaction leading to killing of the tumour cell and the observed side effects in patients treated with cisplatin. It is evident that on the route to the tumour cells, the Pt compounds can interact with e.g. blood proteins and with cell-wall components. These reactions might also contribute to the toxic side effects observed upon treatment of patients with cisplatin and analogs. However, these reactions have hardly been studied and therefore will not be discussed here.

Binding of Platinum Compounds

The recognition that binding of cisplatin to DNA is important, has directed many studies towards the area of metal-nucleobase interactions. For details the reader is referred to numerous reviews. In Figure 4 a tetranucleotide with the four common nucleobases of DNA is drawn with the recommended numbering system.

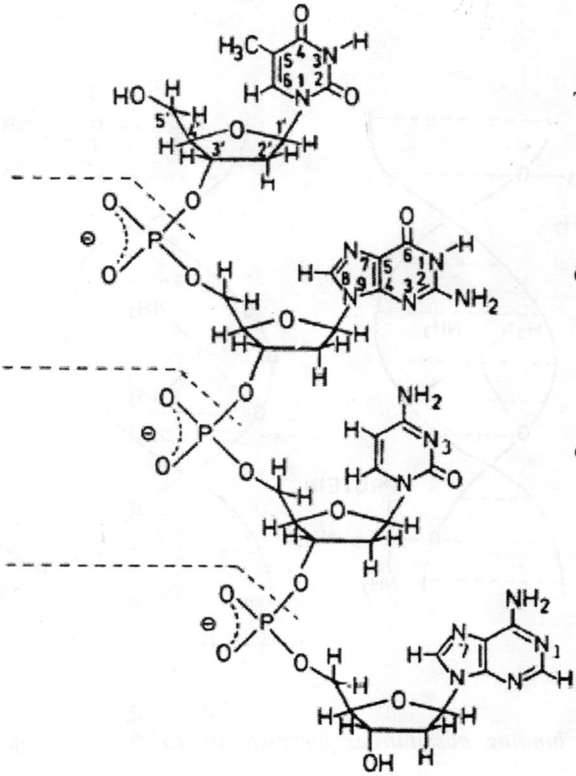

Fig 4: Tetranucleotide showing all the four nucleobases with the used ring numbering.

Metal binding can occur in principle at all sites having a lone pair of electrons. However, in practice metal binding occurs only to a relatively small number of sites, i.e. cytosine-N3, guanosine-N7, adenosine-Ni and adenosine-N7. Binding of metal ions to the sugar ring oxygen atoms has rarely been observed in nucleotides and nucleosides. Binding of metal ions to the phosphate groups in nucleotides is quite common, especially for class A metal ions such as Mg(II), Ca(II), but also binding of Zn(II), Cu(II) and Ni(II) has been observed. In case of Pt(II)—a class B metal -, however, only binding to the nitrogen atoms of the nucleobases has to be considered, although Pt-phosphate interactions do occur in certain cases. The role of phosphate under physiological conditions could be only secondary for coordination, but quite important or hydrogen bonding.

Especially the various studies that make use of NMR techniques have contributed a lot to a better understanding of the manner in which DNA fragments react with metal ions. High-field NMR spectroscopy is a very useful tool to determine the platinum binding sites in nucleic acid fragments, since crystal-structure determinations have been reported only for a limited number of cases. The proton NMR evidence for platinum binding to nucleobases can be summarised as follows:

— A downfield chemical shift of about 0.2—1.0 ppm for a nearby proton.

— The absence of shifts due to (de)protonation proves platination at that site, i.e. when a metal ion is coordinated at e.g. N7 of guanine, protonation cannot occur anymore at this site, resulting in the absence of a protonation PKa at about pH = 2. Metal binding can also alter pKa's of sites where no metal is bound, e.g. the decrease of the PIKa at Ni for N7-platinated gunanine residues.

— A characteristic 3J(^{195}Pt-1H) coupling, manifested as "satellites" around H8 (34% of isotope ^{195}Pt) is often observed at low magnetic fields. At high fields (i.e. above 1.5 Tesla) this coupling disappears because a scalar chemical shift anisotropy relaxation becomes important.

In biological systems, all kinds of possibilities are present, but only the kinetically preferred ones are expected to be selected. It should be realised that in case of platinum binding, the kinetic effects dominate the binding, since ligand-exchange reactions of Pt(II) are quite slow. In other words: when platinum is attached to a certain site, it will take a long time (hours, days) before the Pt-nucleobase bond is broken. From competition studies, it became apparent that the guanine-N7 binding is strongly preferred under a variety of conditions, followed by adenine-N7, adenine-Ni and cytidine-N_3. In addition, the presence of a 5'-phosphate group in a mononucleotide results in a faster reaction than in the presence of a 3'-phosphate group.

In the past the role of the oxygen atom of guanine, 06, has been the subject of much debate. It had originally been assumed that an intrabase chelate with cis-$Pt(NH_3)_2^{2+}$ and the N7-06 groups of guanine (Figure 4) was an important end product, possibly explaining the difference between the cis- and trans-isomers. However, this binding is now generally accepted to be highly improbable and irrelevant, though a hydrogen-bond accepting role of the 06 group has been demonstrated from several X-ray studies of solid products.

An important observation, reported by Cramer and by Marcelis, was that in certain cases the rotation about the Pt-N7 bond is slow on the NMR time scale, as a result of steric hindrance in the coordination sphere of platinum. From the fact that DNA reacts with the .a available coordination sites of cisplatin, the species formed upon the first binding step should be able to undergo fast rotation about the Pt-N bond in a "search" for a second binding position on the DNA. The fact that with small amine ligands on platinum, the anti-tumour active compounds, such a rotation is always fast, agrees with the earlier observation (vide supra) that amine ligands with at least one N-H group yield active compounds. Inactive compounds with bulky amine groups not only have different dynamics (rotation about the Pt-N bond), but also yield different structures of the formed adducts.

Binding to Single-stranded Oligonucleotides

The knowledge that cisplatin has two binding sites available and that binding to neighboring nucleobases in DNA may occur, initiated investigations on Pt-binding to dinucleotides. In these compounds the second step of the binding (chelation) becomes possible within the same molecule. Of all 16 possible combinations of nucleobases (GG, AG, GA, CG, GC, TO, GT, AA, AC, CA, AT, TA, CC, CT, TC, TT) and other variations (i.e. ribose or deoxyribose; with or without a terminal 3'- or 5'-phosphate group), only a few have been studied in great detail. In addition to dinucleotides, larger oligonucleotides of defined sequences (up to dodecamers) have been used to study the binding of cisplatin, and other compounds.

Important questions addressed in these studies are:

- Is there a kinetic preference for any of these combinations (i.e. will a certain sequence be recognised by cisplatin)?
- What are the binding sites at the nucleobases?
- Are products with different sugar conformations or base orientations (isomers) formed?
- Can a detailed structure be determined, either in the solid state or in solution?
- What is the effect of the binding on the sugar ring?
- What is the effect of phosphate groups on the kinetics and structure of the formed chelate?

Work in this area is still ongoing in many research groups and will only be briefly summarised here. A very interesting observation, first made by Chottard et al., deals with the occurrence of conformational isomers after chelation of cisplatin to a number of dinucleotides. In certain cases these isomers are rapidly interconverted, but in other cases they can easily be separated by chromatographic techniques, thus allowing characterisation by NMR techniques. Apart from NMR, other techniques—such as CD and IR -may be used to study the structure and conformation of small oligonucleotides.

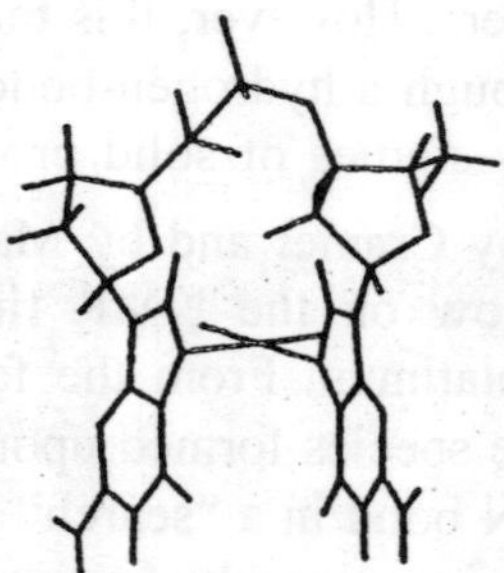

Fig. 5. Projection of a structural model for the $Pt(NH_3)_2(d(GpG))$ adduct, deduced from analysis of the NMR spectra.

Using high-resolution NMR techniques it appeared possible to determine the most likely structure in solution for the adduct cis-$Pt(NH_3)_2(d(GpG)^+$. Complete assignment of the proton resonances, decoupling experiments and computer simulations yielded a structural model, depicted in Figure 5, with the following characteristics:

- — The two bases, coordinated through N7, are oriented head-to-head with a dihedral angle for the bases of about 600;
- — The sugar ring at the 5' side of d(GpG) has changed to the N-conformer (compared to the S conformer in free d(GpG));
- — The other conformational characteristics of d(GpG) are hardly changed upon chelation to · cisplatin.

From this model it was predicted that the distortion of the DNA structure, after chelation of cisplatin to a GG sequence, would be rather small, probably yielding a kink in the helix of about 140-70°.

Recently, Sherman *et al.* have solved the X-ray crystal structure of the adduct cis-$Pt(NH_3)_2(d(pGpG))$. This species differs just one phosphate group with the abovementioned adducts of d(GpG). The role of this phosphate group appears to be a stabilising one, i.e. it is involved in a hydrogen bond with a NH_3 ligand at platinum. This could well be an important explanation for the observation that platinum anti-tumour drugs need an acid N-H group to donate a hydrogen bond. The O_3P-O--H-NH_2 interaction could stabilise and induce the distortion of the DNA, thereby interfering with replication processes (vide infra). The most important finding of this study, however, appears to be that the solidstate geometry is substantially the same as the one in solution. Part of the solid state structure is redrawn in Figure 6, based on the coordinates of Sherman *et ai.*. The analogy with Fig. 5 is clearly observed. Finally it should be mentioned here that such a distortion for a GG-chelate has also been calculated using molecular mechanics.

Four major questions have been adressed in studies dealing with trinucleotides containing:

1. What is the conformational effect of a third terminal base on the chelating properties of a -GG- unit?
2. Knowing the strong preference of cisplatin for guanine-N7 binding, does chelation occur to two guanines separated by a third base?
3. Knowing that chelation to AG-units—but NOT GA—occurs *in vitro*, what are the products of reaction when GAG is treated with cisplatin?
4. Can chelation of the trans-isomer to GNG occur through both guanine-N7 atoms?

To study the first question, the trinucleotide d(CGG) was treated with cisplatin and the formed GG-chelate was investigated by X-ray diffraction techniques. It appears that again the 5'-G has a N-type sugar ring as a result of chelation, whereas the cytosine residue is involved in several hydrogen bonding and stacking interactions; NMR studies had earlier indicated a similar structure for the GO-part. The chelate formed with the G(2)G(3)-unit is roughly the same as the one found with d(GG). A structure projection is depicted in Figure 7. In the earlier findings that cisplatin induces base-pair substitutions in -GCG- and -GAG- base sequences in bacterial DNA (*E.coli*) and has mutagenic properties, has led Jan Reedijk to a study of cisplatin interactions with d(GCG), d(GTG), d(GAG) and d(pGGG). In all cases chelation through the guanines has been observed. Recent investigations on d(GAG) have shown that cisplatin also has a tendency to form a AG-chelate (20%, compared with 80% of a chelate through both guanines).

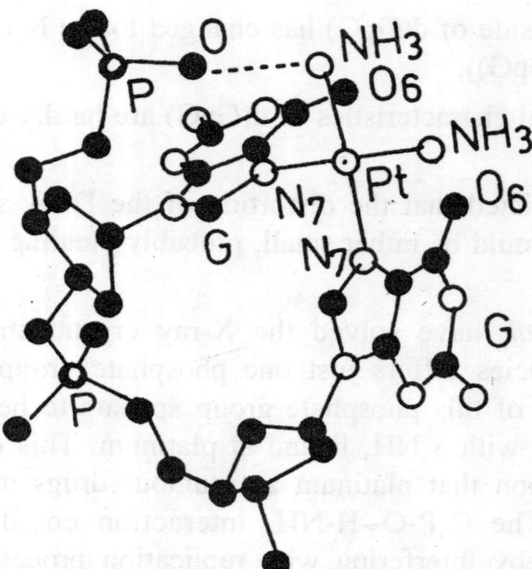

Fig. 6. Projection of the structure of cis—$Pt(NH)_2(d(pGpG))$ as determined from single crystal structure analysis.

The first studies with larger single-stranded oligonucleotide in fact dealt with self-complementary (i.e. forming a duplex with itself) tetranucleotides and hexanucleotides, viz. d (CCGG), d (AGGCCT) and d (TGGCCA).

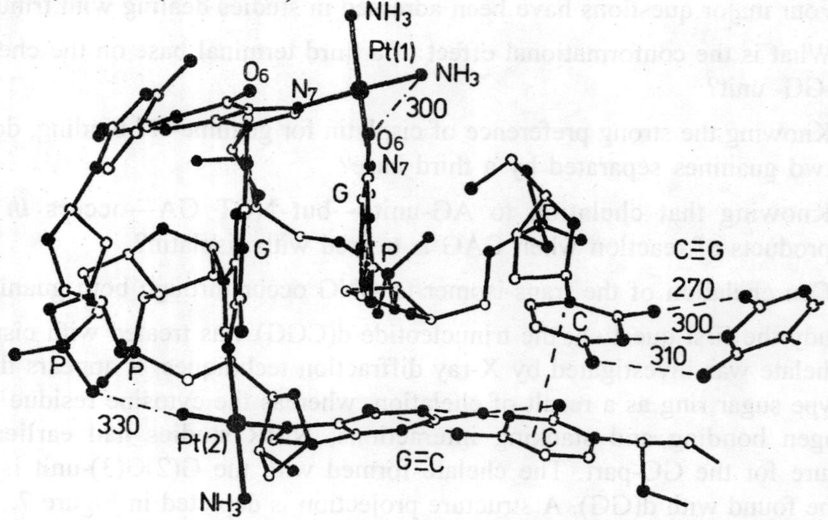

Fig. 7. Projection of one part of the X-ray structure of the adduct cis—$Pt(NH_3)_2(d(CpGpG))$

In all these studies it became immediately clear that the double helix is disrupted upon chelation of the cisPt unit at the GG site. Again NMR proved to be a very useful technique to study these reactions and their products. Another technique that can be applied to determine the binding sites of cis-$Pt(NH_3)_2^{2+}$ in oligonucleotides, is based upon the application of enzymes that digest the NA (or oligonucleotides) to smaller fragments, some of which contain platinum species. He has first introduced this technique for the analysis of cisplatin-treated salmon sperm DNA. Since then it has also been applied for the analysis of platinated oligonucleotides. Because large oligonucleotides are needed to maintain a double-stranded structure after platination, more attention has been given recently to octa, deca, undeca and even larger oligonucleotides. Since these larger fragments are usually studied in connection with a (self)-complementary strand.

Binding to Double-stranded Oligonucleotides

As discussed above, the first investigations on oligonucleotides have made clear that—at least up to hexamers—platination results in disruption of the double-helical structure. One should realise, however, that in these self-complementary oligonucleotides 2 Pt groups will chelate. Therefore, Jan Reedijk has undertaken a study with a duplex fragment, having a centrally located GG-unit in only one strand. The performed reactions are depicted schematically in Figure 8.

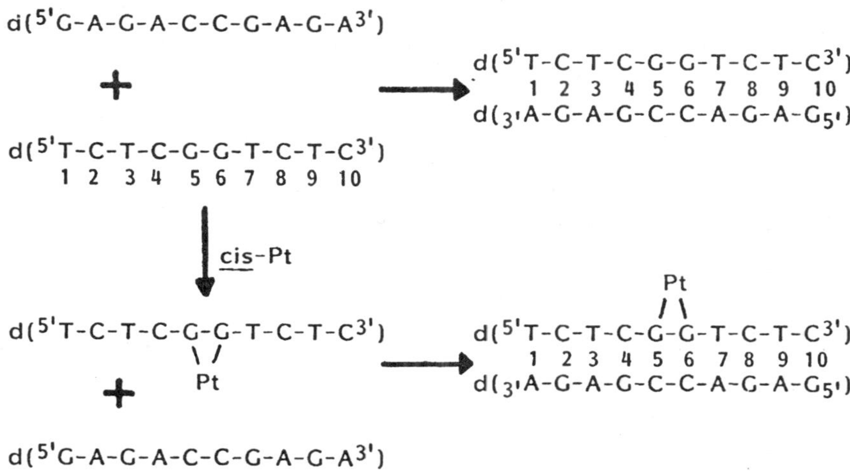

Fig. 8. Representation of the reaction of a decanucleotide with cisplatin.

Upon reaction of cisplatin with the strand d(TCTCGGTCTC) a high yield of the GG chelate is formed, as would be expected; subsequent addition of the complementary strand d(GAGACCGAGA) apparently results in duplex formation. Detailed analysis of proton and phosphorus NMR spectra and consideration of the CD spectra, has resulted in the following conclusions:

1) The double helix is somewhat destabilised, compared with the unplatinated duplex, as seen from the decrease in melting temperature of about 10-20°.

2) Observation of the imino-proton resonances—though somewhat shifted and broadening at low temperatures—of the central G:C base pairs, indicates that hydrogen bonding (Watson Crick type or related) remains possible after platination.

3) Careful analysis of the chemical shifts and coupling constants after platination, suggests a distortion of the double helix which is rather small and which can—at the moment—beat be described as a kink in the helical axis of about 140-70°, without a large change in overall helical unwinding.

4) Comparison of CD spectra and ^{31}P spectra of platinated DNA (from several sources) and the platinated ds decanucleotide, strongly suggests similar distortions in both cases.

In an independent study, using the platinated octanucleotide d(GATCCGGC) and the complementary decanucleotide strand d(GCCGGATCGG), Chottard arrived at similar conclusions. Studies with the adduct of cisplatin with d(GCCGGATCGC) complemented with d(GCGATCCGGC) clearly showed early metling of the inner G:C base pairs.

In a related piece of work Jan Reedijk has platinated the undecamer d(TCTCGTGTCTC) and have concluded that a chelate using both guanosines (a GBG chelate) is again formed, despite the T in between them. Now addition of the complementary d(GAGACACGAGA) again results in duplex formation. However, the CD spectrum has significantly changed compared with the unplatinated duplex and this change is different from the change in the case of the decanucleotide. Detailed study of the proton NMR spectrum in the imino region indicates that two H-bridges between the bases have disappeared and that the helix melting temperature has decreased by about 25°C. Apparently the central T:A base pair cannot be formed at all, since T is bulged out as a result of chelation of the guanines to cisplatin. Addition of the almost complementary decanucleotide d(GAGACCGAGA) did not result in a better duplex formation.

Recently also Kidani has shown—by using enzymatic digestion and HPLC—that cisplatin binding to self-complementary d(GGTCGACC) and to d(CGGATCCG) results mainly in *cis*Pt-GG adducts.

From all the above-mentioned results the following general conclusions can be drawn:

1) When adjacent guanines are present in an oligonucleotide, cisplatin chelates to these dinucleotides at the N7 positions.

2) When GNG sequences are present in an oligonucleotide, cisplatin chelates also to both guanines; when the central base is adenine then also an AG chelate can be formed.

3) Double-stranded oligonucleotides may be formed with the above-mentioned platinated GO- and GNG-containing oligonucleotides (at least 8 units seem to be required). The distortion of the double helix is rather small in case of GO-chelation by cisplatin. A larger distortion occurs in the case of GTG-chelation.

5) Melting of the double helix appears to start from the central part where the GO-Pt unit is located.

Binding of Platinum Compounds to DNA and RNA in Vitro

As soon as it became apparent that the cisplatin-DNA interaction is an important event in the cytostatic process, investigations were started using DNA from several sources. In this short review only attention will be given to recent spectroscopic studies dealing with DNA and RNA binding, with special emphasis on the molecular details. Other aspects have been treated in earlier reviews

Upon binding of cisplatin to DNA in vitro one may monitor the differences in spectroscopic properties. A very sensitive technique has been shown to be circular dichroism spectroscopy. In fact, early observations already dealt with the enhancement of the ellipticity at 275 nm, indicating the occurrence of intrastrand cross linking. The results on the decanucleotide show the same CD effects after platination, suggesting similar distortions in the decamer nucleotide and in DNA.

More recently it has become possible to use 31P NMR to study DNA and—under comparable conditions—double-stranded oligonucleotides. As already discussed above, the first results indicate that similar distortions occur in DNA and oligonucleotides. It is even possible now, applying special proton NMR techniques, to use imino proton signals to study the distortions induced by cisplatin. This has been nicely shown for ribo-poly(I):poly(C) by Marzilli. In all these studies one should realise, however, that under *in vivo* conditions the loading of the DNA by cisplatin is much lower than in the DNA samples, which are suitable for spectroscopic studies. Therefore, one should be careful in extrapolating such results to in vitro mechanisms.

A different approach to study the cisplatin binding positions in DNA is based upon degradation of high-molecular weight DNA after platination, and analysing the fragments—after separation—by a special technique, such as NMR. In the laboratories, a method has been developed to degrade cisplatin-treated DNA with enzymes to ((P1=98/0)) mononucleotides and Pt-containing mono- and dinucleotides. This mixture was subsequently separated on the basis of their net charge by anion-exchange chromatography. Four reaction products of cisplatin with salmon sperm DNA were isolated, together comprising at least 90% of the total platination. The structures of these adducts have been unambiguously proven by proton NMR spectroscopy and Pt-analyses with atomic absorption spectroscopy (AAS). In all cases the Pt was linked to the N7 atom of guanine or adenine and no indications have been found for binding at adenine-Ni or cytosine-N3 at the level of platination studied (up to 6 Pt atoms per 100 nucleotides; rb = 0.06).

The formation of intrastrand crosslinks on GG and AG base sequences in DNA could be deduced from the presence of the adducts cis-$Pt(NH_3)_2$(d(pGG)) and cis-$Pt(NH_3)_2$(d(pAG)) in digests of platinated DNA. However, no indications for the presence of GA adducts could be found, and therefore crosslinks with a GA sequence are unlikely.

The formation of interstrand crosslinks in DNA, also demonstrated with alkaline elution techniques, could be related to the presence of the adduct cis-$Pt(NH_3)_2(dGMP)_2$. This product might originate from two guanines on different strands. The same pt-digestion product, however, would also be obtained from intrastrand crosslinks on two guanines separated by one or more nucleobases. The occurrence of this type of crosslinks in DNA was deduced from mutation studies in bacteria, and proven in oligonucleotides.

Already some years ago he observed that the results of enzymatic disgestion of cisplatin-treated DNA from several sources, result in larger amount of GG-adducts than would be expected from a random approach of the G-sites of the DNA by the cisplatin unit. In other words, when approaching the DNA, the cis-$Pt(NH_3)_2$ unit already "knows" which G is part of a GG-fragment (preferred) and which one is not. The molecular basis of this selectivity is not yet understood. It has also been observed under *in vivo* conditions.

Reactions of Platinum Amine Compounds *in vivo*

Apart from the binding of cisplatin and analogs to DNA in the tumour cell—which is generally believed to be the origin of the antineoplastic activity—several other events will take place after administration of the drug into the body (usually through infusion in the blood stream). A large variety of reactions may take place during the transport through the body via the blood. As discussed above, hydrolysis reactions are severely suppressed because of the high Cl^- concentration in the blood. Nevertheless, reactions with e.g. blood proteins or peptides may occur, although systematic studies about the molecular basis of these reactions are not yet available. Likely candidates for binding may be the thiol and thioether groups of proteins and peptides, which are known to have a high affinity for platinum.

During the first 24 hours after the application of the drug, already 30-70% of the administered platinum is excreted from the body, mainly through the kidneys. Initially this was leading to a severe problem, because of the resulting kidney damage. Nowadays, the patients receiving cisplatin are simultaneously treated with a high dose of fluid that prevents this damage. Other toxic effects observed in the early days, when high doses of cisplatin were administered, were probably due to the formation of dinuclear hydroxo-bridged Pt-species. In later administration protocols, the platinum concentrations were lowered, so that only monomeric species are supposed to enter the body. With the new reagent paraplatin (figure 1) higher doses seem possible.

The circulating platinum species is transported—most likely as a neutral species—into both tumour cells and normal cells, and then undergoes hydrolysis inside the cells (Figure 2). Subsequently, the hydrolysed Pt-species can react with all kinds of cell components, such as peptides, proteins, DNA and RNA.

Although the binding to DNA appears to be responsible for the ultimate cell killing, it cannot be excluded that reactions of the Pt species with other molecules are leading to toxic side effects. These effects can be classified as acute, subacute, and chronic. The molecular basis of these toxicities are far from understood, although medical research has resulted in protocols that reduce the effects enormeously; (e.g. the pretreatment hydration of the patient, which almost avoids nephrotoxicity).

Recently, it could be established that in living cells the same types of Pt-adducts are formed as in DNA treated with cisplatin in vitro. Also the strong preference of cisplatin to form intrastrand crosslinks on GG base sequences is found upon interaction with DNA *in vivo.*

The nature of the difference in response to Pt-chemotherapy between tumour and normal cells is still unknown. It is possible that rapidly dividing (tumour) cells are more susceptible, because of the inhibitory effect of Pt-DNA adducts during the DNA replication—a prerequisite for cell devision. Another explanation may be a difference in the capacities of the various cell types to remove Pt-DNA adducts. The influence of DNA repair systems on the survival of cisplatin-treated cells has clearly been demonstrated in *E.coli* bacteria and in cultured human cells.

The question which of the induced Pt-DNA adducts is (are) responsible for the antitumour activity is still unanswered. Many investigators have tried to correlate interstrand DNA and DNA-protein crosslink formation with the cytotoxic action of cisplatin, but conflicting results were obtained. Untill now a biological role inside the cell has been found for only one type of Pt-DNA adduct. Brouwer et al. showed that cisplatin can induce base-pair substitutions in *E.coli* bacteria at GAG and GCG base sequences. This strongly suggests that the intrastrand crosslink of cisplatin on GBG is responsible for this effect.

Studies are ongoing to establish not only the induction but also the repair of the various Pt-DNA adducts, because it is assumed that incorrectly repaired and non-repaired, persistent, lesions will be responsible for the antitumour effects of the drug. For the study of the Pt-adduct formation and repair in DNA of cells treated with biologically relevant doses, e.g. from patients treated with cisplatin, very sensitive detection methods are required. For this purpose immunochemical techniques have been developed with which, at the moment, femtomol (10^{-15} mol) amounts of the Pt-adducts in DNA can be detected with anti-Pt-adduct antibodies.

Differences between Cis and Trans Isomers

It has been known for some time that also *trans*-$PtCl_2(NH_3)_2$ binds to DNA. It is likely that the first binding step does not differ very much from that of cisplatin. With mononucleotides also *trans*-PtCl2(NH)2 easily forms 1:2 adducts, and these differ hardly from those derived from cisplatin. Also the kinetic differences of the cis- and the *trans*-isomer hardly differ. The biological effect of *trans*-$PtCl_2(NH_3)_2$ is quite interesting and DNA binding studies deserve detailed study. Since it is clear that the binding affinities to the nucleobases are quite comparable, the different anti-tumour activity must originate from differences related to the bifunctional binding.

Comparative studies with *trans*- and cis-platin on single-stranded and double-stranded oligonucleotides are beginning to contribute to the understanding of these differences. It has recently been shown that also *trans*-$Pt(NH_3)_2^{2+}$ can chelate to GNG sequences through both guanine N7 atoms. The distortion of DNA after such a *trans* binding is likely to be very large. This has put forward the hypothesis that repair enzymes recognise and remove the *trans* compound easier than the cis compound. Replication, however, would not proceed when a cis-$Pt(NH_3)_2^{2+}$ unit is attached to a GG unit.

The above mentioned influence of the N-H group in the amine ligand of the Pt-compound may have an indirect role in the binding to nucleic acid fragments and to DNA. On one hand it could have a directing effect on the approach to the DNA, by having a H-bond interaction with guanine-06; this possibly explains the J<inetic preference of platinum compounds for guanine-N7. On the other hand, after formation of a bis-nucleotide adduct, it may have a Hbond interaction with a phosphate group, as has recently been found in a crystal structure determination of cis-$Pt(NH_3)_2$(pGG-N7,N7) and was expected from molecular mechanics calculations.

Chapter 12

Impact of Anticancer Drugs on Microtubules

Microtubules are central to a number of cellular processes including the formation of the mitotic spindle. Indeed it was the destruction of the spindle, leading to a loss of chromosome segregation with consequent inhibition of cell division and cell death that gave rise to the term 'antimitotic agent'. Despite the spindle being an obvious target for further drug development, research in this area was unfashionable until the exciting clinical results of the taxane drugs were reported in the late 1980s.

The lead compound is a natural product isolated initially from the Pacific Yew. The taxane drugs again disrupt the tubulin-microtubule equilibrium, however these agents, in contrast to the vinca alkaloids, push the equilibrium towards the assembled microtubule. Tubulin-related functions within the cell rely on the dynamic nature of the tubulin-microtubule equilibrium, and microtubule stabilisation can lead to the same end result as microtubule disruption. Research in the field has again increased following the observation that agents that bind to tubulin can selectively destroy the vasculature within tumours, causing extensive tumour necrosis. It is likely that the vinca alkaloids and the taxanes also exert part of their antitumour activity via an antivascular mechanism. The destruction of the tumour vasculature also arises via a tubulin-related mechanism. Microtubules are involved in the maintenance of cell shape, and the disruption of the microtubular network leads to a rapid rounding up of cells. The tumour selectivity arises from the leaky and disorganised nature of tumour blood vessels which are weak owing to the lack of supporting smooth muscle cells and pericytes seen in normal blood vessels.

The endothelial cells lining tumour microvessels are therefore likely to be more susceptible to the effects of agents that target tubulin. This, coupled with the relatively high levels of exposure to drug experienced by all vascular endothelial cells, results in the rounding up of these susceptible cells with consequent blockage of microvessels, leading to necrosis of those tumour cells that were dependent on the microvessels. One problem seen in these studies is the survival of tumour cells at the periphery of the tumour. These are fed from the vasculature of the normal surrounding tissue and are therefore not affected by the destruction of blood vessels within the tumour. These tumour cells form the nucleus for re-growth, and it is therefore unlikely that these agents will be effective unless given in combination with other therapies.

Synergy between combretastatin A-4 and other therapies has been reported in experimental tumours.

CLINICAL USE

Clinical use of agents targeted at tubulin is currently restricted to the vinca alkaloids (vincristine, vinblastine, vinorelbine) and the taxanes (paclitaxel, docetaxel), although a number of agents are undergoing early clinical trials. Those can be broadly classified according to their effect on tubulin-microtubule equilibrium. It is interesting to note that some of these trials (e.g. combretastatin A4 phosphate) are using blood flow and tumour permeability as pharmacodynamic end points.

A number of agents have been evaluated clinically. Results are available for amphethinile, maytansine, cemadotin, rhizoxin, 1069C85, CI-980 and dolastatin 10. It is noteworthy that the trial of amphethinile reported two deaths due to vascular causes. No blood flow measurements were performed in this early study. Tumour pain was also noted, a toxicity also seen for cemadotin. The toxicity profile of these agents shows neurotoxicity to be a common observation. However this is perhaps to be expected as neural tissue is the most abundant source of tubulin. Neutropenia, nausea and vomiting are also seen. The vinca alkaloids and taxanes are amongst the most effective anticancer drugs. Both series of drugs are based on large, complex, natural products. The failure of the synthetic agents to achieve clinical success so far has been disappointing. However the realisation that these molecules may possess potent antivascular activity has raised interest in this area. It is likely that these agents will be more active when given in combination with therapies that will destroy those tumour cells that are supported by vasculature from normal tissue surrounding the tumour.

Microtubule Network

Microtubules (MTs) are polar, non-covalent assemblies of the protein tubulin found in all dividing eukaryotic cells. During cell division, a large dynamic array of microtubules, the mitotic spindle, functions to physically segregate the duplicated chromosomes and to orient the plane of cleavage (Figure 1).

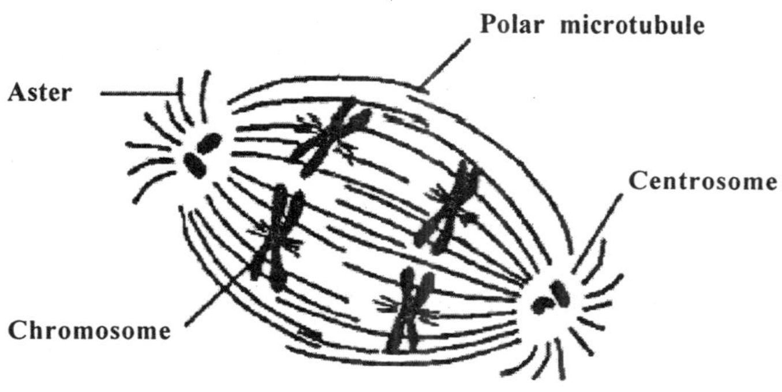

Figure 1. The mitotic spindle.

In non-dividing cells, microtubules organise the cytoplasm, position the nucleus and organelles and serve as the principal element of flagella and cilia. Agents in Clinical Development. filaments, microtubules constitute the cytoskeleton which provides shape and strength to the cytoplasm.

Microtubules (Figure 2) are composed of two structurally similar protein subunits, namely a- and b-tubulin arranged head-to-tail at 80 Å intervals to form a linear protofilament. A single microtubule is composed of thirteen protofilaments, forming a hollow structure of ca. 240 Å diameter. Each tubulin monomer is composed of approximately 440 amino acids (50 kDa) forming a core of two b-sheets surrounded by a-helices and binds a guanine nucleotide, which is non-exchangeable when bound in the a-subunit (N-site) and exchangeable when bound to the b-subunit (E-site).

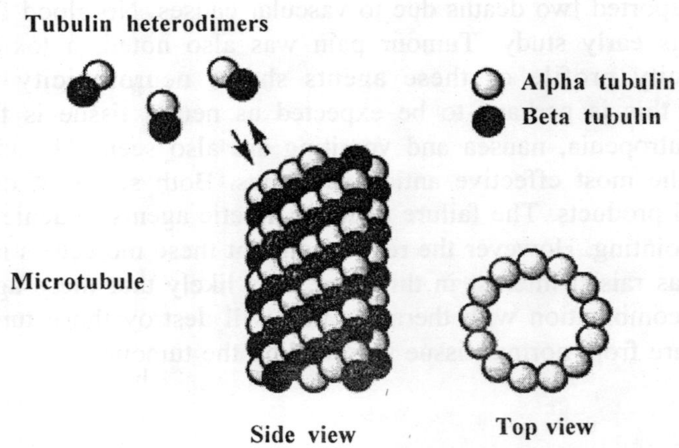

Figure 2. Structure of microtubules.

To carry out their tasks properly, microtubules assume a protean array of different stability states and degrees of organisation and they respond rapidly to requirements of the cell by modification of their organisation and stability. This organisation is regulated by numerous factors including the intrinsic ability of microtubule subunits, tubulin heterodimers, to form non-equilibrium, dynamic polymers. This process of polymerisation/depolymerisation is referred to as dynamic instability.

GTP hydrolysis at the exchangeable E-site is required to establish a flux of subunits through the polymer, adding tubulin heterodimers to the "plus" end and dissociating heterodimers from the "minus" end of the microtubules, resulting in microtu-bule destabilisation. Magnesium ions (Mg^{2+}) are also required for assembly in part because GTP binds as a Mg-GTP complex. The extent of dynamic behaviour depends upon the competing role of both stabilising proteins, microtubule-associated proteins (MAPs: tau, MAP1, MAP2, MAP4, XMAP215) and dynamic regulatory proteins that destabilise microtubules (stathmin, XKCM1, XKIF2, katanin).

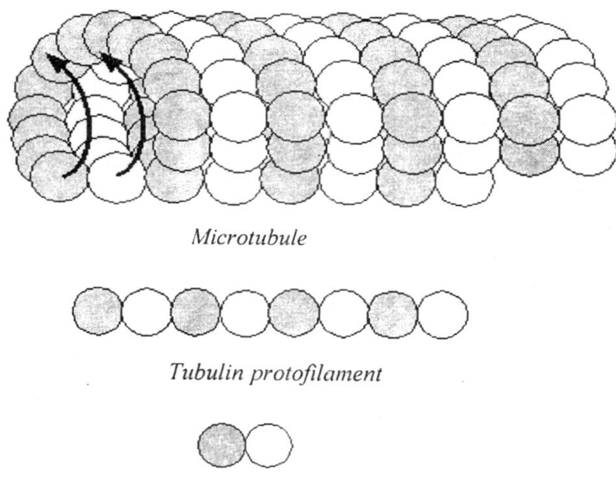

Figure 3. Tubulin subunits

This dynamic process of assembly and disassembly can be blocked by various agents that bind to distinct sites in the β-tubulin subunit. By interfering with microtubule function in vitro, these agents arrest cells in mitosis, eventually leading to cell death, by both apoptosis and necrosis. So far, three binding domains have been identified, the colchicine site close to the α/β interface, the area where the vinca alkaloids bind, and the taxane binding pocket. Colchicine binds to a site near the intra-dimer interface and alters lateral contacts within the microtubule, blocking microtubule polymerisation.

Vinca alkaloids inhibit microtubule assembly by cross-linking at the inter-dimer interface, they sterically distort the protofilament and induce tubulin to form alternate spiral polymers. The mechanism of action of taxanes is quite different from that of the other two, for it promotes the assembly of microtubules, resulting in highly stable, non-functional polymers. Taxanes bind at the M loop on the β-subunit, stabilising lateral contacts between protofilaments.

Antimitotic agents that interact with microtubule components are of interest for the insights they can provide into the roles of microtubules in cells and the subtleties of tubulin structure and also for their potential activity in the treatment of human neoplastic diseases. A variety of bioassays have been used to identify new antitubulin agents and new techniques have been developed to further understand their biological potency and mechanistic basis at the molecular level.

Effects on culture human cancer cell growth. Cell proliferation assays offer quantitative methods for evaluating a cell population's response to candidate drugs, whether they cause an increase in cell growth, no effect, or a decrease in growth due to necrosis, apoptosis or cytostasis. The MTT and sulforhodamine B assays are of particular importance for routine screening of anticancer drugs as they reflect cellular growth and viability. Antimitotic agents inhibit the proliferation of cancer cell lines.

Cell cycle analysis. Flow cytometry provides an accurate method for determining the proportions of cells in G_1, S and G_2/M stages of the cell cycle by measuring the DNA content of eukaryotic cells after propidium iodide staining. Flow cytometric studies show that cancer cells treated with antitubulin agents arrest in the G_2/M phase of the cell cycle.

Indirect Immunofluorescence. The intracellular distribution of microtubules following treatment with drugs can be observed by using an indirect immunofluorescence approach (Figure 4). Cells treated with paclitaxel-like agents develop characteristic extensive microtubule bundles that arise randomly throughout the cytoplasm while colchicine-like agents uniformly cause the disappearances of microtubules leaving a granular cytoplasm (Figure 4).

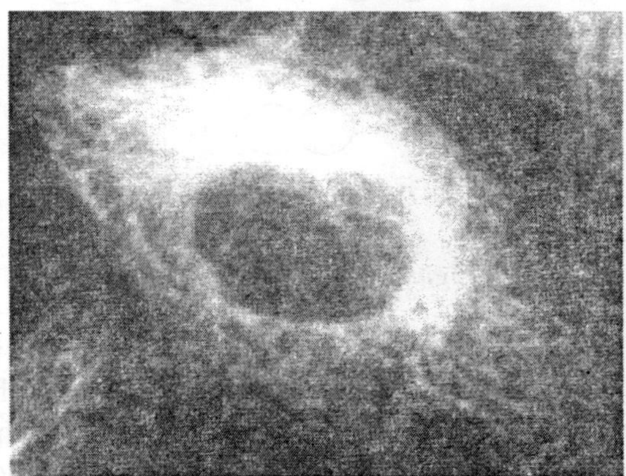

Figure 4. Indirect immunofluorescence of microtubules.

Tubulin isolation and purification. Tubulin has been isolated from porcine or bovine brains using procedures that rely on the ability of tubulin to assemble at 37°C in the presence of GTP to give microtubules and disassemble in the cold (4°C) in the presence of calcium ions to regenerate α/β-tubulin heterodimers. This property is also used to purify tubulin by cycles of assembly and disassembly, collecting the tubulin and microtubules whilst reducing the content of other proteins.

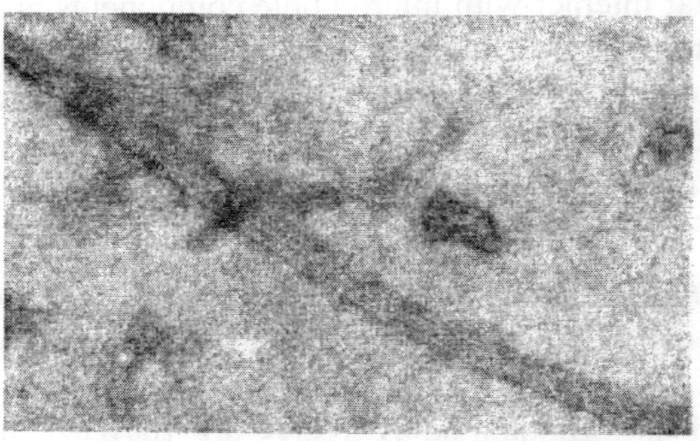

Figure 5. Electron Microscopy of a Microtubule.

Tubulin assembly assay. Antitubulin agents fall into two classes: promoters and inhibitors of microtubule assembly. The in vitro polymerisation of microtubules is associated with turbidity which can be quantified in a spectrophotometer by measuring increase in optical density of a tubulin solution at 350 nm. This intrinsic process is very much a function of the in vitro experimental conditions.

For example, the presence of GTP, magnesium ions, glycerol and/or a physiological temperature of 37°C are known to promote the assembly of tubulin into microtubules while the presence of calcium ions and a low temperature of 4°C trigger disassembly of microtubules into tubulin heterodimers. Microtubule-stabilising agents have been shown to initiate the polymerisation of tubulin in the absence of polymerisation promoters and stabilise microtubules against cold- and calcium-induced depolymerisation. Microtubule-destabilising agents have been reported to inhibit tubulin polymerisation against heat- and GTPinduced polymerisation.

Sedimentation assay. The quantity of microtubules formed during the tubulin assembly assay has been determined by comparing the supernatant protein concentration before and after centrifugation. This was then compared to the quantity of microtubules formed in the presence of GTP.

Competitive binding assay. Drugs that bind tubulin have been subdivided into three main classes depending on which area of the tubulin they populate. The ability of antitubulin agents to compete with [^{3}H]-radiolabelled colchicine, vinblastine or paclitaxel for binding to tubulin has been examined and quantified by scintillation counting.

Electron microscopy. Tubulin polymers formed in the presence of microtubule-stabilising agents have been observed by electron microscopy (Figure 5). This powerful tool is useful to determine whether the increase in turbidity measured during polymerisation experiments is caused by the formation of microtubule-like polymers or the formation of other structures.

Cytoblot assay. This phenotype-based screen is a chemical genetic approach to identify cell permeable compounds that affect the cell cycle. The assay uses TG-3, a monoclonal antibody that recognises a form of the protein nucleolin specifically phosphorylated during mitosis, to report indirectly on the progress of the cells through mitosis.

Interaction of the antimitotic Agents

The interaction of the different classes of antimitotic agents with their binding sites on tubulin. The binding regions discussed are the colchicine, vinca, paclitaxel and unknown/uncharacterised sites. Some agents have been described as binding to more than one site and this will be mentioned where appropriate. There are more than thirty agents in clinical or pre-clinical development and new antimitotic agents are discovered regularly.

Colchicine Binding Site

Colchicine 1

Colchicine 1 (Figure 6), a natural product isolated from Colchicum autumnale (meadow saffron), is the classic tubulin binding agent. Colchicine 1 was one of the first antimitotic agents investigated and although used in the treatment of gout, its high toxicity prevents its use in other therapies.

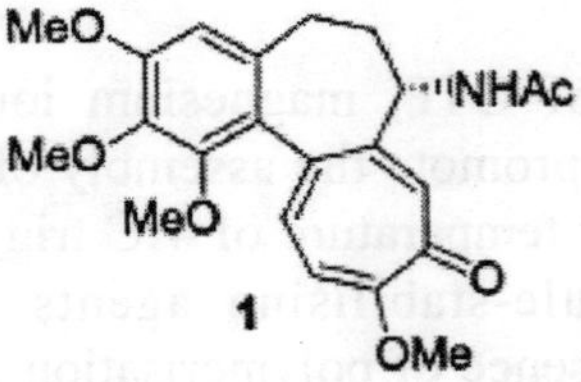

MeO MeO MeO NHAc O OMe 1

Figure 6. Structure of Colchicine 1.

In vitro, the major effect of colchicine 1 is to cause a change in the secondary structure of tubulin by bin-ding with a high affinity site on the heterodimer. This causes a change in the structure/conformation of the heterodimer which hinders the formation of microtubules. This binding is temperature dependent and irreversible. Colchicine 1 has also been shown to bind with a second, lower affinity site on tubulin.

Combretastatin A-4 2

The natural product combretastatin A-4 2 (Figure 7) was isolated from the South African bush willow Combretum caffrum. Combretastatin A-4 2 is highly cytotoxic to several cancer cell lines and shows potent antimitotic activity. Stilbene 2 inhibits in vitro tubulin polymerisation, stimulates tubulin-dependent GTP hydrolysis and competitively inhibits the binding of colchicine 1 to the protein.

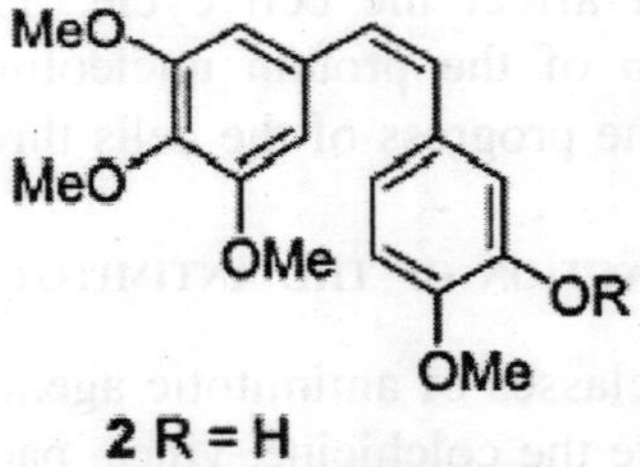

MeO MeO OMe OR OMe

2 R = H
3 R = P(=O)(ONa)$_2$

Figure 7. Structures of combretastatin A-4 2 and phosphate 3.

Several structure-activity studies have been carried out on combretastatin A-4 in order to identify the structural motifs important for bioactivity. A number of agents have shown equipotency as combretastatin A-4, although this latter agent was the one chosen for clinical studies. Like several other antimitotic agents, combretastatin A-4 2 can selectively damage tumour vasculature. However, whereas most antimitotic agents are effective only at or close to their maximum tolerated dose (MTD) combretastatin A-4 2 is effective at 10% of its MTD.

Although combretastatin A-4 2 is a simple cis-stilbene, its original 5-step synthesis produced a mixture of cis- and trans-isomers. A convenient two-step stereoselective synthesis which can provide multigram quantities for clinical use, has recently been published. Combretastatin A-4 2 itself is insoluble and several attempts have been made to produce an active water-soluble derivative. The most effective is the disodium phosphate salt 3 (>20 mg/ml) and is the form of combretastatin A-4 presently in Phase II clinical trials.

ZD6126 4

ZD6126 4 (Figure 8) is a prodrug phosphate derivative of N-acetylcolchinol which has been synthesised in 4 steps from colchicine. ZD6126 4 is a tubulin binding agent; however it is also a vascular targeting agent that selectively damages tumour vasculature. The antivascular effect of ZD6126 4 occurs within fifteen minutes of administration and can be administered at 20% of its MTD in mice. ZD6126 4 shows antitumour effects when administered alone, but enhanced activity has been noted when used in combination with cisplatin.

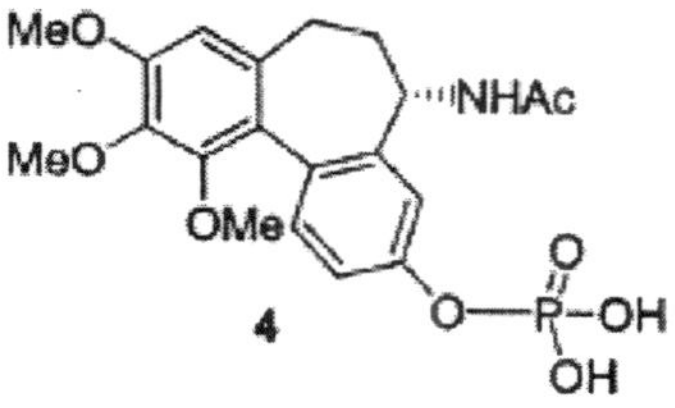

Figure 8. Structure of ZD6126 4.

Phenstatin 5

The ketone 6 (Figure 9) showed moderate cytotoxicity to several cancer cell lines and the ability to inhibit the assembly of tubulin. A phenolic derivative named phenstatin 5 was obtained unexpectedly from the Jacobsen oxidation of a combretastatin A-4 derivative.

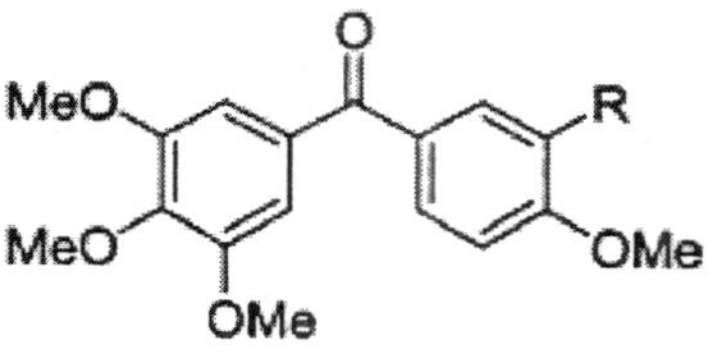

Figure 9. Structures of phenstatins 5-7.

A water-soluble phosphate derivative 7 has been synthesised and has shown equivalent potency to combretastatin A-4 2 in its ability to inhibit of tubulin polymerisation and to displace colchicine from its binding site on tubulin.

E7010 8

E7010 8 (Figure 10) is a synthetic diaryl 2,3-diaminopyridine showing a broad spectrum of activity against various murine and human tumour cell lines and against human tumour xenografts. E7010 8 is an antimitotic agent which inhibits microtubule assembly and inhibits the binding of colchicine to tubulin. This agent 8 can be administered orally and preliminary data from Phase I clinical trial shows promising efficacy and tolerability.

Chalcones

The chalcones are a series of biaryl enones which show potent toxicity to several cancer cell lines and interact with tubulin at its colchicine-binding site. The most potent agents are those possessing a trimethoxy unit on one aryl ring.

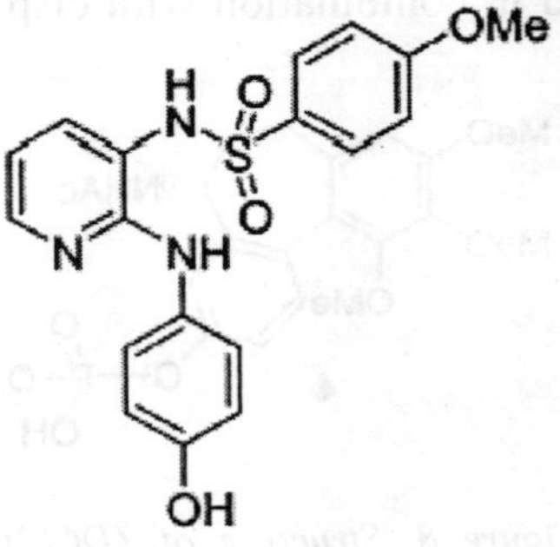

Figure 10. Structure of E7010 8.

Substitution analogous to that of combretastatin A-4 has provided the most potent agent 9 (Figure 11). However substitution with a methyl group a to the carbonyl increases the cytotoxicity several-fold eg chalcone 10. This increased cytotoxicity is thought to arise from geometric considerations. The a-methyl chalcones can adopt an s-trans conformation similar in shape to combretastatin A-4 2 whereas with a lack of methyl substitution, the chalcones prefer to adopt a less active s-cis conformation. Some chalcones have also been shown to bind to type II oestrogen receptors.

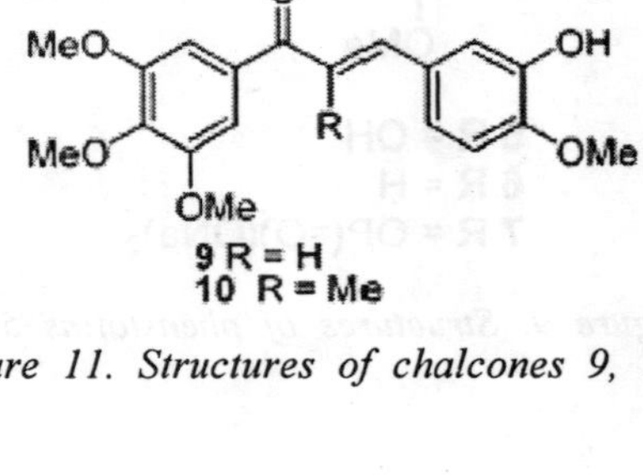

Figure 11. Structures of chalcones 9, 10.

Indanocine 11

Indanocine 11 (Figure 12) is potently cytotoxic to several tumour cell lines, but unlike other antimitotic agents, induces apoptotic cell death in stationary-phase multidrug-resistant (MDR) cancer cells at concentrations that do not impair the viability of normal non-proliferating cells. Three MDR cell lines tested were more sensitive to growth inhibition on treatment with indanocine 11 than in the parental cell lines. This sensitivity to MDR cancer cells suggests that indanocine 11 should be considered as a lead compound in the search for agents to combat drug-resistant disease.

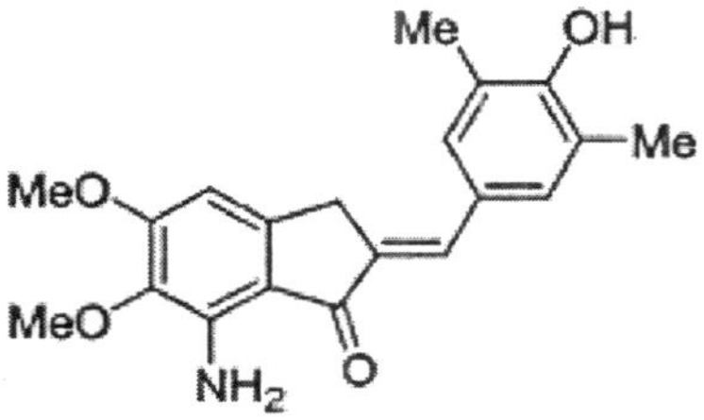

Me OH Me MeO MeO NH_2 O

11

Figure 12. Structure of indanocine 11.

T138067 12

T138067 12 (Figure 13) is a synthetic sulfonamide which disrupts tubulin polymerisation by binding covalently to the βII- and βIV-tubulin isotypes. Importantly, this agent 12 exhibits cytotoxicity against tumour cell lines that show substantial resistance to vinblastine, paclitaxel, doxorubicin and actinomycin D. T138067 12 is presently in Phase II clinical trial, and a similar amide, T900607, is also in Phase I clinical trials.

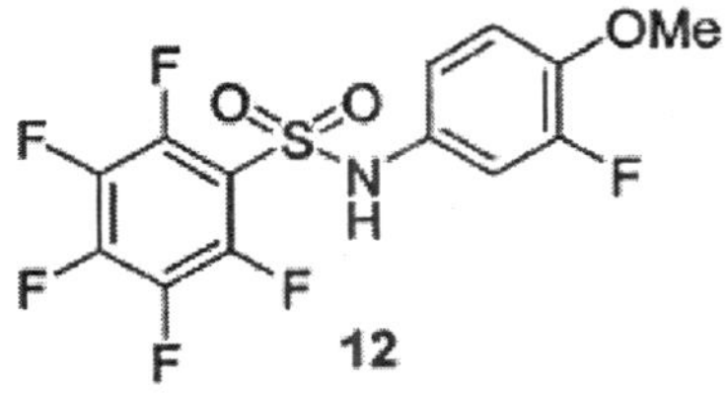

OMe F O S O F N H F F F F

12

Figure 13. Structure of T138067 12.

Indolyloxazolines

The synthetic oxadiazoline A-105972 13 (Figure 14) was recognised as a colchicine site binder and structural modifications have led to the discovery of indolyloxazolines as more stable and orally active antimitotic agents. A-289099 14 has been identified as the most active antimitotic indolyoxazoline against various cancer cell lines including those that express the MDR phenotype.

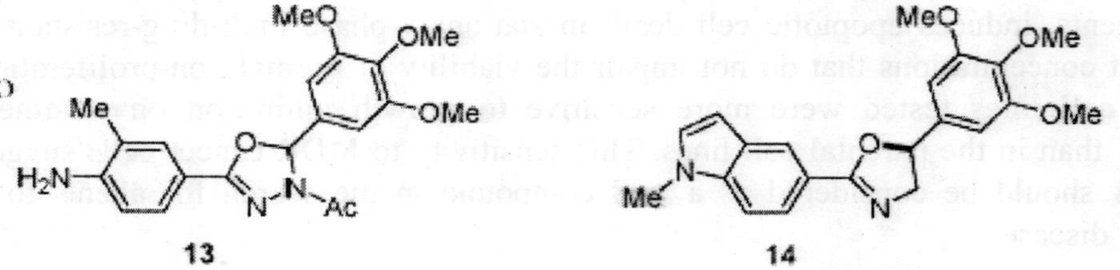

Figure 14. Structures of A-105972 13 and A-289099 14.

Oncocidin 15

Oncocidin 15 (Figure 15) is a simple diaryl ether developed as a thyroxin analogue. Oncocidin 15 is thought to bind to the colchicine site on tubulin and can adopt a 3D structure similar to combretastatin A-4. The agent 15 is orally bioavailable, has a rather low systemic toxicity in mice, and can induce apoptosis in breast carcinoma cells. These biochemical properties and the simple structure of Oncocidin 15 lend this type of agent suitable for the design of a new class of antimitotic agent.

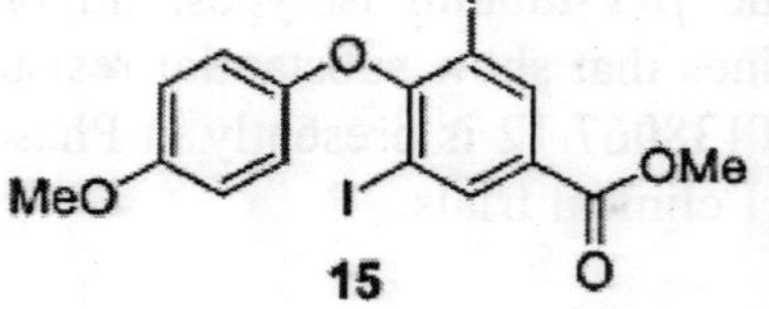

Figure 15. Structure of oncocidin 15.

Alkaloid Binding Sites

Vincas

Vincristine 16 and vinblastine 17 (Figure 16) are two complex bisindole alkaloids isolated from Catharanthus roseus G. Don (Madagascan periwinkle). Both agents inhibit the assembly of tubulin and are clinically used to treat various malignancies such as acute lymphoblastic leukaemia and Hodgkin's disease respectively.

Vincristine 16 and vinblastine 17 can tolerate only small changes to their structure without loss of activity. Vinorelbine 18 (Figure 17), a semi-synthetic vinca, has shown significant antitumour activity in diseases previously known to respond to vinca alkaloids. Several Phase II studies with vinorelbine 18 have been carried out. A more recent modification comprised the introduction of fluorine to give vinflunine 19.

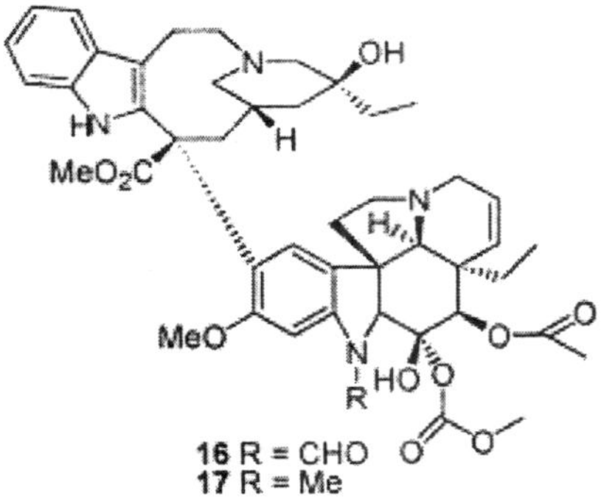

Figure 16. Structures of vincristine 16 and vinblastine 17.

Vinflunine 19 was identified in preclinical studies as having marked antitumour activity in vivo against a large panel of experimental tumour models, with tumour regressions being recorded in human renal and small cell lung cancer tumour xenografts. Overall its level of activity was superior to that of vinorelbine 18 in many of the experimental models and the agent is presently in Phase II clinical trials.

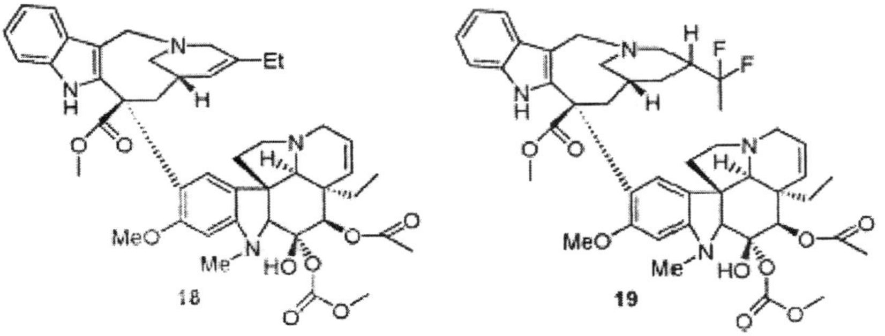

Figure 17. Structures of vinorelbine 18 and vinflunine 19.

Halichondrins

The halichondrin B 20 (Figure 18) is one of a group of complex polyether macrolides isolated from the marine sponges Halichondria okadai, Axinella carteri, Phakellia carteri and Lissodendoryx n. sp. 1. Halichondrin B 20 is a non-competitive inhibitor of the binding of vinblastine 17 and is thought to bind in the vinca domain of tubulin. The agent 20 has been successfully synthesised.

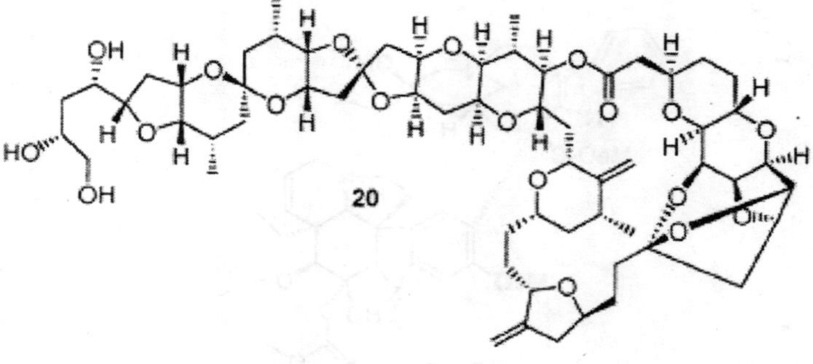

Figure 18. Structure of halichondrin B 20.

A related polyether, isohomohalichondrin B 21 (Figure 19) (isolated from Lissodendoryx n. sp. 1) has been shown to possess cytotoxic activity against multiple tumour types. Isohomohalichondrin B 21 has shown high antiproliferative activity in prostate cancer cell lines and is presently in preclinical evaluation. ER-86526 22 (Figure 20), a structurally simpler halichondrin analogue, is also in preclinical investigation.

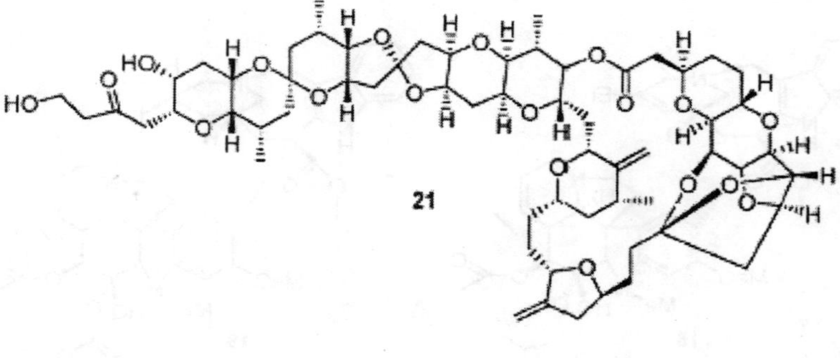

Figure 19. Structure of isohomohalichondrin B 21.

Pironetin 23

Pironetin 23 (Figure 20), a pyrone derivative isolated from Streptomyces sp. which acts as a plant growth regulator, has been successfully synthesised. Pironetin 23 inhibits tubulin polymerisation by binding to the vincristine site while promoting the binding of colchicine to tubulin. It induces the disassembly of microtubules in a dose-dependent and reversible manner. The agent 23 has also been shown to induce apoptosis via microtubule disassembly.

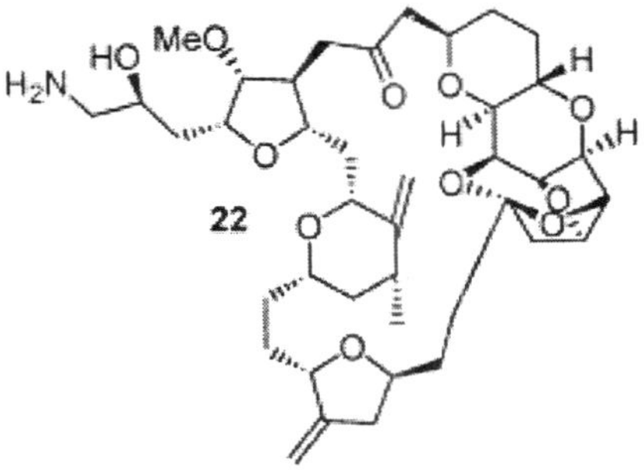

Figure 20. Structure of ER-86526 22.

Spongistatin 1 24 /Spiket P 25

The spongistatins are a group of complex macrocyclic lactones isolated from the sponges Spirastrella spinispirulifera and Hyrtios erecta. Spongistatin 1 24 (Figure 23) has no effect on the binding of colchicine to tubulin, but was a potent inhibitor of the binding of vinblastine and GTP to tubulin. Spongistatin 1 24 has shown particularly potent activity against several solid tumour cell lines including melanoma, lung cancer and colon cancers.

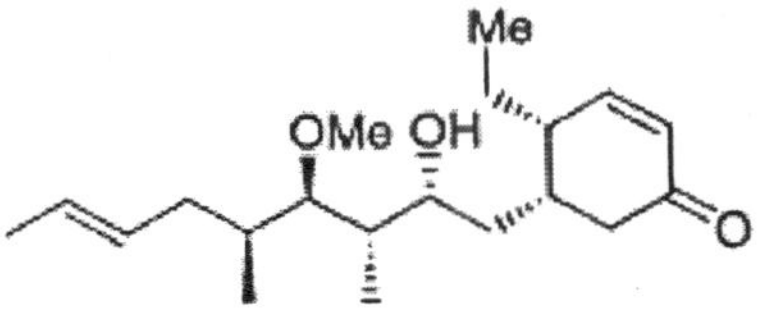

Figure 21. Structure of pironetin 23.

Even though the spongistatins are structurally complex the total synthesis of spongistatin 1 has been accomplished by several groups. Recently a structurally simpler, rationally designed spiroketal subunit of the spongistatins, Spiket P 25 (Figure 23), has been synthesised.

Spiket P 25 causes tubulin depolymerization in cell-free turbidity assays and exhibited potent cytotoxic activity against cancer cells as evidenced by destruction of microtubule organisation and prevention of mitotic spindle formation in human breast cancer cells. The discovery of simpler antimitotic agents derived from complex natural products such as Spiket P 25 should lead to synthetically more accessible second or third generation agents.

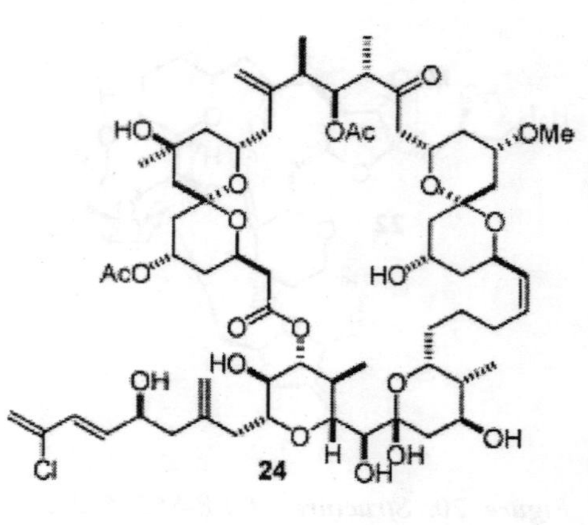

Figure 22. Structure of spongistatin 1 24.

Cryptophycins

The cryptophycins are a class of cyclic depsipeptides whose parent cryptophycin 1 26 (Figure 24) was isolated from the cyanobacterium Nostoc sp. Cryptophycin 1 26 suppresses tubulin dynamics and induces apoptosis. The cryptophycins are thought to bind at the vinca site on tubulin or a site that overlaps with the vinca site. Several synthetic analogues have been prepared and cryptophycin 52 27 has been discovered as an extremely potent antiproliferative agent showing activity in several drug resistant cell lines. These exciting properties have led to cryptophycin 52 27 entering Phase II clinical trials.

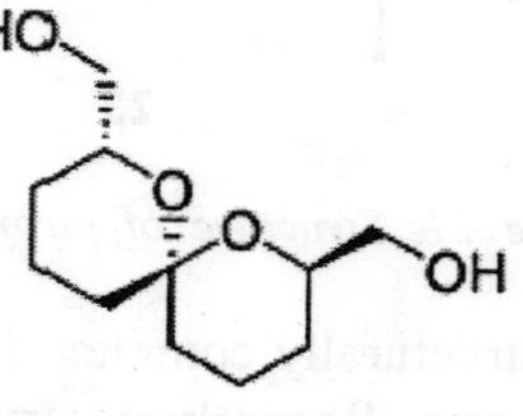

Figure 23. Structure of spiket P 25.

Dolastatins

The dolastatins are a group of pseudopeptides isolated from the Indian Ocean shell-less mollusc Dolabella auricularia. These agents inhibit microtubule formation by binding to β tubulin near the vinca domain and induce apoptosis. The lead compound dolastatin 10 28, has been synthesised. Several dolastatins and derivatives are currently under clinical investigation. These include dolastatin 10 28, cematodin 29 (a water-soluble pentapeptide analogue of dolastatin 15), and

TZT-1027 30 (Figure 25). This latter agent 30 has shown activities similar or superior to vincristine 16, the ability to damage tumour vasculature and the ability to induce apoptosis.

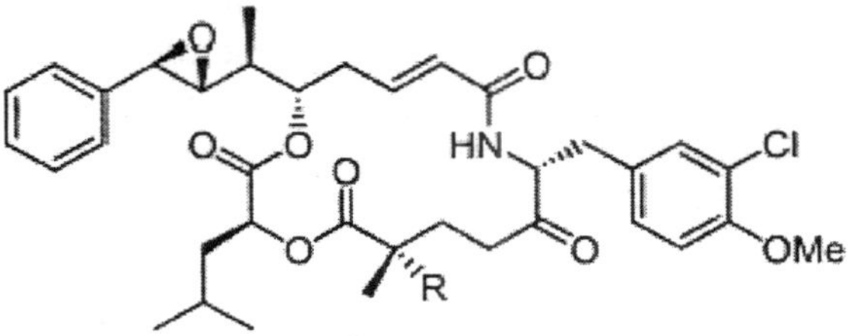

26 R = H
27 R = Me

Figure 24. Structures of cryptophycin 1 26 and 52 27.

Rhizoxin 31

Rhizoxin 31 (Figure 26) was identified as an antifungal agent from Rhizopus chinensis. This compound 31 interacts at the vinca site on tubulin, although it has also been considered to have its own binding site which overlaps the vinca site. The total synthesis of rhizoxin 31 has been accomplished. The agent 31 is presently in Phase II clinical trials.

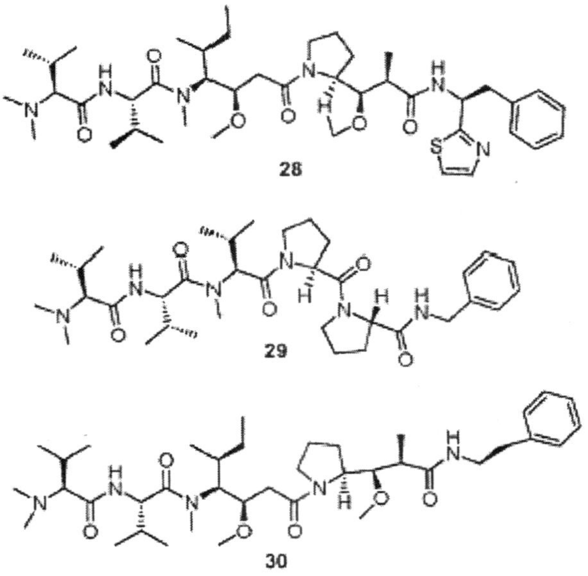

28

29

30

Figure 25. Structures of dolastatin 10 28, cematodin 29 and TZT-1027 30.

Paclitaxel Binding Site

Drugs that target the paclitaxel-binding site are known to act as microtubule-stabilising agents. Paclitaxel 32 was the first compound found to promote the assembly of tubulin heterodimers into microtubules and to stabilise them. The success of paclitaxel 32 stimulated the search for other natural products that could share the taxanes ability to inhibit the disassembly of microtubules and the extended family of microtubule-stabilising agents now includes the epothilones, discodermolide, laulimalide, the sarcodictyins, eleutherobin, GS-164 and FR182877.

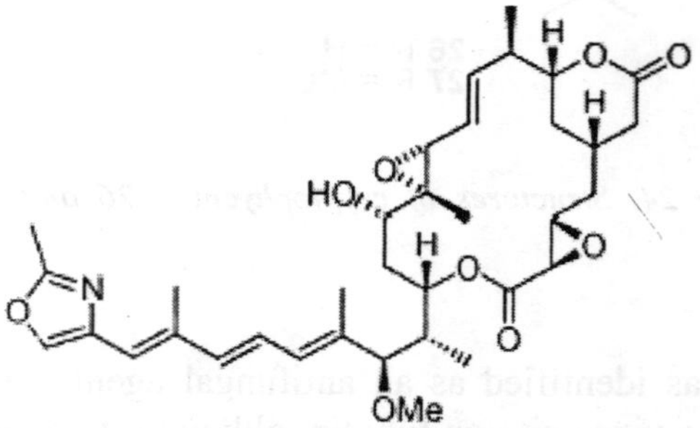

Figure 26. Structures of rhizoxin 31.

Taxanes

Paclitaxel (Taxol®) 32 (Figure 28) was discovered in 1964 when Wall and Wani isolated the compound from the Pacific yew tree, Taxus brevifolia, and noted its antitumour activity. By 1971, the two scientists had determined its structure and in 1979, Horwitz reported that paclitaxel 32 had the unique mechanism of action of stabilising microtubules, making it the prototype for a new class of chemotherapeutic agents. This discovery renewed interest in paclitaxel 32 and there have been substantial research programmes in taxane chemistry and biochemistry for almost twenty years now, directed at synthesising the challenging natural product, understanding how it works at the molecular level and designing improved microtubule-stabilising agents.

Paclitaxel 32 was first approved by the FDA for the treatment of ovarian cancer in 1993. Paclitaxel 32 and its semi-synthetic derivative docetaxel 33 (Figure 28) are now considered established drugs for the treatment of breast cancer and other malignancies. In more recent years, activities in the taxane area have focused on the major downsides of paclitaxel 32 and docetaxel 33 by improving their poor aqueous solubility and overcoming their multi-drug resistance.

Epothilones

The cytotoxic macrolide natural products, epothilone A 34 and B 35 (Figure 29), were isolated in 1993 from the myxobacterium Sorangium cellulosum as potent antifungal agents. It was not until 1995 that the paclitaxel-like mechanism of action of these compounds was discovered through induction of tubulin polymerisation and microtubule stabilisation. The epothilones 34, 35 were also claimed to be 30 to 50 times more water soluble than paclitaxel 32 and in contrast to paclitaxel, they inhibited the growth of MDR cell lines. Several total syntheses of epothilone A 34 and B 35 have been reported allowing the exploration of the various regions of the molecule and providing insight into the pharmacophore. The very detailed structure-activity relationship studies led to the discovery of desoxyepothilone B 36, which lacks the epoxide ring. Its biological superiority relative to epothilone B 35 and paclitaxel 32 has been demonstrated in a variety of competitive in vitro and in vivo settings and as a result deoxyepothilone B 36 has recently entered a Phase I clinical trial.

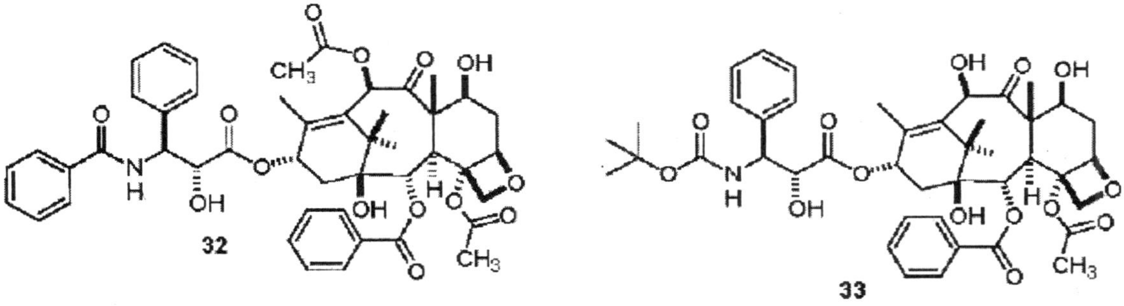

Figure 27. Structures of paclitaxel 32 and docetaxel 33.

Discodermolide 37

The polyhydroxylated lactone discodermolide 37 (Figure 29) was isolated from the marine sponge Discodermia dissoluta in 1990 and initially identified as a potential immunosuppressive agent. Total syntheses of the natural product 37 ascertained its structure and supply and further biological testing resulted in the discovery that discodermolide 37 was a microtubule-stabilising agent Discodermolide 37 competes with paclitaxel 32 for the same binding site with higher affinity, is more potent than paclitaxel 32 in stabilising microtubules and like the epothilones 34, 35 retains its growth inhibitory properties against MDR cell lines. Since discodermolide 37 is proving to be an interesting candidate for drug development, efficient total syntheses have been reported and SAR studies are currently underway.

Sarcodictyins and Eleutherobin

Sarcodictyin A 38 and B 39 (Figure 30) were first isolated from the Mediterranean soft coral Sarcodictyon roseum and found to enhance tubulin assembly. Later, eleutherobin 40 (Figure 32) was isolated from an Eleutherobia sp. and also found to stabilise microtubules. Their biological activities resemble those of paclitaxel 32 with slightly reduced potencies and eleutherobin 40 has been shown to be a substrate for the MDR pump. Total syntheses of the natural products have been reported, and very detailed SAR studies confirmed the importance of the urocanic side-chain and showed that some simplified analogues retained their microtubule-stabilising properties.

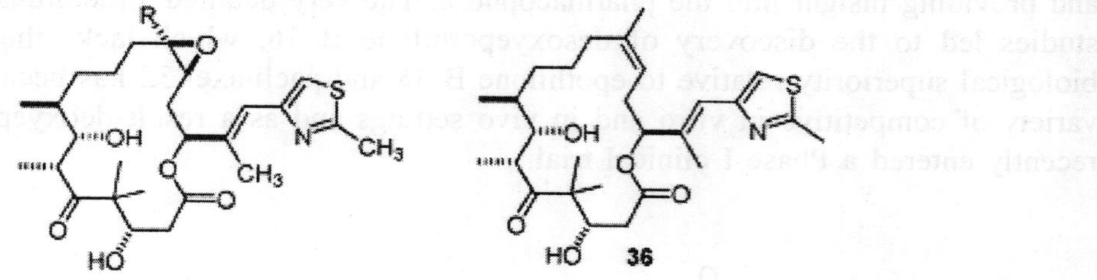

Figure 28. Structures of epothilone A 34, B 35 and desoxyepothilone 36.

Laulimalide 41

Although isolated in 1988 from the Pacific sponge, Cacospongia mycofijiensis, laulimalide 41 (Figure 32) was only recently identified as a microtubule-stabilising agent. Like the epothilones, it was also found to retain potency in MDR cell lines. Total syntheses of the macrolide 41 have been reported. GS-164 42 The synthetic molecule GS-164 42 (Figure 33) has been shown to cause cellular effects associated with microtubule-stabilising agents, although its potency is significantly lower than paclitaxel 32. This example demonstrates that paclitaxel-like behaviour towards microtubules can also be achieved with structurally simple molecules.

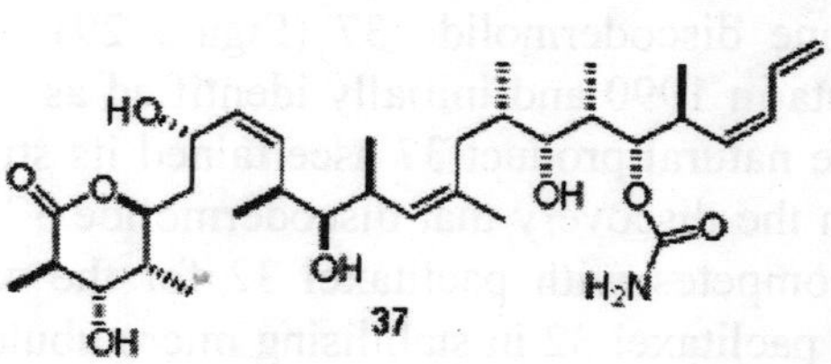

Figure 29. Structure of discodermolide 37.

FR182877 43

FR182877 43 (Figure 34) is the most recent addition to the microtubule-stabilising agents. It induces G2/M phase arrest and promotes microtubule assembly. Efforts to synthesise the complex metabolite 43 are in progress.

The ability of these structurally diverse molecules to interfere in the cell cycle by a similar mode of action, i.e., microtubule stabilisation, has led to speculation concerning the existence of a common pharmacophore that is recognised by the paclitaxel binding domain of β-tubulin or an overlapping binding site. Models have been proposed regarding this possibility and it is thought that these could lead to the design of better binding and more efficacious microtubulestabilising drugs in the future.

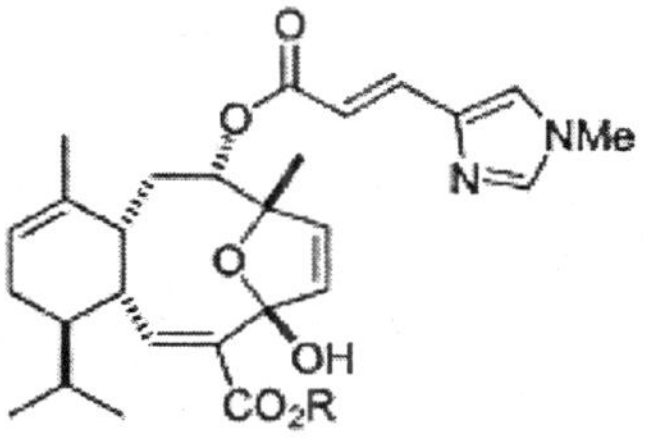

38 R = Me
39 R = Et

Figure 30. Structures of sarcodictyin A 38, B 39.

Unknown site binders

Cobras

The cobras 44, 45 (Figure 35) are rationally designed anticancer agents targeting a leucine-rich domain on a-tubulin. Treatment of human breast cancer and glioblastoma cells with cobra 1 45 causes destruction of microtubule organisation and apoptosis. The cobras are relatively simple tetrahydrofuran derivatives and would lend themselves to structure-activity studies.

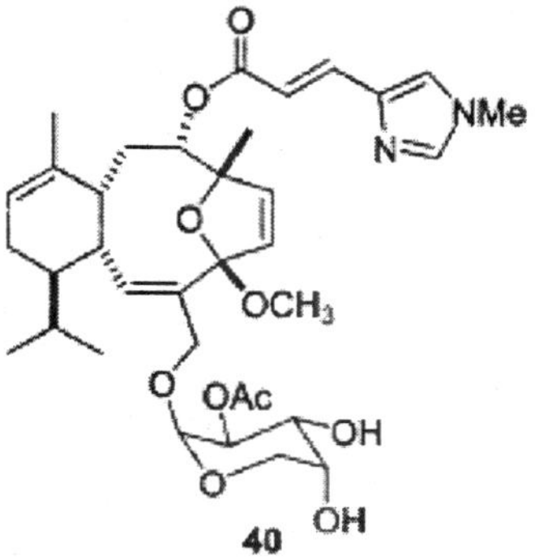

40

Figure 31. Structure of eleutherobin 40.

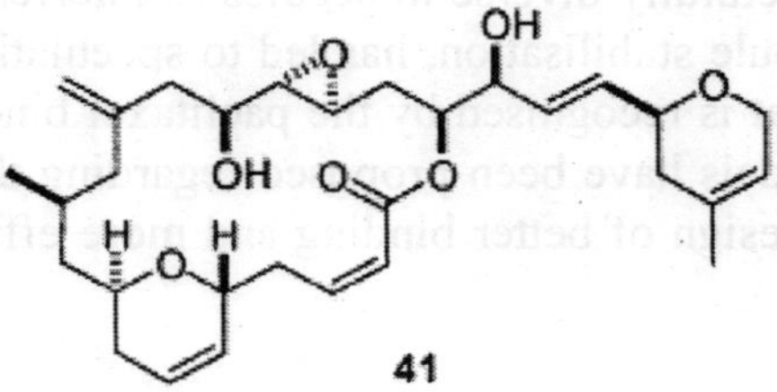

Figure 32. Structure of laulilamide 41.

D-24851 46

The indole D-24851 46 (Figure 36) is a synthetic agent which destabilises microtubules by interaction at an unknown site on tubulin. The indole 46 shows potent cytotoxicity to several human tumour cell lines and caused complete tumour regression in tumour-bearing rats. D-24851 46 is orally available, is effective against MDR tumours and lacks neurotoxicity. These factors indicate that this agent 46 or one derived from it may have significant potential for the treatment of various malignancies.

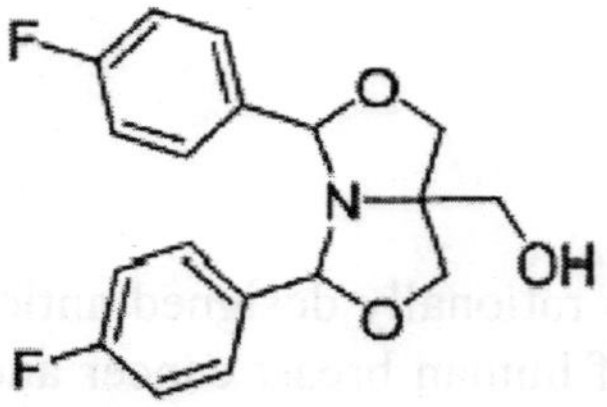

Figure 33. Structure of GS-164 42.

Other Biological Activities of Agents

Vascular Damage

Tumour vasculature currently represents a new target for the development of cancer therapy since a functional vasculature is critical to both the survival and growth of the tumour. The concept was pioneered by Folkmann who in 1972 identified the vasculature as a target for solid tumour therapy. Until recently, the majority of the research in this area focused on the discovery of antiangiogenic agents, that is agents which prevent the growth of new blood vessels within the tumour. More recently various strategies have been employed in antivasculature targeting. In contrast to the antiproliferating effects of antiangiogenic therapy, antivasculature approaches aim to cause a rapid and extensive shutdown of the tumour vasculature, leading to extensive tumour death.

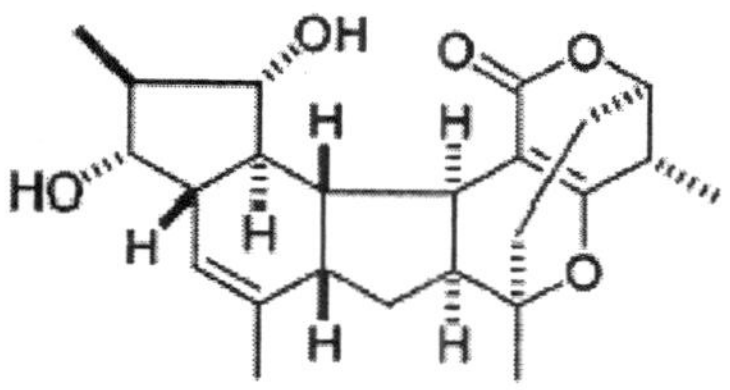

Figure 34. Structure of FR182877 43.

Differences in the architecture, cellular and biochemical composition between normal and tumour blood vessels provide the foundations for specific vascular targeting of tumours. Tumour vasculature endothelial cells, in contrast to normal endothelial cells, are highly proliferative. In addition, tumour vessels are usually thin-walled capillaries or sinusoids with little more than an endothelial lining backed by a basement membrane, and are prone to spontaneous haemorrhage and thrombosis. The tumour vasculature also contains vessels that are leaky to macromolecules, which, combined with poor lymphatic drainage, can lead to increased interstitial pressure, further impairing tumour perfusion.

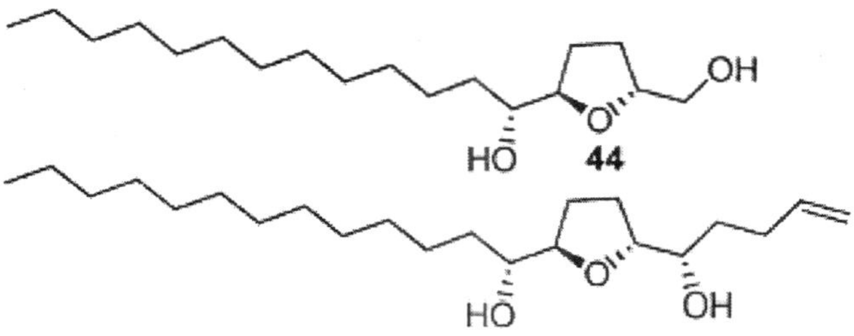

Figure 35. Structures of cobras 44 and 45.

Various approaches have been employed in antivascular targeting for cancer therapy. Targeting of the tumour vasculature directly was demonstrated using flavone acetic acid (FAA) 47 (Figure 37), a synthetic flavonoid, which has been shown to cause vascular shut-down and reduction in tumour blood flow in experimental tumours. These observations were found to be mediated by the production of necrosis factor alpha (TNF-alpha). Although the results of flavone 47 were never repeated in clinical trials, its related analogue, 5-6-dimethylxanthenone-4-acetic acid (DMXAA) 48 is currently under investigation in the clinic.

Considerable interest now lies with the tubulinbinding agents as potential tumour selective vasculature targeting drugs. Colchicine 1 was the first tubulin-binding agent noted to show antivasculature action, producing haemorrhagic necrosis in experimental tumours which resembled that produced by bacterial toxins. Furthermore, it appeared that endothelial cells of growing capillaries are sensitive to its toxic actions. Unfortunately the therapeutic window was very limited due to the drug's toxicity.

Chaplin and co-workers have reported that several of the tubulin-binding drugs could induce blood flow reduction in an experimental tumour system. Agents such as vincristine 16 and vinblastine 17, both potent anticancer drugs used in the clinic, also induce vasculature damage in animal tumours close to the MTD. Furthermore, newly identified vinca alkaloid derivatives, such as vinflunine 19, have demonstrated antivascular effects in the MAC 15A transplantable murine colon adenocarcinoma. Morphological changes, particularly the appearance of haemorrhagic necrosis, accompanying tumour growth delay, were suggestive of antivascular activity. However, vasculature shutdown of these MAC 15A tumours was obtained with vinflunine 19 at doses greatly below the optimum effective single dose.

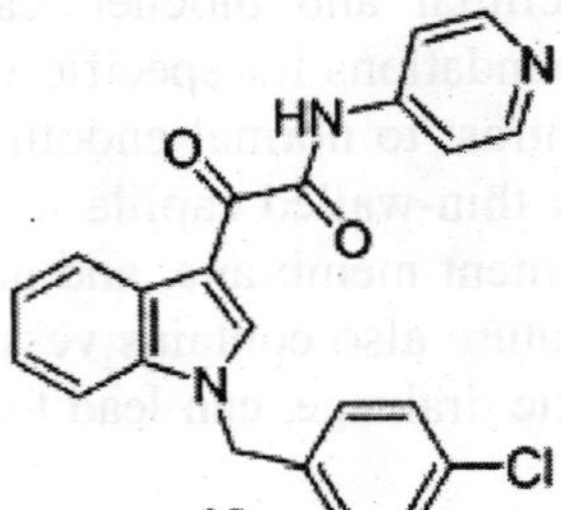

Figure 36. Structure of D-24851 46.

Much interest has recently been generated in the family of combretastatins. The most studied of these are combretastatin A-4 2 and its soluble phosphate prodrug 3. The prodrug 3 has been shown to induce vascular shutdown within tumours at doses less than onetenth of the MTD and to be more cytotoxic to rapidly proliferating rather than quiescent endothelial cells in vitro. It is not yet understood why proliferating endothelial cells show such sensitivity to the action of combretastatin A-4 2 and its prodrug 3 in vitro. Indeed the mechanisms governing the tumour-selective effects of the tubulin-binding agents are largely unknown, but it is likely that the inhibition of tubulin polymerisation affects endothelial cell shape. These cells are highly structured. They form the lumen of the blood vessels and the destruction of their cytoskeleton brings about the collapse of the vessel which is then blocked, starving the tumour of oxygen and nutrients.

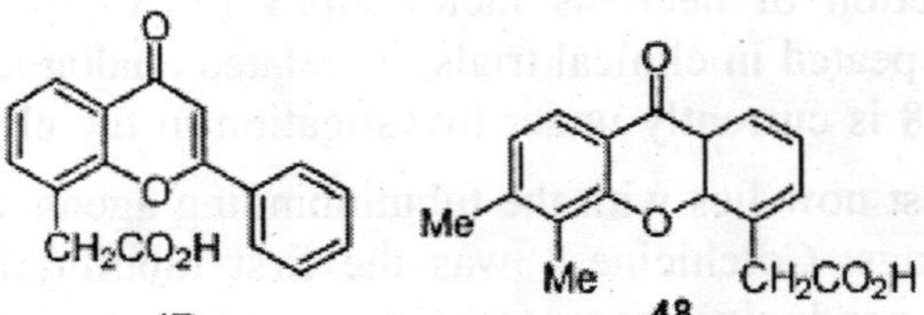

Figure 37. Structures of FAA 47 and DMXAA 48.

Topoisomerase Inhibition

Podophyllotoxin 49 (Figure 38) is another natural product (from Podophyllum peltatum) that has an affinity for tubulin. Podophyllotoxin 49 has been shown to bind to part of the colchicine site and to act as a competitive inhibitor. The agent 49 has been used for several hundred years in conditions ranging from sclerosis of the liver through constipation, rheumatism and cancer. As podophyllotoxin 49 proved to be too toxic in clinical trials, two less toxic analogues, etoposide 50 and teniposide 51 have been developed. Etoposide 50 is currently in clinical use in the treatment of cancers, particularly small cell lung carcinoma and testicular cancer.

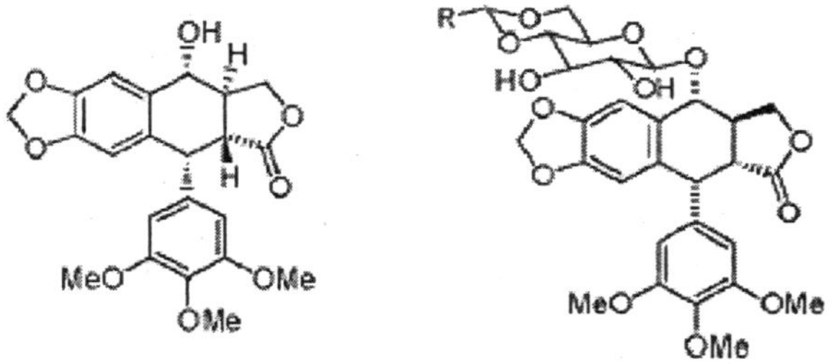

49

50 R = Me
51 R = 2-Thienyl

Figure 38. Structure of podophyllotoxin 49, etoposide 50 and teniposide 51.

The mechanism by which podophyllotoxin 49 blocks cell division is related to the inhibition of microtubule assembly in the mitotic spindle. However, etoposide 50 and teniposide 51 were shown not to be inhibitors of microtubule assembly which suggested that their antitumour properties were due to another mechanism of action. They have been found to interact with DNA and inhibit topoisomerase.

DNA topoisomerases are nuclear enzymes which make transient DNA strand breaks allowing the cell to manipulate the topology of its DNA. DNA topoisomerases are essential for DNA replication, transcription, chromosomal segregation and DNA recombination. Topoisomerases are divided into two groups by whether they transiently break only one DNA strand (type I enzymes) or both strands (type II). Etoposide 50 was the first agent recognised as a topoisomerase II inhibiting anticancer drug and its continued study has greatly helped the understanding of mechanisms by which drugs poison topoisomerase II.

Multi-Drug Resistance

A major problem in the treatment of cancer is the development of resistance to chemotherapy and subsequent relapse. Ultimately, patients may fail to respond to any therapy and exhibit clinical resistance to a wide range of structurally diverse drugs. Despite the appearance of clinical MDR in these patients, little is known about the underlying mechanisms. In contrast to the clinical situation, there is a broad cross-resistance to "natural product" anticancer drugs.

These include the vinca alkaloids, anthracyclines, epidophyllotoxins and other antimitotic agents such as colchicine and paclitaxel.

Mammalian cells exposed to a single cytotoxic natural product drug, such as vincristine or dactinomycin, can develop resistance to the selective agent and cross-resistance to a broad spectrum of structurally and functionally distinct antibiotics and alkaloids. This phenomenon, termed multidrug resistance has been widely studied. The most consistent feature of cells with high level MDR is amplification and overexpresion of genes encoding an integral plasma membrane protein known as P-glycoprotein.

The P-glycoprotein belongs to a family of proteins called the ABC (ATP-binding cassette) transporters. These proteins act as potent ATP-dependent drug efflux pumps, reducing the net intracellular concentration of natural product drugs. The drugs, almost all of which are amphipathic compounds that are preferentially soluble in lipids, interact with a specific binding domain on the P-glycoprotein molecule. This interaction facilitates their expulsion from the cell.

Typically, there is a large variety of cytotoxic agents to which multidrug cells are resistant. Vinca alkaloids, whilst very useful in the initial course of chemotherapy, are extremely good substrates for transport by P-glycoprotein and thus are of limited usefulness against drug resistant tumours. In contrast, cryptophycin 1 25 is an antimicrotubule agent which appears as a poorer substrate for P-glycoprotein than the vinca alkaloids. Hence it has reduced susceptibility to P-glycoprotein-mediated MDR, in comparison with other natural product anticancer drugs. Similarly combretastatin A-4 2 is active in cell lines overexpessing the MDR pump.

Paclitaxel 32 is another antimitotic agent which despite its preclinical and clinical success, cannot overcome the ability of tumours to develop an acquired resistance to its action. Although resistance to paclitaxel 32 has been shown to be mediated by P-glycoprotein, it could also result from alterations in its intracellular target, tubulin. Although precise molecular alterations in microtubules have not been identified, several mechanisms have been proposed. These include decreased tubulin, point mutations as evidenced by altered migration of a- or β-tubulin on two-dimensional SDS polyacrylamide gel electrophoresis, differential expression of β-tubulin isotypes, and acetylation of α-tubulin.

Chapter 13

Assay of Thyroid Hormones

In the 1950s, only one thyroid test was available - an indirect estimate of the serum total (free + protein-bound) thyroxine (TT4) concentration, using the protein bound iodine (PBI) technique. Since 1970, technological advances in radioimmunoassay (RIA) and immunometric assay (IMA) methodologies have progressively improved the specificity and sensitivity of the methods. Currently, serum-based methods are available for measuring both total (TT4 and TT3) and free (FT4 and FT3) thyroid hormone concentrations. In addition, measurements can be made of the thyroid hormone binding proteins, Thyroxine Binding globulin (TBG), Transthyretin (TTR)/Prealbumin (TBPA) and Albumin, as well as for the pituitary thyroid stimulator, thyrotropin (thyroid stimulating hormone, TSH) and the thyroid hormone precursor protein, Thyroglobulin (Tg).

The recognition that autoimmunity is a major cause of thyroid dysfunction has led to the development of tests for thyroid autoautoantibodies - thyroid peroxidase antibodies (TPOAb), thyroglobulin antibodies (TgAb) and TSH receptor antibodies (TRAb). Currently, thyroid testing is performed on serum specimens using either manual or automated methods employing specific antibodies. Methodology is still evolving as performance standards are established by the professional organisations and new technology and instruments are developed by manufacturers.

Hormone Measurements

Thyroxine (T4) circulates ~ 99.97% bound to the plasma proteins: TBG (60-75%); TTR/TBPA (15 -30%) and Albumin (10%); whereas triiodothyronine (T3) is ~ 99.7% bound, primarily to TBG. Technically, it has been easier to develop methods to measure the concentrations of total (free + protein-bound) thyroid hormones (TT4 and TT3) that circulate at nanomolar concentrations, in contrast to the free hormones (FT4 and FT3) that are measured in the picomolar range. Serum TT4 measurement has evolved through a variety of technologies over four decades. The PBI tests of the 1950s were replaced first by competitive protein binding methods in the 1960s, and later superseded by RIA methods in the 1970s.

Table 1. Agents Inhibiting Thyroid Hormone Synthesis and Secretion

	Results		*Likely Cause*	*Action*
Misleading Test	*TSH*	*FT4*		
FT4	↑	N	1. Untreated - mild hypothyroidism	1. Measure TPOAb.Confirm TSH after 6 weeks.
			2. Treated - inadequate L-T4 dose or noncompliance	2. Increase L-T4 dose/ councel compliance
FT4	↓	N or ↓	1. Mild (subclinical) hyperthyroidism	1. ? Autonomous functioning goiter
			2. Overtreatment with T3-containing prep.	2. Measure FT3 to rule out T3-toxicosis
FT4	N	↑	1. Common during L-T4 treatment	1. Expect higher FT4 in L-T4 Rx. hypo.
			2. Abnormal Binding proteins (i.e. FDH)	2 & 3. Check FT4 by alternate FT4 method ideally one using physical separation i.e. equilibrium dialysis or ultrafiltration
			3. Antibody interferences (T4 antibody, HAMAor rheumatoid factor	
FT4	N	↓	1. Bindingprotein competitor drugs	1. Check FT4 by method using minimal dilution
			2. Pregnancy	2. Check FT4 by albumininsensitive method. Use method- & trimesterspecific refrerences ranges
TSH	↑	N	1. Dysequilibrium (first 6-8 weeks of L-T4 Rx. for 1o hypothyroidism)	1. Recheck TSH before adjusting L-T4 dose. High TSH persists for months after Rx. severe hypo.
			2. HAMA & other interferences	2. Check TSH (new spec.) by alternate method
TSH	↓	N	1. Dysequilibrium (first 2-3 months post Rx. for hyperthyroid)	1. Use FT4 and FT3 during early Rx. of hyper to monitor thyroid status.

			2. Medications, i.e. glucocorticoids, dopamine	TSH may take months to normalise after starting Rx. for severe hyperthyroidism
TSH	N or ↑	↑	1. TSH-secreting pituitary adenoma	1. Check TSH (new spec.) by alternate method 2. TRH-stim or thyroid hormone suppression test 3. TSH alpha subunit 4. Pituitary Imaging
TSH	N	↓	1. Central hypothyroidism	1. Reduced bioactivity of immunoreactive TSH 2. ? other signs of pituitary deficiency 3. ? blunted (<2 fold) TRH response

Currently TT4 and TT3 concentrations are primarily measured by non-competitive immunometric assay (IMA) methods that use radioactivity, enzymes, fluorescence or chemiluminescent molecules as signals. Total hormone methods require the inclusion of inhibitors, such as 8-anilino-1-napthalene-sulphonic acid, to block hormone binding to serum proteins in order to facilitate hormone binding of to the antibody reagent.

Serum TT3 method development has paralleled that of TT4, the ten-fold lower TT3 concentration presents a sensitivity and precision challenge despite the use of higher specimen volumes. The inter-method variability for total hormone measurements (TT4 and TT3) is surprisingly large (10-17% and 20-30%, TT4 and TT3, respectively) considering that highly purified preparations of crystalline L-thyroxine and L-triiodothyronine are readily available and a reference technique has been established. This variability most likely relates to matrix differences between calibrators and the efficiency of the blocking agent employed by different manufacturers.

Clinical Performance

The diagnostic accuracy of total hormone measurements would equal that of free hormone if all patients had similar binding protein concentrations. Unfortunately, serum TBG abnormalities that distort the total: free hormone relationship, are commonly encountered in clinical practice (Table 2). Additionally, some patients have abnormal thyroid hormone binding proteins or autoantibodies that render total hormone measurements diagnostically unreliable (Table 1). Consequently, TT4 and TT3 measurements are rarely used as stand-alone tests, but are employed in conjunction with a binding protein estimate test [i.e. thyroid hormone binding ratio, THBR,] to form a free hormone index (FT4I or FT3I). The index approach corrects for common abnormalities in the proteins that bind thyroid hormones in the circulation. Total T4 reference ranges vary to some extent between methods but in general have approximated 58 to 160 nmol/L (4.5-12.6 μg/dL) for more than two decades. There is a new trend to use TT4 measurements in preference to FT4 estimate tests to monitor pregnancy, provided that the TT4 reference range is adjusted by a factor of 1.5 to accommodate the effects of the predictable TBG elevation. Likewise, serum TT3 reference values are also method dependent, with ranges approximating to 1.2 - 2.7 nmol/L (80-180 ng/dL).

Free Hormone Tests (FT4 and FT3)

In accord with the free hormone hypothesis, it is believed that the minute free fraction of hormone (0.02% versus 0.2%, FT4 versus FT3, respectively) is responsible for the biologic activity of thyroid hormones at the cellular level. It follows that free hormone measurement will better reflect the physiological effects of the hormone than total hormone measurements, especially when binding proteins are abnormal. Unfortunately, techniques for physically separating the exceeding small amount of free hormone from the dominant protein-bound moiety are too technically demanding, inconvenient and expensive for routine clinical laboratory use. Such physical separation methods, which employ equilibrium dialysis, ultrafiltration and gel filtration, are typically only available in reference laboratories.

Clinical laboratories prefer to use one of a number of single-test immunosequestration techniques that are designed to estimate free hormone concentration in the presence of protein-bound hormone. In reality, despite manufacturers claims, all of the FT4 and FT3 estimate tests are binding-protein dependent to some extent. The current trend is to use one of the newer single-test free hormone immunoassay methods in preference to the older two-test (total hormone + THBR) index approach.

Considerable confusion still surrounds the nomenclature of the free hormone tests and controversy continues regarding the technical validity of the measurements themselves, and their clinical utility in conditions associated with binding protein abnormalities of pregnancy and non-thyroidal illness (NTI). Methods that are sensitive to albumin concentrations, the presence of certain drugs, high free fatty acid (FFA) levels or hormone binding inhibitors are reported to give inappropriately abnormal results for such patients.

Nomenclature of Free Hormone Measurements

Free hormone measurements (FT4 and FT3) are made either by two-test index methods or single-test methods that include reference techniques that employ physical separation of free from bound hormone, and immunoassay "sequestration" methods that are usually automated. The single-test methods are either standardised with gravimetric preparations or use calibrators with values assigned by a reference method. Since the reference methods are manual, technically demanding techniques they are too inconvenient and expensive for routine clinical use.

Clinical laboratories typically use either two-test index or single-test immunoassays. Unfortunately, a confusing plethora of terms have been used to distinguish the different free hormone methods, and the literature abounds with inconsistencies in nomenclature of these tests. Currently, the methodologic distinction between terms such as "T7", "effective thyroxine ratio", "one-step", "analog", "two step", "backtitration", "sequential" and "immunoextraction" has become blurred, as manufacturers have modified the original techniques or adapted them for automation.

Following the launch of the original one-step "analog" tests in the 1970s, the term "analog" became mired in confusion. This first generation of hormone-analog tests were shown to be binding-protein dependent and have since been replaced by a new generation of labelled-antibody "analog" tests which are more resistant to the presence of abnormal binding proteins. Unfortunately, manufacturers rarely disclose all assay constituents or the number of steps involved in an automated procedure. Thus, it is not possible to use the method's nomenclature to predict its diagnostic accuracy for assessing patients with the binding protein abnormalities shown in Table 1.

Two-Test Index Methods (FT4I and FT3I)

Free Hormone Indexes (FT4I and FT3I) are based on simple calculations that modify the total hormone value to provide an approximation of the free hormone concentration in the presence of abnormal binding proteins. These indexes have been used to estimate free hormone concentrations for 40 years and require two separate measurements. One test is a measurement of total hormone concentration (TT4 or TT3), the other, an assessment of binding protein concentration using either (a) a TBG immunoassay, (b) a Thyroid Hormone Binding Ratio (THBR) or "Uptake" test, or (c) an estimate of the free hormone fraction determined by isotopic dialysis or ultrafiltration. The purity of the hormone tracer critically impacts the diagnostic accuracy of indexes calculated using isotopic THBR tests or free fraction assessments.

Current THBR tests are usually able to produce normal FT4I and FT3I values when TBG abnormalities are mild (i.e. pregnancy and estrogen therapy). However, these tests often fail to normalise FT4I and FT3I values in euthyroid patients with grossly abnormal binding proteins (congenital TBG extremes, familial dysalbuminemic hyperthyroxinemia (FDH), thyroid hormone autoantibodies and NTI). These tests also have impaired diagnostic accuracy in the presence of certain drugs that influence thyroid hormone protein binding.

TBG Immunoassays

There is no improvement in diagnostic accuracy of free hormone indexes (TT4/TBG) calculated using direct TBG measurement in preference to a THBR ("uptake") test (Table 2). Further, the TT4/TBG index approach is not independent of the TBG concentration, nor does it correct for non TBG-related binding protein abnormalities (Table 1). Thus, despite the theoretical advantages of using direct TBG measurements, indexes employing THBR ("uptake tests") appear to be diagnostically superior to the TT4/TBG index approach that is now rarely used.

Thyroid Hormone Binding Ratio (THBR)

The first "T3 uptake" tests developed in the 1950s employed the partitioning of T3-I tracer between the plasma proteins in the specimen and an inert scavenger (red cell membranes, talc, charcoal, ion-exchange resin or antibody). The scavenger "uptake" of T3 tracer was an indirect, reciprocal estimate of the TBG concentration of the specimen. Initially, T3 uptake tests were reported as percent uptakes (free/total tracer).

T3 binding to TBG is ten-fold lower than T4-TBG binding, resulting in a greater percentage of unbound T3 tracer being available for scavenger pick up, and consequently required shorter gamma counting times. As non-isotopic technology became more widely used, labelled T4 became the preferred hormone for the THBR testing because a T4 "uptake" more appropriately correct for T4-binding protein effects. Current THBR tests usually produce normal FT4I and FT3I values when TBG abnormalities are mild (i.e. pregnancy). However, these tests may not produce normal index values when patients have congenital TBG extremes, familial dysalbuminemic hyperthyroxinemia (FDH), thyroid hormone autoantibodies, nonthyroidal illness or medications that directly or indirectly influence thyroid hormone binding to plasma proteins.

Free Fraction Measurements

The first free hormone tests developed in the 1960s were indexes, calculated from the product of the free hormone fraction, measured isotopically by dialysis, and TT4 measured by PBI and later, by RIA. The early isotopic detection systems were technically demanding and included paper chromatography, electrophoresis, magnesium chloride precipitation and column chromatography. The free fraction index approach was later extended to ultrafiltration and symmetric dialysis, the latter measuring the rate of transfer of isotopically-labelled hormone across a membrane separating two chambers containing the same undiluted specimen. These techniques eliminated the dilution effects that had been shown to influence tracer dialysis values. The free hormone indexes calculated with isotopic free fractions are not completely independent of the TBG concentration and furthermore, are influenced by tracer purity and the buffer matrix employed by the method.

SINGLE-TEST FREE HORMONE METHODS

These methods fall into two categories –methods employing physical separation of free from protein-bound hormone, and immunoassays that employ sequestration.

Physical Separation

These methods are generally considered reference methods. A variety of techniques are employed to physically isolate free from protein-bound hormone before a sensitive immunoassay is used to measure the free hormone concentration in the free hormone fraction using gravimetrically prepared standard solutions. The physical isolation of the free from protein-bound hormone is accomplished with either

(a) dialysis through a semi-permeable membrane,

(b) ultrafiltration, or

(c) Sephadex LH-20 resin adsorption chromatography.

Although these methods are generally considered "gold standard" methods they are prone to dilutional errors that may cause underestimation of FT4 when binding inhibitors are present in the specimen. Further, such methods are too labour intensive and expensive for use in the routine clinical laboratory setting.

Equilibrium Dialysis and Ultrafiltration Methods. Ellis and Ekins first reported the direct measurement of FT4 in dialysates in 1973. Subsequently, the technique has been applied to serum ultrafiltrates. In the last decade, such physical separation methods have been refined by the adoption of more physiologic buffer diluents and improvements to the design of dialysis cells and ultrafiltration devices.

Gel Absorption Methods. These methods are based on the ability of free hormone to penetrate the internal compartments of Sephadex LH-20 columns. Protein-bound hormone is excluded and eluted before the free fraction, which is assayed for hormone by sensitive RIA. Commercial kits have been based on this methodologic approach but currently are not widely used.

An exquisitely sensitive T4 RIA method is needed to measure the picomolar concentrations of FT4 in free fraction isolates as compared with total hormone that circulates at nanomolar concentrations. The requisite sensitivity has been achieved by using higher affinity T4 antibodies (>1x10 ^{11}L/mol) and higher specific activity T4-I tracers. It has been more difficult to apply this approach to FT3 measurement since T3 RIA sensitivity has been limited by the requirement for T3 antibodies with affinities >3x10 ^{11}L/mol.

A number of direct absolute FT4 methods employing equilibrium dialysis and ultrafiltration have been developed for both research and commercial use. It is generally considered that the direct absolute methods are least influenced by binding proteins and by inference, provide free hormone values that best reflect circulating free hormone concentrations. When available, they may be useful for evaluating the thyroid status of the unusual patient in whom routine FT4 test results and serum TSH values are diagnostically discordant. These methods are also important for assigning values to the calibrators used for some of the immunoassay methods.

There is closer agreement between the reference ranges of methods employing physical separation than between the FT4 immunoassays and tracer dialysis techniques. The reference ranges for FT4 immunoassay methods approximate to 9-23 pmol/L (0.7 –1.8 ng/dL). In contrast, the upper FT4 limit for methods such as equilibrium dialysis that employ physical separation extends above 30 pmol/L (2.5 ng/dL). Reference ranges for FT3 immunoassay methods approximates to 3.5-7.7 pmol/L (23 –50 ng/L). FT3 methods that employ physical separation are only available as research assays.

FT4 and FT3 Immunoassay Methods

Free hormone immunoassays use a specific hormone antibody to sequester a small amount of the total hormone. The antibody occupancy, which is proportional to the free hormone concentration, is quantified using gravimetric standards or calibrators with free hormone values assigned by a reference equilibrium dialysis method. The actual proportion of total hormone sequestered (up to 5%) varies with the methodologic design, but greatly exceeds the actual free hormone concentration. The key to the validity of these methods is the use of conditions that maintain the free to protein-bound hormone equilibrium and minimise dilution effects that weaken the influence of any endogenous inhibitors present in the specimen. Three general approaches have been used to develop comparative FT4 and FT3 immunoassay methods:

(a) Two-step Labelled-Hormone;

(b) One-step Labelled-Analog; and

(c) Labelled Antibody.

Two-Step, Labelled-Hormone

This approach was first developed by Ekins and colleagues in the late 1970s and subsequently adapted to form the basis of many current FT4 and FT3 methods. Two-step methods typically employ anti-hormone antibody bound to a solid support. The solid-phase, high affinity (>1x10 ^{11}L/mol) antibody sequesters a small proportion of total hormone from a diluted serum specimen during the first incubation step. Unbound serum constituents are washed away before the addition of labelled hormone that is taken up by the unoccupied antibody-binding sites during a second incubation step. After washing, the amount of labelled hormone bound to the solid-phase antibody is quantified relative gravimetric standards or calibrators that have free hormone values assigned by a method employing physical separation. Following the introduction of the less labour-intensive one-step labelled hormone-analog methods two-step techniques lost popularity, despite comparative studies showing that they were less affected by the binding protein abnormalities that were problematic for most one-step methods. Currently, some manufacturers still base their FT4 and FT3 methods on the two-step approach.

One-Step, Labelled Hormone-Analog

Classic "one-step" methods use hormone analogs having molecular structures that claim to be totally non-reactive with serum proteins, but still able react with unoccupied hormone antibody sites. If these methods were physiochemically valid, the hormone-analog, coupled to a signal-generating molecule such as an isotope or enzyme, could compete with free hormone for a limited number of anti-hormone antibody binding sites in a classic competitive immunoassay format. Unfortunately, although conceptually attractive, this approach was technically difficult to achieve in practice, despite early claims of success.

The hormone-analog methods were principally engineered to give normal FT4 values in high TBG states (i.e. pregnancy). They were found to have poor diagnostic accuracy in the presence of abnormal albumin concentrations secondary to Familial Dysalbuminemic Hyperthyroxinemic (FDH), NTI, high FFA concentrations or thyroid hormone autoantibodies. During the 1980s, attempts were made to overcome the albumin dependence of these methods by adding propriety chemicals to block analog binding to albumin or by empirically adjusting calibrator values to correct for protein-dependent biases.

Labelled Antibody

Most clinical laboratories now use free hormone methods based on this approach. Labelled antibody methods are competitive immunoassays that measure free hormone as a function the fractional occupancy of hormone-antibody binding sites. The methods use specific immunoabsorbants to quantify the unoccupi ' antibody binding sites in the reaction mixture. The physiochemical theory of analog-based labelled antibody methods would suggest that they would be as susceptible to the same errors as the labelled hormone-analog methods described above. However, the physicochemical differences arising from the binding of analog to the solid support confer kinetic differences that result in decreased analog affinity for endogenous binding proteins and more reliable free hormone measurements. The labelled antibody approach is becoming the favoured technique for automated free hormone testing.

Clinical Performance Free Hormone Tests

The impetus for free hormone test development has been the high frequency of binding-protein abnormalities encountered in clinical practice. The only reason to select a free hormone method (FT4 or FT3) in preference to total hormone is to improve the diagnostic accuracy for detecting hypo- and hyperthyroidism in patients with thyroid hormone binding abnormalities that compromise the diagnostic utility of total hormone measurements. Situations where current FT4 methods may give misleading results are listed in Table 1. Unfortunately, the diagnostic accuracy of the free hormone methods cannot be predicted either from their methodologic classification (Table 2) or an in vitro test of technical validity, such as dilution.

The indirect index tests (FT4I) and the direct comparative methods are all protein dependent to some extent, and most are prone to underestimate FT4 in patients with binding protein abnormalities. These binding protein abnormalities cause discordances between the total and free hormone concentrations. Unfortunately, no current FT4 method is universally valid in all clinical situations. However, pre-analytical or analytical assay artifacts may arise in a number of situations associated conditions not related to abnormal TBG concentrations: (a) when the binding of the assay tracer to albumin is abnormal; (b) in the presence of medications that displace T4 from TBG and (c) during critical NTI. Assay artifacts causing paradoxically abnormal FT4 values are more common than assay problems causing inappropriate TSH as shown in Table 1.

Anomalous FT4 values are suspected when there is gross FT4/TT4 discordance or a history of medications affecting FT4 methodology. In these situations it is useful to check the FT4 result with another manufacturer's FT4 test. Gross discordance between two methods suggests that the FT4 abnormality is a technical artifact and that TSH is the more reliable test, unless the discrepancy is due to central hypothyroidism (Table 1).

Common conditions that decrease the diagnostic accuracy of current free hormone tests in ambulatory patients include;

a) TBG abnormalities (TBG excess or deficiency);

b) Familial Dysalbuminemic Hyperthyroxinemia, FDH;

c) T4 and T3 autoantibodies and

d) interfering substances such as Rheumatoid Factor and Heterophilic antibodies (HAMA).

There is growing concern that the lack of method specific, trimester-specific reference ranges for current FT4 tests renders these tests unreliable for assessing thyroid status during pregnancy when there are progressive changes in binding proteins. TT4 measurements may be a better option if the reference limits are adjusted for increased TBG by a factor of 1.5.

Additionally, free hormone tests have not received adequate testing in hospitalised patients in whom NTI and drug therapies decrease the diagnostic accuracy of both thyroid hormone and TSH testing. Ideally, three classes of hospitalised patients should be tested: a) patients without thyroid dysfunction but with low or high TT4 due to NTI; b) patients with hypothyroidism associated with severe NTI and, c) patients with hyperthyroidism associated with NTI. However, since it is not feasible for the manufacturers to obtain such pedigreed patient specimens, it will continue to be difficult to ensure the diagnostic reliability of a FT4 test made in a sick hospitalised patient.

A combination of FT4 plus TSH is recommended for assessing the thyroid status in this setting. In most clinical situations involving discordant FT4 and TSH results, the TSH test is usually the most reliable, provided that the patient is not receiving medications such as glucocorticoids and dopamine that directly inhibit TSH secretion, or the condition involves pituitary failure (Table 1). When the diagnostic accuracy of the FT4 test is questionable, the demonstration of gross discordance between the relationship between FT4 and TT4 measurements may suggest problems with the FT4 and more rapidly available than an FT4 measurement made by a reference method employing physical separation.

TSH Assays

Over the last thirty years, the sensitivity and specificity of Thyroid Stimulating Hormone/ Thyrotropin (TSH) assays have under- gone dramatic improvements. The first clinical TSH assays developed in the 1960s used RIA methodology. These methods were too insensitive to detect TSH in all euthyroid subjects and were limited to the diagnosis of primary hypothyroidism. In 1970, the hypothalamic tripeptide, Thyrotropin Releasing Hormone (TRH) (Thyroliberin) was synthesized, and shown to stimulate serum TSH into the measurable range in all euthyroid subjects but not patients with hyperthyroidism or hypopituitarism. These observations led to the practice of measuring serum TSH 15-30 minutes following a 200-500μg IV TRH dose as a way to overcome the insensitivity of the TSH RIA methodology.

The practice of TRH testing fell into decline after the development of the inherently more sensitive TSH immunometric assay (IMA) methods (also called "sandwich" or "noncompetitive" assays) that became available in the mid-1980s. These IMA techniques were based on the excess antibody approach of Miles and Hales, originally reported in the 1960s. This new approach could not be universally adopted until the 1980s when monoclonal antibody technology facilitated the large-scale production of specific antibodies.

IMA methods were quickly found to be ten-fold more sensitive than the older RIA methods and as such were able to distinguish between the basal TSH levels of hyperthyroid and euthyroid subjects without TRH testing. By 1990, IMA methodology had replaced most RIA methods and TSH euthyroid reference ranges became narrower as a result of reduced glycoprotein hormone cross-reactivity and improved precision.

In the last decade, TSH assay sensitivity has been further enhanced by the adoption of non-isotopic signals such as chemiluminescent and fluorescent molecules that are inherently more sensitive than isotopes and, in addition, are easier to automate. Most current TSH methods are automated and report sensitivities down to 0.01 mU/L. With this level of sensitivity, TSH measurement can be used to diagnose both overt and mild (subclinical) degrees of thyroid hormone excess without TRH stimulation.

Methodology and Nomenclature

TSH assay "quality" has historically been defined by its clinical sensitivity – the ability to discriminate between hyperthyroid and euthyroid values. During the RIA era considerable efforts were made to develop methods that could achieve this clinical discrimination. Although some success was reported using modified RIA procedures employing a pre-treatment lectin affinity extraction or a long pre-incubation period before tracer addition, these were principally research techniques. The TSH IMA methods launched in the mid-1980s rapidly replaced the insensitive RIA tests because they could distinguish hyperthyroid from euthyroid values after only a relatively short incubation. Mechanistically, these IMA methods employed an excess of TSH monoclonal antibody, bound to a solid support to capture TSH from the serum specimen during a 20 to 120 minute incubation period.

A different poly- or monoclonal TSH antibody, targeted to a different TSH epitope(s) and labelled with a signal, was included in the reaction mixture. After removing unbound constituents by washing, the signal bound to the solid support is directly proportional to the serum TSH concentration. Later modifications to this basic concept have included the use of fragmented monoclonal antibodies to reduce interference by heterophilic antibodies and the use of Avidin-Biotin and magnetic particle separation techniques. The first IMA methods used a radioisotopic signal (I-) and were designated "immunoradiometric assays", or IRMAs. Current TSH IMA methods use a variety of non-isotopic signals that have given rise to a lexicon of terminology to distinguish between assays using different signals. Immuno enzymometric assays (IEMA) use enzyme signals; immunofluorometric assays (IFMA) use fluorophors as signals, immunochemiluminometric assays (ICMA) use chemiluminescent molecules as signals and immunobioluminometric assays (IBMA) use bioluminescent signal molecules.

Initially, TSH IMA methods were designated as "sensitive", "highly sensitive", "ultrasensitive" or "supersensitive" assays, terms used to distinguish the new IMA methodology from the older insensitive RIA methods then still in use. This descrip-tive nomenclature was confusing and led to a debate concerning the meaning of "sensitivity". It is now accepted that TSH between-run precision should be the determinant of TSH assay sensitivity.

A new parameter "functional sensitivity", defined as the TSH value associated with 20 percent CV established from assays run over a 6 to 8 week period, is now generally accepted as the determinant of the lower reporting limit for TSH assays. During the last decade, a generational nomenclature system based on functional sensitivity (FS) has been widely adopted.

This nomenclature defines each generation as having a ten-fold difference in functional sensitivity. For example, the older RIA methods are designated "first generation" (FS = 1-2 mU/L); IRMA methods "second generation" (FS = 0.1 – 0.2 mU/L) and the even more sensitive current non-isotopic methods (IFMA, ICMA and IBMA) referred to as "third generation" (FS = 0.01 –0.02 mU/L).

Serum TSH Values

Serum TSH values are higher in neonates and children. Although serum TSH exhibits a diurnal variation with a peak between midnight and 0400, diagnostic information is not significantly influenced by the time of the blood draw. The reference range for euthyroid adults has progressively contracted over the last two decades. Improvements in the sensitivity of TSH methodology over the last three decades, from first-generation (1G) radioimmunoassays to current 3G immunometric assays (IMA), have facilitated the determination of the lower TSH reference limit. This contraction also reflects the adoption of IMA methodology employing monoclonal capture antibodies that eliminate crossreactivity with other glycoprotein hormones (a problem with the 1G RIAs) and the use of more sensitive and specific Thyroid Peroxidase Antibody (TPOAb) immunoassays that have replaced insensitive antimicrosomal antibody (AMA) tests for excluding individuals with autoimmune thyroid conditions, the inclusion of whom skew TSH reference limits.

There is a high degree of confidence that the mean TSH value for truly euthyroid subjects lies in the 1.0-1.5 mIU/L range, and that the lower 2.5% confidence limit approximates to 0.4 mIU/L. This being the case, the upper limit of the Gaussian distribution would be expected to approximate 2.5 mIU/L. However, a higher upper reference limit is cited for most current TSH assays (3.5 to 4.0 mIU/L) based on population studies. Possible reasons for the persistent 'skew' in the TSH upper limit include:

It is difficult to reliably exclude all individuals with thyroid dysfunction when determining the TSH reference range - given the high prevalence of thyroid dysfunction in the general population exaccerbated by the sensitivity and/or specificity limitations of current TPOAb tests. Although a number of possible reasons for the persistent 'skew' in the TSH upper limit are listed above, the dominant cause is likely to be the inclusion of TPOAb-negative individuals with occult autoimmune thyroid dysfunction that is not detected as a humoral response.

In fact, it has now been shown that the inclusion of TPOAb-positive persons in the cohort of subjects used for the TSH reference range skews the upper but not the lower limit. The need for rigorous exclusion of thyroid antibody-positive individuals is reinforced by the new NACB consensus guideline #22 that states, "TSH reference intervals should be established from the 95 % confidence limits of the log-transformed values of at least 120 rigorously screened normal euthyroid volunteers who have; (a) No detectable thyroid autoantibodies, TPOAb or TgAb; (b) No personal or family history of thyroid dysfunction; (c) No visible or palpable goiter and, (d) Who are taking no medications except estrogen". Figure 1 contrasts the new reference range for euthyroid subjects and the target range for L-T4 replacement therapy with the typical laboratory reference range (0.4-4.0 mIU/L).

The new TSH upper limit of 2.5 mIU/L is supported by the data from the Whickham follow-up study that found an increased risk of hypothyroidism in individuals with serum TSH > 2.0 mU/L, especially if TPOAb was present. This report suggests that when longitudinal study data are available, the TSH upper reference limit may fall further.

Clinical Performance

A number of current guidelines recommend the use of TSH as the first-line test for detecting both hypo- and hyperthyroidism in ambulatory patients with stable thyroid status and intact hypothalamic/pituitary function. This strategy is considered more cost-effective than a panel approach (TSH + FT4 or FT4 + FT3) but necessitates the use of a TSH assay with a functional sensitivity below 0.02 mU/L (i.e. third generation). However, this TSH-first strategy can miss patients with central hypothyroidism or TSH secreting pituitary tumours. As shown in Figure 1, third generation sensitivity (< 0.02 mU/L) is critical for the detecting subnormal TSH values, since less sensitive second generation assays are prone to produce falsely negative (normal range) values for sera with subnormal TSH concentrations.

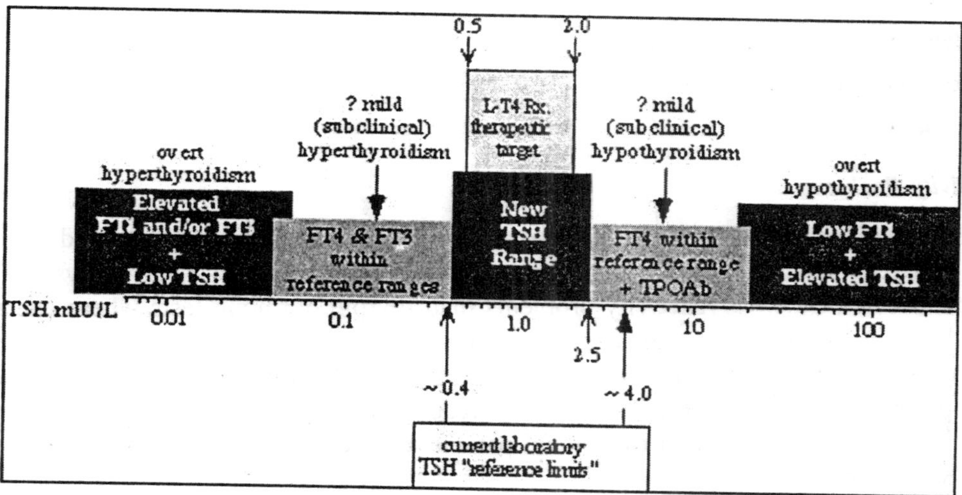

Figure 1. TSH Reference Ranges

The reliable detection of subnormal TSH is critical for evaluating both ambulatory and hospitalised patients. In older ambulatory patients, mild hyperthyroidism is a risk factor for atrial fibrillation. In hospitalised patients, it is important to be able to distinguish the mild TSH suppression typical of euthyroid patients with NTI (0.02 –0.4 mU/L) from the profound TSH suppression typical of sick hyperthyroid patients. Additionally, the use of a wider reference limit (0.02 –10 mU/L) is thought to improve its positive predictive value of TSH measurement for evaluating sick hospitalised patients.

Thyroid Test Methodologies

It is difficult for the laboratory to proactively detect interference from a single measurement i.e. a first-line TSH test. Interference is more often suspected by the physi-cian who observes that a reported value is inconsistent with the clinical status of the patient. Classic laboratory checks of analyte identity, such as dilution, may not always detect an interference problem.

The most practical way to test for an interference is to measure the specimen by a different manufacturers method and check for discordance between the test results reflecting that methods vary in their susceptibility to interfering substances. In addition, in the case of suspected TSH interferences, the physician can perform a TRH-stimulation or thyroid hormone suppression test as a biological check. Interferences producing a falsely elevated basal TSH values will usually be associated with a blunted (<3-fold increase) TRH response, or a less than the 90 percent suppression by 48 hours following the oral administration of 1mg of L-T4 or 200μg L-T3.

Interferences in competitive immunoassays and non-competitive IMAs fall into four classes:

(a) crossreactivity problems,

(b) endogenous analyte antibodies,

(c) heterophilic antibodies and

(d) drug interactions.

Crossreactivity Problems

The specificity of an assay depends on the ability of the antibody reagent to discriminate flawlessly between analyte and it's structurally related target ligand. TSH assays are more likely to be affected by cross-reactivity problems than thyroid hormone tests where chemically pure iodothyronine preparations are available for selecting antibody specificity. The use of monoclonal antibodies for current TSH IMA methods has virtually eliminated cross-reactivity with glycoprotein hormones (LH or hCG) that plagued the early TSH RIA methods. However, even when using current assays an occasional patient will be identified with an unusual cross-reacting TSH isoform.

Endogenous Antibodies

In 1956 Robbins and colleagues first reported an unusual thyroxine binding protein globulin in serum. Subsequently, both T4 and T3 as well as TSH autoantibodies have been identified in sera from patients with autoimmune thyroid and nonthyroid disorders. Although there are a number of reports of anomalous total and free thyroid hormone, and TSH test values caused by T3, T4 or TSH autoantibodies, these autoantibodies rarely cause interference with current methods. Endogenous antibody interferences are characterised by either falsely low or falsely high values, depending on the type and composition of the assay.

Heterophilic Antibodies

HAMA affect IMA methods more than competitive immunoassays by forming a bridge between the capture and signal antibodies, thereby creating a false signal, resulting in a high value artifact. The inappropriate value may not necessarily be abnormal but may be inappropriately normal. Since antibodies cross the placenta, they have the potential to influence the neonatal screening result. Heterophilic antibody interference falls into two classes:

HAMA: These are relatively weak multispecific, polyreactive antibodies that are frequently IgM rheumatoid factor are often called human anti-mouse antibodies (HAMA). HAMA has been reported to interfere with both total and free T4 and T3, as well as TSH measurements. Approaches to reduce this form of interference have included the use of Fab fragments and hetero-species assay configurations.

HAAA: Specific human anti-animal (immunoglobulins) antibodies (HAAA) are produced against well-defined specific antigens following exposure to a therapeutic agent containing animal antigens (i.e. murine antibody) or by coincidental immunisation through workplace exposure (i.e. animal handlers). Although assays for HAMA have been developed, inter-method differences are so large that these tests are currently not reliable for detecting heterophilic antibody interference.

Drug Interferences

Drug interferences may relate to in-vitro or in-vivo effects. In-vitro affects result when the specimen contains a sufficient concentration of certain therapeutic and diagnostic agents to produce methodologic interference. Alternatively, following heparin administration, the heparin in the specimen can cause in-vitro stimulation of lipoprotein lipase that liberates free fatty acids that inhibit T4 binding to serum proteins. In-vivo effects on thyroid test results are caused by the therapeutic agents themselves (i.e. furosemide) competitively inhibiting thyroid hormone binding to serum proteins in the specimen, thereby causing abnormal (usually low) thyroid hormone test results. In certain pathologic conditions such as uremia, abnormal serum constituents such as indole acetic acid may accumulate and interfere with thyroid hormone binding. Thyroid test methods employing fluorescent signals may be sensitive to the presence of fluorophor-related therapeutic or diagnostic agents in the specimen.

Thyroid Specific Autoantibodies

Tests for antibodies against thyroid-specific antigens, anti-thyroid peroxidase (TPO), thyroglobulin (Tg) and TSH receptors are used in the diagnosis of autoimmune thyroid disorders. Over the last four decades, antibody measurement techniques have evolved from semi-quantitative agglutination and complement fixation tests and whole animal bioassays to specific ligand assays using recombinant antigens and cell culture systems transfected with the human TSH receptor. Unfortunately, the diagnostic and prognostic value of these thyroid autoantibody measurements is hampered by differences in the sensitivity and specificity of current methods. Although autoantibody tests have inherent clinical utility in a number of clinical situations, these tests should be selectively employed.

Thyroid Peroxidase Autoantibodies (TPOAb)

TPOAb antibodies were initially detected as antibodies against thyroid microsomes (antimicrosomal antibody, AMA) using semi-quantitative complement fixation and tanned erythrocyte hemaagglutination techniques. Recent studies have identified the principal antigen in the AMA tests as the thyroid peroxidase (TPO) enzyme, a 100kD glycosylated protein present in thyroid microsomes. Currently, manual agglutination tests are being replaced by automated, more specific TPOAb immunoassay or immunometric assay methods, based on purified or recombinant TPO. Despite the use of the same International Reference Preparation (MRC 66/387), inter-method variability of current TPOAb assays is significant. It appears that both the methodologic principles of the test, and the purity of the TPO reagent used, may influence the sensitivity, specificity and normal reference range of these methods.The variability in the sensitivities (range <0.3 to >20 kIU/L) and reference ranges of the different methods has led to different concepts regarding the "normality" of a detectable TPOAb. Specifically, assays characterised by a low detection limit (<10 kIU/L) typically report unmeasureable TPOAb values for normal subjects, suggesting that the detection of this autoantibody is a pathologic finding.

Clinical Use of TPOAb Tests

High between-method variability precludes the numeric comparison of serum TPOAb values measured by different tests.

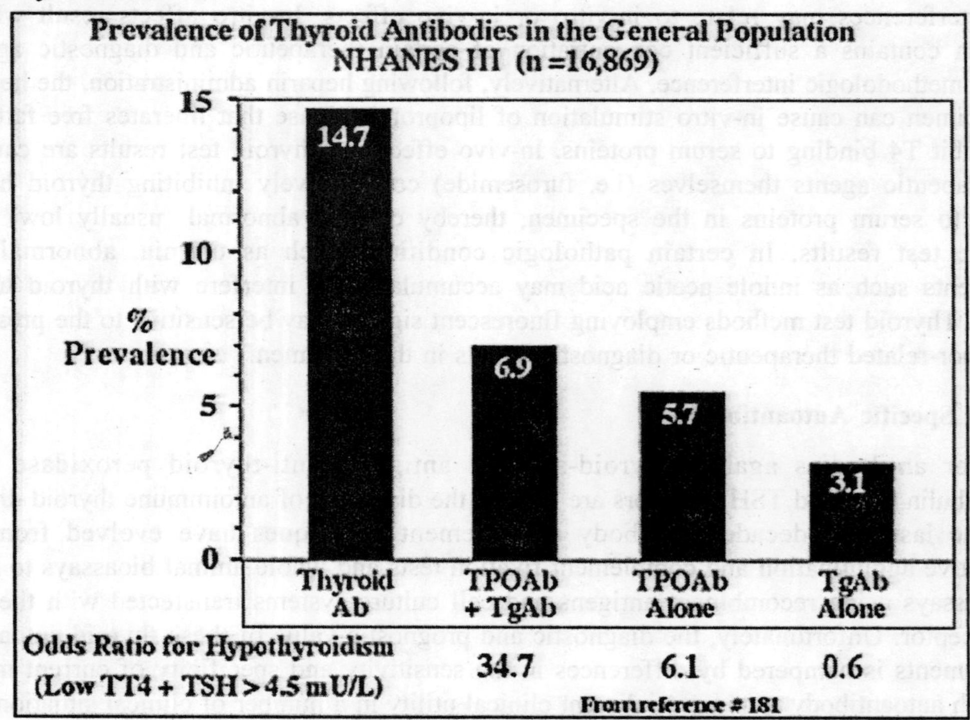

Figure 2. Thyroid Autoantibody Prevalences and Associations with Hypothyroidism

An abnormal TPOAb is detected in 12 to 14 percent of "healthy" euthyroid subjects and even higher percentages of patients with various non-thyroid autoimmune disorders. Approximately 70-80 % of patients with Graves' disease and virtually all patients with Hashimoto's, atrophic thyroiditis or post-partum thyroiditis have TPOAb detected. In fact, TPOAb is implicated as a cytotoxic agent in the destructive thyroiditic process.

TPO antibodies are involved in the tissue destructive processes associated with the hypothyroidism observed in Hashimoto's thyroiditis. Some studies suggest that TPOAb itself may be cytotoxic to the thyroid. Estimates of TPOAb prevalence depend on the sensitivity and specificity of the method employed. The recent NHANES survey of ~17,000 subjects without apparent thyroid disease, reported that TPOAb was detected in 12.6 % of subjects using a competitive immunoassay method. As shown in Figure 2, the odds ratio for hypothyroidism was strongly associated with the presence of TPOAb and not TgAb.

TPOAb as a Prognostic Indicator

In the future, TPOAb measurement may be used as a prognostic indicator for thy-roid dysfunction. Although the appearance of TPOAb usually precedes the development of thyroid dysfunction, recent studies suggest that a hypoechoic ultrasound pattern may precede a biochemical TPOAb abnormality, as shown in Figure 3.

The paradoxical absence of TPOAb in some patients with unequivocal TSH abnormalities likely reflects the suboptimal sensitivity and/or specificity of current TPOAb tests or non-autoimmune thyroid failure (atrophic thyroiditis). Although changes in autoai. ibody concentrations often reflect a change in disease activity, serial thyroid autoantibody measurements are not recommended for monitoring treatment for AITD. This is not surprising since treatment of AITD addresses the consequence (thyroid dysfunction) and not the cause (autoimmunity) of the disease. The presence of TPOAb has now been established as a risk factor for developing thyroid dysfunction in patients taking Amiodarone, Interferon-alpha, Interleukin-2 or Lithium therapies.

TPOAb prevalence is increased in patients with non-thyroid autoimmune diseases such as type 1 diabetes and pernicious anemia. Aging is associated with an increasing prevalence of TPOAb that parallels the increasing prevalence of both subclinical (mild) and clinical hypothyroidism. As shown in Figure 3, the detection of TPOAb in a euthyroid subject is a risk factor for the future development of hypothyroidism as shown by the 20-year follow up of the Whickham survey.

Longitudinal studies suggest that even low levels of TPOAb, that could not be detected by the insensitive AMA agglutination tests used in the 1970s, are likely reflect the presence of occult thyroid dysfunction and are a risk factor for the development of clinical hypothyroidism over a timescale of years or decades. Specifically, as shown in Figures 2 and 3, a detectable TPOAb is a risk factor for hypothyroidism and typically precedes the development of an elevated TSH.

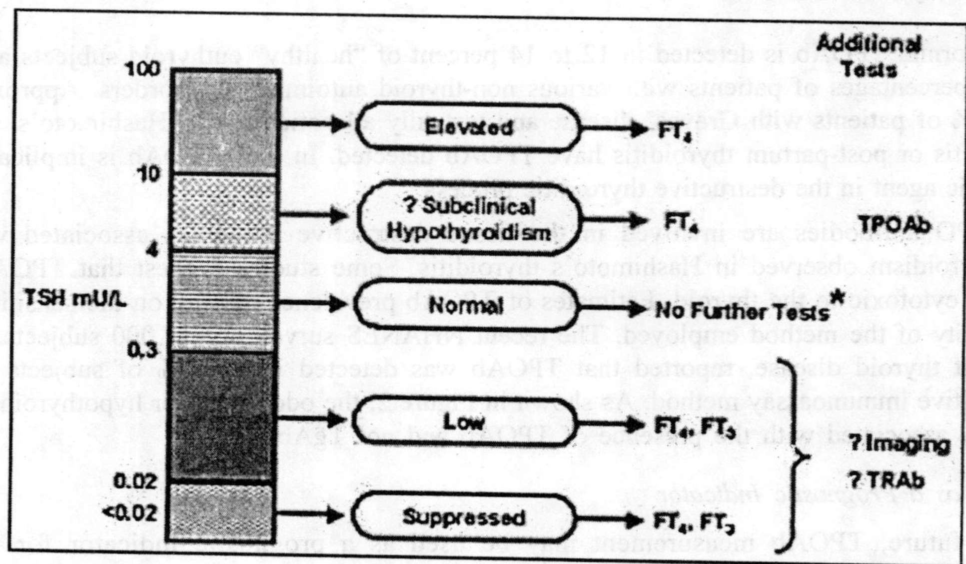

Figure 3. Developing Autoimmune Thyroid Dysfunction (Hashimoto's Thyroiditis)

In addition, the presence of TPOAb has been linked to reproductive complications such as miscarriage, infertility, IVF failure, fetal death, pre-eclampisa, pre-term delivery and postpartum thyroiditis and depression. It is generally considered cost-effective to replace semi-quantitative AMA agglutination tests by TPOAb immunoassay methods, since the enhanced sensitivity and specificity of the new immunoassays obviates the need for additional TgAb measurements in the routine diagnosis of autoimmune thyroid disorders.

Thyroglobulin Autoantibodies (TgAb)

Antithyroglobulin autoantibodies (TgAb) were the first thyroid antibodies to be recognised to circulate in patients with autoimmune thyroid disorders. The first TgAb methods were based on tanned red cell hemagglutination. Subsequently, methodologies have evolved in parallel with TPOAb methodology from semi-quantitative techniques, to more sensitive ELISA and RIA methods and more recently chemiluminescent immunoassays). This variability likely reflects both the purity and the epitope specificity of the Tg protein reagent, as well as the inherent heterogeneity of the antibodies present in different patient sera.

As with TPOAb methods, TgAb tests report a wide range of sensitivity limits (<0.3 to >20 kIU/L) as well as reference ranges, despite the use of the same International Reference Preparation (MRC 65/93). There is some data to suggest that low levels of TgAb may be present in normal individuals. However, studies suggest that the TgAb epitope specificity expressed by normal individuals and patients with either differentiated thyroid carcinoma or thyroid autoimmune disorders may be qualitatively different. These differences in test specificity can impact the suitability of a TgAb method for screening sera for prior to serum Tg measurement.

Clinical Use

Autoantibodies against Tg are encountered in autoimmune thyroid conditions, usually in association with TPOAb. However, the recent NHANES III study found that 3 % of subjects with no risk factors for thyroid disease had detectable TgAb without TPOAb. In these subjects with only TgAb detected, no association with TSH abnormalities was found so that the clinical significance of an isolated TgAb abnormality remains to be established. This suggests that it is unnecessary to measure both TPOAb and TgAb for a routine evaluation of thyroid autoimmunity. TgAb is primarily used as an adjunctive test for serum Tg measurement.

Current guidelines recommend that all sera be prescreened for TgAb by a sensitive immunoassay method, prior to Tg testing. Immunoassay methods detect TgAb in approximately 20% of patients with differentiated thyroid carcinoma (DTC) compared with 10 % of normal subjects. There appears to be no threshold TgAb concentration above normal that precludes TgAb interference. When patients with differentiated thyroid carcinoma have TgAb detected, serial measurements of TgAb concentrations serve as an independent parameter for detecting changes in tumour mass. For example, TgAb-positive patients who are rendered athyreotic by surgery display progressively declining TgAb concentrations during the first few post-operative years. Such patients usually become TgAb-negative after four years. In contrast, a rise or appearance of serum TgAb concentrations is often the first indication of recurrence.

When using serial TgAb measurements as a surrogate marker for changes in tumour burden, between-method differences necessitate the use of the same manufacturer's TgAb test kit. Usually, serial serum Tg RIA and TgAb changes in such patients are concordant. However, a disparity between these two parameters may indicate recurrence, since as tumour mass increases and more Tg is secreted more Tg-TgAb immune complexes may form, which may be cleared more rapidly. Early data suggesting that quantitation of Tg mRNA in peripheral blood might be an additional tumour parameter for detecting the presence of thyroid tissue in TgAb-positive patients with DTC has now been discredited. Recent studies report that false positive results may arise from assay artifacts or illegitimate transcription. Additionally, false negative results are seen in patients with metastatic disease.

TSH Receptor Autoantibodies (TRAb)

TSH Receptor Antibodies (TRAb) were first recognised as a long-acting thyroid stimulator (LATS) using mouse bioassays. These autoantibodies are directed against epitopes on the ectodomain of the TSH receptor. Studies carried out by several laboratories suggest that TRAb epitopes, as well as TSH binding domains, are distributed, throughout the extracellular parts of the receptor. However, the complex structure of the TSH receptor makes it difficult to localise the epitopes targeted by these autoantibodies. Methods for measuring TRAb are even more varied than for TPOAb and TgAb.

Two classes of TRAb can be associated with autoimmune thyroid disorders – (a) thyroid stimulating autoantibodies (TSAb) that cause Graves' hyperthyroidism and (b) thyroid stimulation-blocking antibodies (TBAb) which block receptor binding of TSH. Attempts to distinguish stimulatory from inhibitory TRAb epitopes have not provided clear results. Each class of TRAb (TSAb and TBAb) may be detected alone or in combination in Graves' disease and Hashimoto's thyroiditis.

The relative concentrations of the two classes of TRAb may modulate the severity of Graves' hyperthyroidism and may change in response to therapy or pregnancy. Two different methodologic approaches have been used to quantify TRAb. Bioassays have been used to measure the stimulating class of antibodies (TSAb). In contrast, radioreceptor assays using isolated, solubilised or more recently, recombinant human TSH receptors have been used to develop thyroidbinding inhibition immunoglobulin (TBII) assays that detect both classes of TRAb.

TSAb Methods

These bioassay techniques use the cAMP second messenger system as the biological endpoint of activity. Assays have been based on both homogeneous (human) and heterogeneous (animal) cells expressing TSH receptors. Early homogeneous assays used surgical human thyroid specimens, whereas heterogeneous assays used mice, guinea pigs and later rat FRTL-5 cell lines. Recently, assays have been developed using cell lines transfected with the human TSH receptor.

TBII Methods

These radioreceptor assays measure the ability of the serum or an immunoglobulin preparation to inhibit the binding of 125I- labelled TSH to solubilised TSH receptors. Typically, animal tissues have been used for TSH receptor preparation. Recently, methods using human receptor expressed in CHO cells or recombinant receptor protein have been developed Cloning of the TSH receptor greatly benefits bioassay development, as TSH receptor transfected cell lines are more useful tools than thyroid cells.

Currently, inter-method differences are wide and the interassay precision of the different tests is very variable so that is difficult to compare values from different methods. New chimeric assays are becoming available, which may be able to target the loci of TRAb epitopes and TSH binding sites may better link assay responses to clinical outcome. New "second generation" assays employing human recombinant TSH-Receptor, are reported to offer improved sensitivity for diagnosing Graves' disease without a loss of specificity. However, preliminary studies of disease suggest that such methods do not appear to offer improved predictability for the response to anti-thyroid drug (ATD) treatment.

Clinical Use of TRAb Tests

TRAb tests are used in the differential diagnosis of hyperthyroidism, the prediction of fetal and neonatal thyroid dysfunction due to transplacental passage of maternal TRAb and prediction the course of Graves' disease treated with antithyroid drugs. Most measurements are made with TBII tests because these methods are available in commercial kit form. Although TBII assays do not directly measure the stimulating antibodies, these tests have comparable diagnostic sensitivity to TSAb bioassays (70-95%) for diagnosing Graves' hyperthyroidism or detecting a relapse or response to therapy. The second generation assays employing human recombinant

TSH receptor are now becoming available and are reported to have superior diagnostic sensitivity for Graves' disease. Current tests are manual and expensive and vary in precision, sensitivity, specificity and reference ranges.

TBII test use is geographically sensitive to differences in practice patterns and reimbursement. In the United States, where reimbursement restricts biochemical testing, TRAb tests are rarely used and thyroid hormone and TSH measurements are preferred for diagnosing or monitoring treatment for Graves' hyperthyroidism. However, the TBII tests are important for evaluating pregnant patients with a history of autoimmune thyroid disease, in whom there is a risk of transplacental passage of TRAb to the infant. Both stimulating and inhibiting activities should be tested since the expression of thyroid dysfunction may be different in the mother and the infant.

The lack of specificity of the TBII methods is actually an advantage in this clinical situation, since a TBII test will detect both the stimulating and blocking classes of TRAb that can produce transient hyper- or hypothyroidism, respectively, in the fetus and newborn. TRAb plays an uncertain role in thyroid-associated opthalmopathy (TAO), which appears to be exacerrbated by radioiodine therapy. Since TRAb and other thyroid antibodies levels increase acutely significantly after radioiodine therapy, a TRAb measurement prior to radioiodine therapy may be useful to predict risk of TAO. However, prospective studies are needed to establish the clinical utility of TRAb measurement in this context. Patients with very high circulating concentrations of hCG due to choriocarcinoma or hydatiform mole, as well as a small number of pregnant patients, may have misleading positive results using TSAb assays.

Thyroglobulin (Tg) Measurement

Serum Tg measurement is used as a tumour marker in the management of patients with differentiated thyroid carcinomas (DTC). Current Tg methods are based either on IMA or RIA techniques. There is a trend for non-isotopic IMA methods to replace RIA methods because IMA methods are easier to automate, have shorter turn around times, wider working ranges and use reagents with a longer shelf life.

Methodologic Problems

Unfortunately, serum Tg measurement is technically challenging and there are five methodologic problems that impair the clinical utility of the test:

(a) between-method biases;

(b) suboptimal sensitivity;

(c) suboptimal interassay precision;

(d) "hook" problems (IMA methods) and

(e) Tg autoantibody (TgAb) interference.

Between-Method Biases

Current methods should be standardised on the International Reference Preparation CRM-457.

Even when methods are standardised against CRM-457 there can be a twofold difference between serum Tg values as shown in Figure 4. This magnitude of between-method bias necessitates that the same Tg method be used for the serial monitoring of patients. Laboratories considering changing to a Tg method having a bias of >15% relative to their current method should alert physicians and allow them sufficient time to re-baseline patients. The causes of Tg method variability are multifactorial and include standardisation differences, matrix effects and Tg epitope recognition differences. The typical serum Tg reference range for a CRM-457 standardised method approximates 3 to 40 µg/L.

Suboptimal Sensitivity

Serum Tg should be detectable in the sera of all normal euthyroid subjects. As shown in Figure 4, current Tg methods can barely discriminate between the lower limit of the euthyroid reference range and the assay detection limit. TSH suppression, which reduces serum Tg by ~50%, exaccerbates assay sensitivity limitations. Insensitive assays are unable to detect small amounts of tumour, especially when TSH is suppressed. A recent metaanalysis reports that 20% of patients with undetectable serum Tg (<1µg/L) during thyroid hormone suppression therapy (THST) display detectable serum Tg values 72 hours after recombinant human TSH (rhTSH) administration. Recent reports also suggest that more sensitive Tg assays will have improved clinical sensitivity for detecting disease without the need for TSH stimulation. Recommendations are that in the case of Tg assays the human serum pools used for precision assessment should not contain any detectable TgAb and the precision should be assessed across the typical clinical interval used for the test, which in the case of serum Tg is 6-12 months.

Suboptimal Between-Run Precision

Normal euthyroid individuals with stable TSH status exhibit intra-person biologic variability of ~14%across a 4-month period. When serum Tg is being monitored in a DTC patient on suppression, any serum Tg increase in excess of 15% should prompt an evaluation for recurrence. Unfortunately, Tg assay precision erodes over the long interval typically used for monitoring DTC patients (6 to 12 months). Unfortunately, the between-run precision of most Tg tests exceeds 15% across this interval when measuring Tg values below the lower reference limit. This precision problem has the potential to mask the detection of clinically important changes. One approach to eliminate interassay imprecision is to remeasure the patient's archived specimen from the previous evaluation alongside the current specimen in the same assay.

"Hook" Problems

Tumour marker IMA methodologies are prone to "high-dose hook effects". This is characterised by inappropriately low values in sera with high analyte concentrations. The problem is caused by an excess of analyte in the specimen overwhelming the binding capacity of the solid-phase capture antibody and inhibiting the ability of endogenous analyte to form a bridge between the two antibodies, resulting in an inappropriately low value. Some Tg IMA methods are prone to this hook problem.

Manufacturers are generally aware of this problem and attempt to overcome it by designing assays using either a two-step process, or recommend that each specimen be run at two dilutions. When an unexpectedly low serum Tg value is reported for a patient with known metastases, the physician should request the laboratory to repeat the measurement at multiple log dilutions to rule out a "hook" problem.

TgAb Interference

Thyroglobulin autoantibody (TgAb) interference with serum Tg measurements remains the most serious problem limiting the clinical value of serum Tg measurement. It is important that the laboratory use a sensitive TgAb immunoassay for identifying sera containing TgAb. Serial TgAb measurements can be used as an inde-pendent prognostic test for the presence of Tg-secreting thyroid tissue. The interpretation of a Tg value depends on the TgAb status of the patient and the class of Tg method used (RIA or IMA). The direction and magnitude the interference relates to the method and the concentration and affinity of the TgAb in the specimen. Typically, unidirectional interference is typical of IMA methods, which display a disproportionately high percentage of low and undetectable serum Tg values when patients are TgAb-positive (Figure 4b).

Some Tg IMA methods claim to overcome TgAb interference by using monoclonal antibodies directed against epitopes not involved in thyroid autoimmunity. Although conceptually attractive, this approach does not appear to overcome interference problems in practice, possibly because less restricted TgAb epitopes are associated with thyroid carcinoma than with the autoimmune thyroid conditions. In contrast, TgAb can cause bidirectional interference with serum Tg RIA measurements. The magnitude and direction of the interference is determined by the factors that affect the partitioning of the Tg tracer between Tg antibody reagent and the endogenous TgAb. In general, fewer low and undetectable values are seen for TgAb-positive sera measured by RIA techniques that appear to represent the measurement of both free Tg and Tg complexed with TgAb.

Figure 4 shows serum Tg measurements made in TgAb-negative and TgAb-positive sera, measured by both classes of method (IMA and RIA). The between–method differences evident in the absence of TgAb (~30% CV) (Figure 4a) reflect the speci-ficity differences between methods. When TgAb is present, sera typically have detectable Tg RIA values and lower or undetectable IMA values. This RIA/IMA discordance appears to be a better indicator of TgAb interference than exogenous recovery tests.

In fact the new NACB consensus guidelines recommend abandoning recoveries and recommend the measurement of TgAb concentrations directly. The mechanism responsible for the low IMA values appears to be an inability of Tg complexed with TgAb from participating in the two-site reaction. When a patient has detectable TgAb, clinical judgment should not be based on an undetectable serum Tg IMA value. It is difficult to prove which class of method (RIA or IMA) most accurately reflects the Tg status of a TgAb-positive patient. It is generally considered that serum Tg underestimation is more problematic than overestimation, since inappropriately low values may mislead the physician into believing that the patient is free of disease, even when metastatic disease is present. In contrast, inappropriately high values that are typical of some RIA methods may prompt unnecessary imaging studies but will have the advantage of promoting a higher level of patient surveillance.

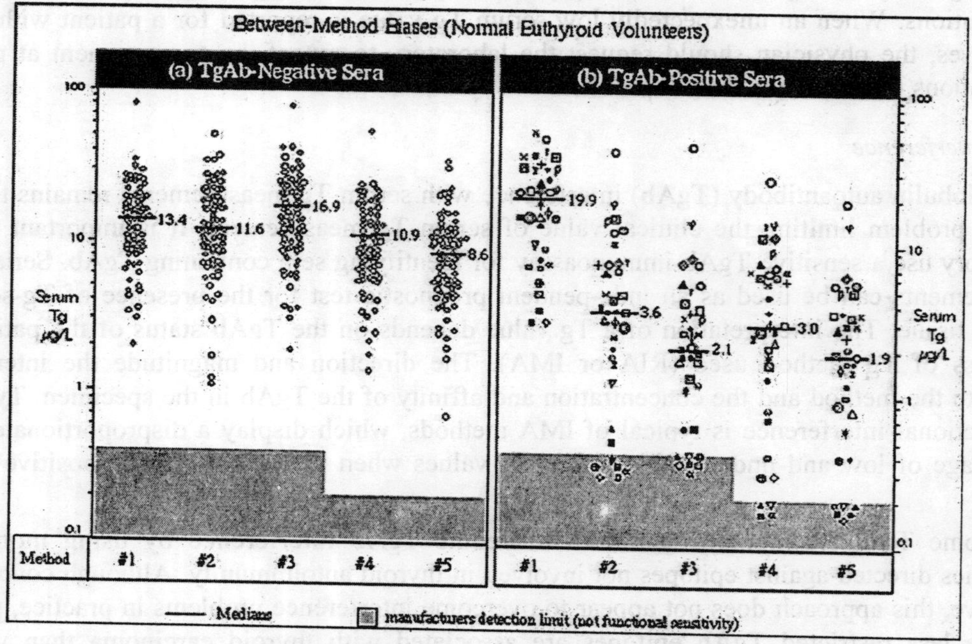

Figure 4. Comparisons of TgAb-negative and TgAb-Positive Subjects

A number of laboratories now believe that TgAb-positive patients are best monitored by serum Tg RIA measurements that appear to correlate better with clinical progression or regression of disease. Many laboratories using IMA methods are currently still reporting falsely undetectable serum Tg values for TgAb-positive patients, albeit with a cautionary comment. Other laboratories are adopting a dual strategy for serum Tg measurement – i.e. restricting the use of the IMA methods to TgAb-negative sera and retaining RIA methodology for TgAb-positive sera. Serum Tg concentrations measured in 88 TgAb-negative (panel a) and 36 TgAb-positive (panel b) euthyroid volunteers (TSH 0.5-2.5 mIU/L) using different methods. Method #1=University of Southern California RIA, Los Angeles, CA, USA; Method #2=Nichols Bead ICMA, Nichols Institute Diagnostics, San Juan Capistrano, CA, USA Method #3= Nichols Advantage ICMA, Nichols Institute Diagnostics ICMA, San Juan Capistrano, CA, USA; Method #4= Access ICMA, Beckman-Coulter, Fullerton, CA USA. Method #5= Immulite ICMA, Diagnostic Products Corporation, Los Angeles, CA, USA. The shaded area indicates the manufacturers recommended sensitivity (not functional sensitivity). Lines indicate median values. The symbols in panel b indicate the values for an individual patient measured across the methods.

Clinical Use of Tg Methods

The technical problems with serum Tg measurement discussed above require a strong physician-laboratory interface for the optimal management of DTC patients. Physicians need to understand the basic technical issues concerning serum Tg measurement, and develop laboratory partnerships that provide an appropriate level of support. It is important to recognise that the serum Tg concentration integrates three principal factors:

(1) the mass of differentiated thyroid tissue present (normal tissue + tumour);

(2) any inflammation of, or injury to thyroid tissue, such as follows fine needle aspiration biopsy, surgery, radioiodine therapy or thyroiditis; and

(3) the degree of stimulation of TSH receptors (by TSH, hCG or TSAb).

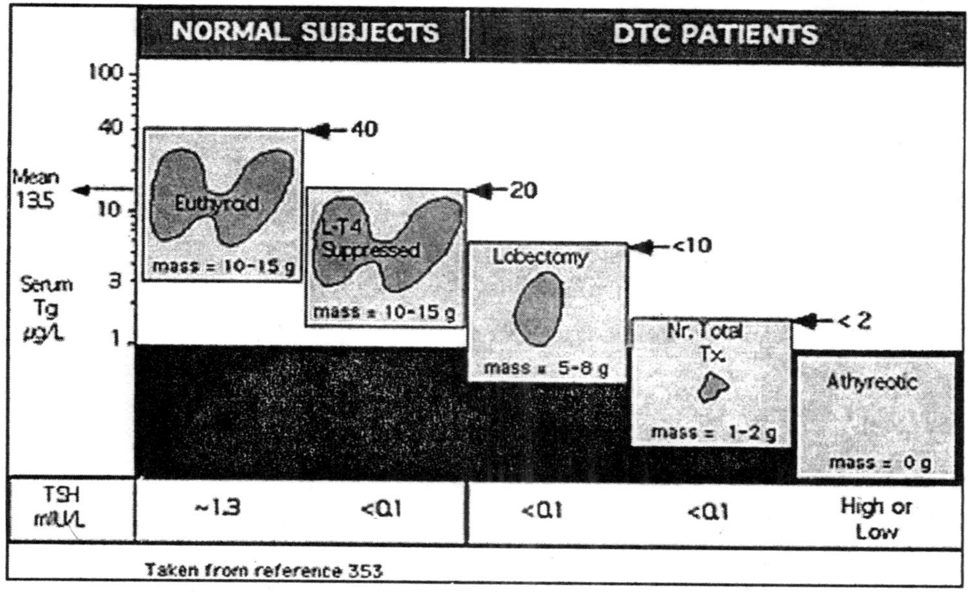

Figure 5. Title

Bibliography

Anstee, J. R.: "Chemical Structure and Analgesic Activity". *Australasian Journal of Pharmacy,* 15 August 1951, *32:* Brisbane Jubilee Conference Issue, p. 820-826

Batterman, R. C.: "The Importance of Addiction to the Newer Synthetic Analgesics in Human Therapy". *Annals New York Academy of Sciences,* 1948, *51:*1,p. 123-129

Beckett, A. H.; Casey, A. F.: "Stereochemistry of Certain Analgesics", *Nature,*1954, *173*:4417,p. 1231-1232

Beecher, H. K.: "Study of Analgesics". *Bulletin on Narcotics,* 1951, *3*:4, p. 9-15

Braenden, O. J.; Wolff, P. O.: "Synthetic Substances with Morphine-like Effect". *Bulletin of the World Health Organization,* 1954, *7*:10, p. 1-25

British Medical Association: "The Relief of Pain". *The Lancet,* 1952, *263:*6725 ,p. 129-130

Chapin, M.: "The Newer Drugs". *Journal of the Maine Medical Association,* 1949, *40:2,*p. 40-42

Costello, C.; Mcdonald, J.: "The Role of a New Analgesic in Surgery". *Journal of the Missouri State Medical Association,* 1949, *46:5,*p .333-337

Cullen, S. C.: "Use of Depressant Drugs". *Journal of the American Association of Nurse Anesthetists,* 1952, *20:*3 ,p .167-174

Dodds, E. C.: "The Search for Morphine Substitutes". *British Medical Bulletin,* 1946, *4*:2, p. 88-91

Douglas S. Johnson, Jie Jack Li (eds.) , *The Art of Drug Synthesis*, Wiley, 2007.

Ebbing, D. D. and Gammon, S. D. *General Chemistry,* Houghton Mifflin, New York,2002.

Eddy, N. B.: "The Relation of Chemical Structure to Analgesic Action". *Journal of the American Pharmaceutical Association,* sc. ed., 1950, *39*:5, p. 245-251

Farmilo, C. G.; Levi, L.; Oestreicher, P. M. L.; Ross, R. J.: "Studies on the Identification of Narcotics. III. Microcrystal and Colour Tests for the New Synthetic Narcotics". *Bulletin on Narcotics,* 1952, *4*:4, p. 16-42

Ferrer, M. I.; Sokoloff, L.: "Antidiuretic Effect of Morphine and Demerol in Congestive Heart Failure". *American Journal of Medical Science,* 1947, *214*:4, p. 372-384

Fraser, H. F.; Isbell, H.; Vanhorn, G. D.; Nash, T. L.: "Use of Measurements of Miotic Effects in Evaluating Analgesic Drugs in Man". *Journal of Pharmacology and Experimental Therapeutics,* 1954, *110*:1, p. 19

Fujimura, Hajime: "Relation between Analgesic Effect of Drugs and the Chemical Structures". *Bulletin of the Institute for Chemical Research, Kyoto University,* 1952, *30*:9, p. 53-54

Govida Rao, A. R.: "The Use of New Synthetic Analgesics in Surgery". *Antiseptic,* 1950, *47*:9, p. 689-691

Hebb, C. O.; Konzett, H.: "Difference between Morphine and Synthetic Analgesics in their Action on Ganglionic Transmission. *Nature* (London) 1949, *163*:4149, p. 720-721

Helliwell, P. J.; Hutton, A. M.: "Analgesia in Obstetrics". *Anaesthesia,* 1949, *4*:1 p.18-21

Houde, R. W.; Wallenstein, S. L.: "Method of Screening Analgetics in Cancer Patients". *Federation Proceedings,* 1953, *12*:1, I, p. 332 (No. 1095)

Irwin, S.; Houde, R. W.; Bennett, D. R.; Hendershot, L. C.; Seevers, M. H.: "The Effects of Morphine, Methadone and Meperidine on Some Reflex Responses of Spinal Animals to Nociceptive Stimulation". *Journal of Pharmacology and Experimental Therapeutics,* 1951, *101*:2, p. 132-143

Irwin, S.; Seevers, M. H.: Neurological changes in the monkey following acute and chronic administration of morphine-like analgesics. *Journal of Pharmacology and Experimental Therapeutics,* 1954, *110*:1, p. 27-28

Isbell, H.: "Methods and Results of Studying Experimental Human Addiction to the Newer Synthetic Analgesics". *Annals New York Academy of Sciences,* 1948, *51*:1, p. 108-122

Kaufmann, H. P.: *Arznelmittel-Synthese* (Synthetic drugs). 1953, Berlin, Springer Verlag, 834 pp.

Keele, C. A.: "Assay of Analgesic Drugs on Man". *Analyst,* 1952, *77*:912, p. 111-117

Levi, L.; Oestreicher, P. M.; Farmilo, C. G.: "Nonaqueous Titration of Narcotics and Alkaloids". *Bulletin on Narcotics,* 1953, *5*:1, p. 15-25

Lu, M. C. In Foye, W. O. (ed.), *Cancer Chemotherapeutic Agents*. Washington DC: American Chemical Society, 1995, 345.

May, E. L.: "The Chemistry of Drugs of Addiction". *American Journal of Medicine,* 1953, *14*:5, p. 540-545

Nathan, P. W.: "Newer Synthetic Analgesic Drugs". *British Medical Journal,* 1952, *2*:4790, p. 903-908

Richard Silverman, *The Organic Chemsitry of Drug Design and Drug Action,* Academic Press, 2004.

Robson, J. M.; Nissim, J. A.: "The New Analgesics". *Chemical Products* (London), 1949, *12*. p. 284-292, 297

Slaughter, D.; Neff, J.; May, M.: "Human Studies on Synthetic Analgesics". *Transactions of the American Therapeutic Society,* 1947, *48*, p. 155-159

Smith, A.: "Safeguards in the Use of New Drugs". *Bulletin New York Academy of Medicine,* 1949, *25*:2, p. 115-124

Tainter, M. L.: *Newer synthetic analgesics,* New York Academy of Sciences, New York, 1948.

Van Dyke, H. B.: "Synthetic Analgesic Drugs", *American Journal of Medicine,* 1949, *6*:6, p. 681-683

Webb, J. L.; Elliott, K. A. C.: "Effects of Narcotics and Convulsants on Tissue Glycolysis and Respiration". *Journal of Pharmacology and Experimental Therapeutics,* 1951, *103*:1, p. 24-34

Index